CAESAR

CAESAR

CHRISTIAN MEIER

*Translated from the German
by David McLintock*

BasicBooks
A Division of HarperCollinsPublishers

Published by BasicBooks,
A Division of HarperCollins Publishers, Inc.

First published in Germany in 1982 by Severin und Siedler.
Published in the United Kingdom in 1995 by HarperCollins
Publishers.

Library of Congress Cataloging-in-Publication Data
Meier, Christian.
   [Caesar. English]
   Caesar : a biography / Christian Meier ; translated from the
German by David McLintock.
     p. cm.
   Includes index.
   ISBN 0-465-00894-1
   1. Caesar, Julius. 2. Rome—History—Republic, 265–30 B.C.
3. Statesmen—Rome—Biography. 4. Generals—Rome—
Biography. I. Title.
DG261.M3713   1996
937'.02—dc20   95-30003
                                             CIP

96 97 98 99 RRD 9 8 7 6 5 4 3 2 1

CONTENTS

CONTENTS

LIST OF ILLUSTRATIONS

LIST OF ILLUSTRATIONS

Due to circumstances beyond their control the Publishers have been unable to acknowledge all the photographic sources for the illustrations used in this book. They will be happy to do so in future editions if they are notified.

Gaul at the Time of Caesar

The Roman World at the Time of Caesar

RICUM
na

Lissus

rrhachium
rundisium · Apolloniae
Tarentum · Oricum · *Epirus*
etapontum · Gomphi

Kerkyra · Pharsalus

MACEDONIA
Thessalonice
Larissa
Thessalia

ACHAIA
Patrae
Athenae
Corinthus

Thrakia

Ilium
Pergamum

Mytelene

ASIA

Ephesus

Miletus

CRETA

Rhodos

I n t e r n u m

P o n t u s   E u x i n u s

BOSPOROS

Sinope
Amisus

BITHYNIA
Nicaea
GALATIA
Zela

Nicopolis

PONTUS

CAPPADOCIA

ARMENIA

LYCAONIA

PISIDIA
CILICIA
Tarsus
Carrhae

Antiochia
SYRIA
PARTHIA

Cyprus
Apamea

Tyrus

Jerusalem

Cyrene

CYRENAICA

Alexandria
Pelusium

Memphis

AEGYPTUS

Rome c. 44 BC

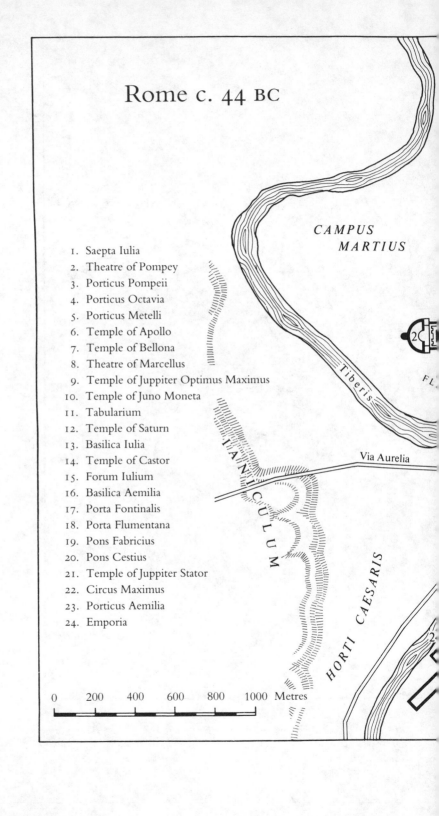

1. Saepta Iulia
2. Theatre of Pompey
3. Porticus Pompeii
4. Porticus Octavia
5. Porticus Metelli
6. Temple of Apollo
7. Temple of Bellona
8. Theatre of Marcellus
9. Temple of Juppiter Optimus Maximus
10. Temple of Juno Moneta
11. Tabularium
12. Temple of Saturn
13. Basilica Iulia
14. Temple of Castor
15. Forum Iulium
16. Basilica Aemilia
17. Porta Fontinalis
18. Porta Flumentana
19. Pons Fabricius
20. Pons Cestius
21. Temple of Juppiter Stator
22. Circus Maximus
23. Porticus Aemilia
24. Emporia

CAMPUS MARTIUS

Tiberis

FL.

Via Aurelia

IANICULUM

HORTI CAESARIS

0   200   400   600   800   1000 Metres

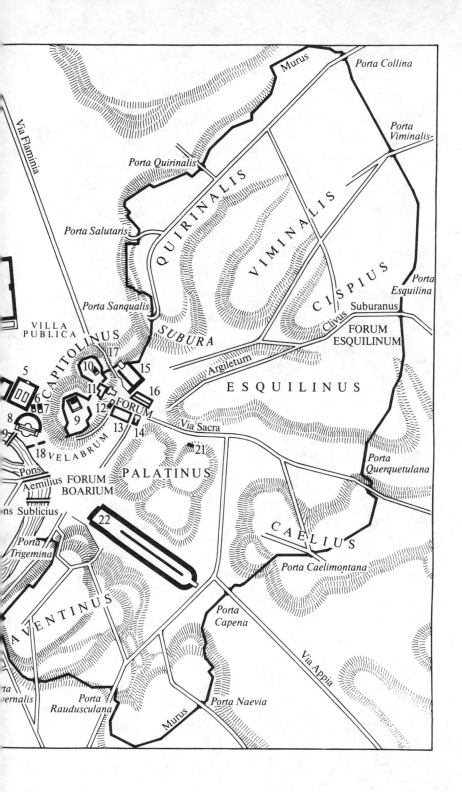

Via Flaminia

Murus

Porta Collina

Porta Quirinalis

Porta Viminalis

Q U I R I N A L I S

V I M I N A L I S

Porta Salutaris

C I S P I U S

Porta Esquilina

Porta Sanqualis

Suburanus

Clivus

VILLA
PUBLICA

S U B U R A

FORUM
ESQUILINUM

17

Argiletum

C A P I T O L I N U S

E S Q U I L I N U S

5

10

15

6

11

16

7

12

FORUM

8

9

13

14

Via Sacra

18

VELABRUM

21

Pons

Porta
Querquetulana

Aemilius

FORUM

P A L A T I N U S

ons Sublicius

BOARIUM

22

C A E L I U S

Porta
Trigemina

Porta Caelimontana

A V E N T I N U S

Porta
Capena

ta
ernalis

Porta
Raudusculana

Murus

Porta Naevia

Via Appia

1

Caesar and Rome – Two Realities

THE SENATE DECLARES A STATE OF
EMERGENCY · CAESAR AT THE
RUBICON · THE MONSTROUS CAUSE OF
THE WAR · THE POSITIONS OF THE
PARTIES · TWO REALITIES

O N I JANUARY 49 BC the consuls began to do everything in their power to remove Caesar from his governorship. He had held it for almost ten years and its term had expired. Caesar now intended to seek the consulship for 48 and return to internal politics. This was just what his opponents wished to prevent. Before he could offer himself as a candidate he was to lay down his command and return to Rome as a private citizen. Once there, he was to be prosecuted for various breaches of the constitution of which he had been guilty during his previous consulship in 59 BC. It was clearly intended that this should take place under military protection, so that he would be unable to put pressure on the court, and also, no doubt, so that the court would not be entirely free from pressure in reaching its verdict. It seems to have been hoped that in this way Caesar's political existence could be destroyed and the power of the

I

Senate régime fully restored. Irrespective of whether or not Caesar really was an opponent of the traditional order, he had persistently interfered with its proper functioning during his previous consulship. It was therefore to be feared that if he succeeded in becoming consul again he would push through various demands against the will of the Senate and thereby become so powerful that repeated conflicts and repeated defeats for the Senate could be foreseen.

For almost two years Caesar's dedicated opponents had sought to persuade the Senate, Rome's central organ of government, to relieve him of his command. They had repeatedly failed because Caesar had won over some of the tribunes of the people, who could use their right of veto to impede any resolution the Senate passed against him. At times they even went on to the offensive and forced it to pass resolutions in his favour. For although the majority of the senators were opposed to the proconsul and desired a speedy end to his governorship, they were even more opposed to a civil war and, knowing that Caesar was not to be trifled with, were inclined to let him have his way.

At the beginning of January Caesar's opponents set all the wheels in motion to oblige the Senate to settle the matter once and for all. Supporters were drummed up, the alarm was sounded, and powerful sentiment aroused. The Senate resolved that if Caesar had not laid down his command by a certain date he would be acting against the republic. The tribunes of the people used their veto. Since they refused to yield, the Senate, on 7 January, passed the 'extreme decree' (*senatus consultum ultimum*); this amounted to declaring a state of emergency.

The Caesarian tribunes of the people thereupon left the city, disguised as slaves, in one of the carriages that stood ready for hire at the city gates. (These conveyances, together with horses and litters, were the normal means of transport for lengthy journeys; the horses could be changed *en route*.) By doing so they indicated that the freedom of the Roman people was in such jeopardy that even its guardians, whom the people had once sworn to protect, were no longer assured of safety.

Caesar was at Ravenna, in the extreme southeast of his province of Gallia Cisalpina. On the morning of 10 January 49 (mid-November according to our calendar) a courier brought news of the Senate's decree and the flight of the tribunes. Without much ado Caesar at once sent a troop of soldiers towards Ariminum (Rimini), the first sizeable town in Italy proper, which lay beyond the Rubicon, the narrow river that bounded his province. It was an immensely bold decision, for Caesar had only one legion with him (five thousand foot and three hundred horse) – the bulk of his army was still in Gaul – but he wanted to take advantage of the element of surprise and hamper his opponents' preparations.

At Ravenna Caesar first attended to routine matters. He inspected a gladiatorial school. After this he took a bath, either at his host's house or in the public baths (bathing had become something of a cult, and Caesar was punctilious in matters of hygiene), then dined in company. At nightfall he took his leave, assuring the others that he would return shortly, and left the city. He did not take a direct route. According to one source he lost his way in the dark; another states that he deliberately set off in a different direction, so that he would not be noticed when he turned south. He had instructed a few friends to follow him, each of them separately. They all met at the Rubicon.

There Caesar halted. He hesitated. Once again, beside the swiftly flowing river swollen by heavy rain, he reviewed the various arguments, then reiterated his decision. For a moment he once again took a detached look at the plan he had already embraced; what he had begun to set in train step by step now appeared to him as an awful vision. All the possible consequences of the enterprise presented themselves in their full horror. Contemplating them, he may well have felt dizzy.

For a long time he stood silent. Then he shared his thoughts with his friends. One of them, Asinius Pollio, recorded the scene in his histories. These have not survived; but Pollio's account has been preserved in slightly differing versions by two other ancient authors. According to Pollio, Caesar was concerned about the misfortune to which he would expose all men if he now took the step leading to war. He considered 'how much misfortune the crossing would bring to all men'. He tried to see how posterity would judge him and his decision. One of our sources shows Caesar's thoughts clearly concentrated on the fateful alternative: 'To refrain from crossing will bring me misfortune; but to cross will bring misfortune to all men.'

3

It is clear, then, that Pollio's account spoke of the misfortune of all men. There is no reason to doubt that Caesar spoke of it too. The forces of his opponents were deployed throughout the whole of the Mediterranean area. It was to be feared that they would mobilize them. He could thus be in hardly any doubt that the war he was about to begin might affect the whole of the known world. He no doubt hoped to get away more lightly; this was why he was anxious to reach a quick decision. Yet if the full implications of the enterprise presented themselves to Caesar's mind with such alarming clarity as he stood at the Rubicon, he probably envisaged the worst.

If the misfortune of all men was in the one scale, was only Caesar's in the other? Was the other side of the alternative as unequivocally clear and as fateful as it appears from our source? Was the war waged solely because Caesar did not wish to be stripped of his command and arraigned before a Roman court? Was he therefore alone against everyone, with only himself to rely on? And if this was really so, could he see it like this without deceiving himself in any way, and speak of it so frankly to his friends at the Rubicon?

At last he roused himself from his anxious ponderings and addressed himself to what lay ahead. With the words 'The die must be cast' he crossed the Rubicon and, having made good headway and reached Ariminum before dawn, led his soldiers into the city. These words are a quotation from a comedy by Menander. The version 'The die is cast' is incorrect, for until then there had been no dicing. The game was just about to begin – the game of war, in which the highest stakes were involved and fortune was an important player. This was clear to Caesar as to few others, but he trusted in the goddess's favour.

At Ariminum he was joined by the tribunes who had fled from Rome. He led them before his soldiers. According to his own account he addressed the troops and explained to them that the Senate had suppressed the legitimate objection of the tribunes by force of arms and quite unjustifiably passed the *senatus consultum ultimum*. He rehearsed 'all the infringements of the law that his opponents had persistently committed against him'. Now they even sought to relieve him of his command. He called upon the soldiers 'to defend the fame and honour of their general, under whose leadership they had for nine years fought so successfully for the commonwealth, won many battles and pacified the whole of Gaul and Germany'. Thus began the civil war, which, with brief intermissions, was to occupy

Caesar for nearly five years, costing much bloodshed and shaking the entire Roman world to its foundations.

If the soldiers were to defend the fame and honour of their general, as he put it, did this not mean that Caesar, solely for his personal interests, was risking a war that, if the worst came to the worst, would affect the whole of humanity? If we consult the relevant sources, of which there are not a few, we can be in no doubt of this. He wanted to deflect the misfortune and danger that threatened his own political existence. By posing as the champion of the tribunes and the liberty of the Roman people, he was merely cloaking his true concerns. The cloak was altogether transparent, and Caesar did not trouble to conceal its transparency by skilful draping. He did not try to deny that the danger to the tribunes, which in any case he grossly exaggerated, resulted solely from their support for him, and he soon abandoned this pretext. From his own statements, and from speeches in which others sought to win his favour, it is evident that the war was waged to preserve Caesar's honour (*dignitas*). 'What else did your armies want,' asked Cicero, 'but to defend you against demeaning injustice?'. On the evening after the battle of Pharsalos, surveying the field strewn with the dead and wounded, Caesar himself said, 'This is what they wanted; after such great deeds I, Gaius Caesar, would have been condemned had I not sought the help of my army.' Such quotations could be multiplied. Caesar had no cause but himself. Cicero wrote: 'This cause lacks nothing but a cause.'

To some extent this is entirely correct: the misfortune that Caesar wished to fend off by embarking upon the war was solely his own. And there is nothing to suggest that this was not clear to him.

It seems monstrous and scarcely credible. How can an individual decide to inflict misfortune on all men rather than suffer it himself? How was it possible to conceive such an idea, to voice it, to act upon it, to persist with it? How could it be justified? Surely anyone who makes such a decision must be a desperado or not in his right mind, not just utterly isolated, but completely out of touch with the world he lives in? Or is it a mark of greatness? But then what is greatness?

We should not see the matter in excessively abstract terms or view it as a purely personal problem. We should rather consider the circumstances, the 'scenario', in which Caesar made his decision.

After all, politicians act not only when viewing situations from outside: they also act within them. In doing so they are not just themselves, but to some extent part of a configuration, and this is probably especially true in extreme situations. We must therefore consider not just the personalities, their general and particular interests and opinions, but the positions they occupy within the configuration, which is of course determined by their action and interaction, but also determines them. The overall situation dictates not only the areas of activity, but perspective and distance. In any situation men are bound to certain positions, and these are defined within the surrounding framework. One must therefore consider not just the participants, but the total situation, which is greater than the sum of its parts. True, such an approach makes it difficult to form one-dimensional judgements with scholarly detachment, but it brings us closer to the reality. So far, all that is clear is who might have been affected in a war, but not yet who was opposed to whom and in what ways.

Caesar rebelled against Rome; this is how it must have seemed to the Senate and to all who acknowledged its authority and its responsibility for the commonwealth – in other words, to Roman society. Caesar's armed invasion was universally condemned, even by some of his prominent friends, relatives and allies.

Modern scholars, on the other hand, have sought to evade the fatal alternative by crediting Caesar with a superior statesmanly vision and a genuine cause, in order to be able to assume that he acted from higher motives: that he was standing up for Rome, Italy, and the peoples of the empire against a blinkered, self-seeking and super-annuated Senate, or that he wished to create a just and effective system of government and fundamentally renew the structure of the Roman empire.

If this was so, Caesar said nothing about it, either at the Rubicon or subsequently, and no other evidence can be cited in support of such a view. On the contrary, it is clear that no one knew anything of it. None of the groupings in the civil war was moved by any such objective considerations. Indeed, there was no split in Roman society corresponding to the division between Caesar and his opponents. Those who went over to him did so because he seemed likely to succeed, or because he was ultimately victorious. On his own side he had only his immediate supporters, whose loyalty – for all the fascination and friendship he inspired – derived chiefly from the hope that his victory would lead to an improvement in their own conditions, and his soldiers, who were similarly motivated. There was no cause that extended beyond his immediate circle. Hence, there can be no doubt that Caesar and his followers were isolated, seemingly alone against Rome.

This was admittedly not how he saw it. His quarrel was not with Rome, but only with his opponents. He did not wish to fight a civil war, but merely to settle 'civil disputes' (*civiles controversiae*). He could not regard the Senate as an independent force. In his view, the resolution it had passed against him did not express the will of the Roman republic, but was engineered by his opponents, to whom he could ascribe no statesmanly motives, but only gross self-interest.

He drew the demarcation lines accordingly. The 'civil disputes' in question should not concern the mass of the citizens. 'What more befits a decent man (*vir bonus*), a decent, peaceful citizen, than that he should remain aloof from civil disputes?' he wrote to Cicero – as though it was the first duty of the citizen in a republican community to keep the peace while the warring parties fought it out. Whoever was not against him he regarded as a friend. The citizens should not involve themselves in what did not concern them. His opponents took the opposite view: anyone who did not join them was an enemy. After all, they represented the cause of the *res publica*, and no one could stand aloof from such a cause. The best evidence for the difference between the two positions was provided by their views on where the neutrals belonged. Caesar had nothing to hope for from their active involvement: they would be for the *res publica* and against him. They were his friends if they were not his enemies. In practical terms they acknowledged that the war was being fought only between Caesar and his senatorial opponents.

In these early weeks of 49 Cicero recalled Solon's law according to which anyone who did not take sides in a civil war was to be punished. This law, designed to ensure that the whole of the body politic brought its influence to bear between the warring factions, dated from the early days of the ancient city states, which were based on the whole of the body politic. For Caesar, at this moment of extreme peril, the body politic as a whole was not involved; consciously or unconsciously he excluded it.

One wonders whether Caesar was being subtle or whether he knew no better. If the former, he was certainly not subtle enough. Who was to accept his view that a senate resolution counted for nothing and that a civil war was merely a quarrel between a few magnates? We may therefore presume that Caesar really saw the matter as he described it. But then we have to ask whether he was so prejudiced, so blind, that he could not perceive the reality of Rome, which was still wholly determined by the *res publica*.

Starting from his own premises, however, he came to conclusions that were sound enough in practical terms. It is true that people thought differently in Rome, but on the whole they behaved as if the war did not concern them. They arrived at a quick and easy accommodation with Caesar. So did many senators. Of the *consulares* barely half joined Caesar's opponents, while the rest remained neutral. And it was not the bulk of the Senate, but only a sizeable minority, that took sides against him. Hence, although Roman society suffered from the war, it did not allow the war to be waged against it. The republic was opposed to the aggressor, but did not defend itself against him. Given this distribution of power and opinion, there was no real republican party – only one that professed allegiance to the republican cause. By wishing to preserve the peace without being able to do so, and hence not committing itself, Rome's 'good society' was effectively in the opposing camp.

It followed that personal supporters played a central role among Caesar's opponents too; these were the followers of Pompey, Rome's leading general. Pompey too had long been opposed by the Senate, as an outsider who did not conform with the discipline of the order and had so much power concentrated in his hands that he seemed to pose a danger to senatorial equality. Yet in the end an accommodation was reached, and Pompey was drawn into the coalition against Caesar. He already commanded many legions and now assembled large additional forces, a 'Pompeian' army recruited in the East, from the cities and princes who were beholden to him. The leading senators were of course also in his camp, or commanded other parts of the joint army and or navy, but they had little power of their own; they had hardly any troops, since the citizens, who were led by the Senate and whose cause was championed by it, had not followed them to arms. Hence, the republican side was essentially represented by Pompey, and it was feared that if he were victorious he would establish an autocracy.

Fundamentally, then, it was a contest between Caesar and Pompey. The republic was present only as a yardstick for a broad, self-assured, but largely ineffectual public opinion. It had no legions. Caesar may or may not have seen through it, but his attention was in any case focused on the effective forces, and he certainly had a better, though by no means complete, appreciation of the military realities than his opponents.

His opponents had only a limited grasp on reality, in that they clearly did not realize what kind of man they were dealing with. For nine years the proconsul had commanded a large army, waged a highly successful war, and made more conquests than any other general in the whole of Rome's glorious history. Yet now he was not only to be denied a triumph – the honour to which he was traditionally entitled – but even punished and robbed of his political existence.

True, ten years earlier he had been guilty of various breaches of the law – committed, incidentally, while asserting important demands by Pompey against Senate opposition. True, he had obtained his Gallic command against the wishes of the Senate, which had no desire for war or the conquests he went on to win. Yet after it had to acquiesce in all this, was it still justifiable, advisable or practical to ignore what had happened, what had been tolerated and was now a *fait accompli*, and revert to the events of 59 in order to threaten Caesar's political existence? Could the victorious leader of nine legions and twenty-two cohorts really be expected simply to hand himself over? Could Caesar's dismissal be realistically presented as the cause of the *res publica* when the Senate had for so long refused to pass a resolution against him? Was the *res publica* in fact conceivable without Caesar?

On the other hand, Caesar's opponents, faced with such a Senate majority, were bound to ask themselves whether it was not incumbent upon them to do all in their power to oppose him. It was traditionally held that the leading circles in the Senate were responsible for its policy, and it had long been taken for granted that men who were too powerful to conform to senatorial discipline must be vigorously opposed. There was every reason to fear Caesar's return to internal politics, and the more powerful he became, the more it was feared.

In such circumstances even civil war was an option that could be contemplated. It need not result in victory for Caesar, though on the other hand it was hardly likely to end in victory for the *res publica*. The shocks it would unleash and the accumulation of military might that was bound to ensue would be so significant that even a victory for Pompey would at the very least result in a palpable restriction and weakening of the Senate régime.

Yet how were Caesar's opponents to appreciate this? How could they gain sufficient detachment to realize that the continuance of the inherited order was in question? Everything we know about Roman society at this period indicates that the inherited order was regarded as the only legitimate one. With it Rome had conquered the world. In it the citizen body had found not only a political form, but a social identity. No one knew anything different. At most Rome might resign itself to making a few concessions. Yet Caesar's opponents believed, as the leading senators had repeatedly believed, that the just order must be strenuously defended, especially in view of the pusillanimity evinced by so many of the citizens. Rome stood at a threshold: if Caesar was allowed to cross it, the worst could be expected.

Today we may wonder whether Caesar's opponents were not acting on the basis of a reality that, though still postulated, was in fact non-existent. Yet at the time this reality was constituted by the whole of society; society saw it in this way and shaped it accordingly, but was unwilling to defend it.

Caesar's opponents, in short, saw Roman reality from within and did not doubt it. Caesar, by contrast, saw it from without. This is why he was able to gauge the power relationships so accurately, though without knowing how firmly rooted the cause of the Senate still was in the public mind. His opponents knew, but deceived themselves by underrating their own weakness.

Yet if a man who had done so much for Rome and led such a great army could see Roman reality from outside and hence stand outside it to this extent, then this reality must have somehow acquired a peculiar character. Its principles, its code of conduct and its fundamental rules no longer had any hold over Caesar; nor were they clearly observed any longer. This meant that his position was neither fortuitous nor wholly alien to Roman conditions – otherwise it is unlikely that any outsider could have become so powerful.

For this very reason Caesar cannot have been simply a desperado who relied solely on his own resources. He had obviously been able to create his own sphere of activity, enjoying so much power and hampered by so little interference that he may be said to have inhabited a world of his own, ruling over his provinces, surrounded by his soldiers, and fully conscious of his immense abilities and achievements. His personal ambition had developed and consolidated itself in a powerful individual position, which was grounded in an admirable, if one-sided, ethos that embodied the old aristocratic ideal of achievement. In this respect only Pompey came close to him. This position acquired a certain independence. It became, as it were, so wide and so powerful that he might feel he had a duty to preserve it. To a certain extent it was a substitute for suprapersonal legitimacy, in other words for the numerous opinions and endeavours that coalesce to objectify and justify a cause. It provided him with a shield against his opponents. This goes some way to explaining Caesar's readiness to embark upon a war that could affect all men. He owed it to himself, even after intense soul-searching. Consciously or unconsciously he may have deceived himself and others in that he could not see – or could not admit – that Roman society was opposed to him. He may have found this impossible to believe after all his services to Rome. And indeed it was true only up to a point.

Admittedly, if a man could see through Rome's republican institutions and perceive only his opponents, his view of Roman reality probably not only differed from that of others, but no longer bore any relation to it. For this reality involved a homogeneity of knowledge regarding the legitimate order; in this respect it allowed no choice. One might oppose the Senate over this or that point, but one could not overlook it.

Caesar and his opponents thus represented two disparate realities: the old reality, which had once been the whole and was suddenly reduced to a part, and the new, which had detached itself from the old and could hardly have been realigned with it even if war had been avoided – so wide was the gap, so great the mutual alienation. It was this disparity that characterized the situation – not just conflicting interests, mistrust, fear, hatred, or the pathological exaggeration of individual pretensions.

While acknowledging the existence of the two opposing realities, we need not refrain from condemning Caesar's crossing of the Rubicon as an act of monstrous presumption committed by an individual against Rome and its empire. Nor need we shrink from criticizing the shortsightedness of his opponents, which led them to overrate their chance of success. Yet we can no longer overlook the strength of the positions that the opposing parties took up.

It is becoming clear that the Roman republic had now reached a point at which not only Caesar and his opponents, but the whole of society, were locked into a situation from which there was no escape. As society was not internally split – but rather united over the need to preserve the *res publica* – its reality was split. An outsider could challenge the whole from a position of power, because he had been able to construct a world of his own.

What kind of a society was this? If outsiders were able to build up so much power within it, against its leading institutions, it can no longer have been properly integrated, but in a state of crisis. Indeed, the old institutions, designed for a city state, had long been over-stretched, as Rome now ruled over a world-wide empire. Yet how could the citizen body remain solid in its adherence to the inherited order? How was it possible, in view of the crisis, that it did not split around great objective oppositions? Why did the deprived populace not rise in revolt? How broad was this society, and what possibilities were there for opposition? Are we dealing here with a phenomenon peculiar to the ancient world? Jacob Burckhardt noted that in the ancient world affairs were ordered in a way that made for rigidity rather than flexibility: the bough was less likely to bend than to break. This is no doubt connected with the fact that the citizen body could be said to *be* a political order rather than to *have* one, and that there was no duality of society and state; rather, the body politic had itself become a political unit. It could therefore have little detachment from itself.

What kind of a crisis was it in which it was not Roman society that fell apart, but Roman reality – the sense of shared security in an order that was essentially unquestioned? It was obviously a crisis that produced outsiders rather than internal opponents, a crisis that – in view of the unanimous adherence to the old – clearly resulted rather from the build-up of the unintended side-effects of action and thus evolved in the form of a process (which looks like a quite modern phenomenon). What did this mean for society in general and for the individuals who grew up and lived in it? How could they feel at home in it – and justifiably feel at home? What kind of reality was it that was at once still valid and clearly no longer valid?

At this period Rome seems to have afforded exceptional opportuni-

ties for the development of personality. 'Where competition between great personalities is concerned,' wrote Jacob Burckhardt,
'this is the foremost period in world history in which we find great
personalities competing with one another. What was not great was
nevertheless characteristic, energetic – even if ruthless – and built on
a grand scale . . . Yet all that was great came together in the
wonderful figure of Caesar.'

Did all this arise because there was special scope for action? If so,
there seems nevertheless to have been no scope for structural
change. If there had been, existing structures would inevitably have
been questioned. It seems then that power to act co-existed with
powerlessness to change anything – power within conditions, but not
over them. However, faith in the traditional afforded a measure of security, just as the failure of the traditional gave rise to special
achievements. There were powerful necessities, strong expectations,
undreamt-of possibilities. In this largely comprehensible and essentially controllable world a great deal was seen to depend on the
individual. And often enough he had to resign himself to failure.
This may have led to the emergence of particular types of character.

Be this as it may, how did Caesar come to occupy the position
from which he was able to launch the war at the Rubicon? How did
he become an outsider? Was it a predisposition from his youth? Was
he as great as he is often said to be – and what does that mean? If he
was indeed great – in whatever sense – can his greatness have
resulted from the coincidence of the universal and the particular,
which Burckhardt discerned in all great individuals? When he crossed
the Rubicon, was he acting solely to save himself, his position, his
honour and his security, as Hegel thought, and thereby doing something timely – something that had to be done – because where great
men are concerned 'their particular aims embody the will of the world
spirit'? Or is this a historical fairy-tale?

Could not Caesar's greatness have amounted merely to the individual make-up of his personality, without predestination or higher
significance? Perhaps he could not act otherwise and simply exploited his innate capacities with exceptional *panache* – and, what is
more, with a charm and intelligence unique in his age – by contending with all the problems and tribulations he encountered, acting
powerfully in his own cause and ruthlessly towards the whole – a
whole that had no power to constrain him because he had freed

himself so completely from its reality? And all this at a time when it was possible for an outsider to become so powerful that he could unleash a civil war for his own sake?

2

Caesar's Fascination

THE EUROPEAN TRADITION · DOUBTS
ABOUT CAESAR'S GREATNESS AND
STATESMANLY ACHIEVEMENT ·
FASCINATION AND FAILURE · SOURCES
OF POSSIBLE DELUSION ·
INDEPENDENCE AND POWER IN A 'BASE
AGE' · 'IT IS OUR CONCERN'

BURCKHARDT'S STATEMENT that everything great came together in the wonderful figure of Caesar is just one expression
of a view that had been universally held in Europe for centuries.
The history of Caesar's fame, written by Friedrich Gu dolf, is long
and contains much evidence of the fascination he has exercised.
Hardly any other figure has left such an enduring impression on
posterity.

The Middle Ages revered him as the first emperor, the founder of the
monarchy, from which the supreme secular power in the west takes
its name (*Kaiser*). Caesar, Rome and the Empire seemed to be one and
to take on mythical dimensions. Then, starting in the Renaissance,
came the discovery of the great personality behind the name, with all

its facets: the general, the conqueror of Gaul and Rome's immense empire, the important writer, the great organizer, credited with having reconstituted Rome after a long crisis, the strategist and military leader, the politician, diplomat and heart-breaker, the victor who showed clemency to the defeated, steadfast and daringly nonchalant, a man who was ceaselessly active and seems never to have known failure, for whom any set-back was merely a prelude to further success, who was swift and brilliant in action – until Brutus and his fellow conspirators brought his life to a seemingly tragic end.

It is true that in the century of the Enlightenment and the French Revolution doubts began to grow, as political opinion ranged itself on the side of liberty, of which Caesar had deprived the Romans, and of the republic he had destroyed. His reputation as a general suffered too. 'We have become too humane', Goethe is said to have remarked, 'not to be repelled by Caesar's triumphs.'

Yet against all these doubts a new form of greatness was conceived – historical greatness. Hegel saw Caesar as the executive arm of the 'world spirit', uniting the universal and the particular, for 'what he gained for himself by attaining his at first negative end – absolute power over Rome – was . . . at the same time a necessary destiny in the history of Rome and the world, so that not only his personal advantage, but all his activity, proceeded from an instinct that brought about what was in itself timely.'

Theodor Mommsen then interpreted Roman history as though its whole course had been leading up to Caesar. It has been said that he had never been so powerfully described because he had never been so longingly awaited. Having shown Rome groping in the dark for so long, Mommsen stages Caesar's appearance on the scene as an epiphany. In Caesar historical necessity was at last made manifest. Even 'where he acted destructively, he did so in accordance with the exigencies of historical evolution.' Roman society had lost control over itself; everything was topsy-turvy; society was drifting towards destruction, unable to find a purchase or defend itself, at the mercy of the train of events in which it was caught up. Then came Caesar. According to Mommsen, he secured a position outside this process of drift and was able to gain power over the whole. Through him the Roman order once more became subject to conscious action; its affairs were once more in human hands.

Mommsen set out from the premise that every situation could be mastered through human action, and of this premise Caesar afforded the finest illustration. He ordered 'the present and future destiny of the world. He thus acted more effectively than any mortal before or since.' Mommsen credits Caesar with a 'rebirth' of Rome and Hellenism. By preventing the Germani from overrunning Rome, he won time for Greek culture to permeate the western half of the Mediterranean. Otherwise the foundations 'of the prouder edifice of subsequent history' could not have been laid there.

The man who fulfilled this historical mission was also a perfect human being. 'In both human and historical terms Caesar stands at the point where the great contrasts of existence meet and combine. Caesar was a man of immense creativity, yet gifted with a penetrating intellect; . . . supreme in will and achievement, imbued with republican ideals, yet born to be king, a Roman to the core, yet with a vocation to reconcile and unite Roman and Greek traditions within himself and communicate them to a wider world. In all this Caesar is the whole and complete man.'

Jacob Burckhardt, though far more sober than Mommsen and somewhat critical of him, nevertheless concurred in his judgment of Caesar, whom he called 'perhaps the most gifted of mortals. Compared with him, all others who are called great were one-sided.' Burckhardt also remarked that 'great individuals represent the coincidence of the universal and the particular.' In times of crisis the existing and the new culminated in them. They belonged to 'terrible times, which provide the only supreme yardstick of greatness and also demand nothing else'.

In his own age Burckhardt observed 'a general decline'. He added, however, that 'we would probably declare the emergence of great individuals impossible, were it not for the feeling that the crisis might suddenly shift from the miserable terrain of "ownership and acquisition" and that the "right man" might suddenly appear overnight – whereupon all would follow him.'

This theme was taken up by Gundolf in the opening passage of his book on Caesar's fame (1924): 'Today, when the need for the strong man is voiced, when we are weary of carpers and cavillers and have to be content with sergeants instead of leaders, when – especially in Germany – we entrust the guidance of the nation to any striking military, economic, bureaucratic or literary talent and confer the name of statesman on social clergymen or antisocial generals, on great tycoons or rabid *petits bourgeois*, we would remind the impatient of the great man to whom the highest authority owes its name and has for centuries owed its ideal: Caesar.' Such a conjuration could not produce a Caesar: history did not repeat itself. 'What the future lord or saviour will look like we shall know only when he holds sway . . . But knowledge can teach us what he will not look like.' The historian 'can help to create a climate in which wise deeds flourish by winning hearts and minds for future heroes.' This was the expectation that Mommsen cherished and that informed his suggestive picture of Caesar.

In our own day, by contrast, greatness seems to have lost all credibility. It is no longer a question of whether it is beneficent, but of whether there can be any such thing. 'Spurious greatness', of which Thomas Mann spoke, with Hitler in mind, seems to have made the very concept of greatness seem spurious. If fascination were to make itself felt across the centuries in his case, it would for the time being meet largely with resistance, if not incomprehension.

Moreover, much that was once ascribed to Caesar has become highly questionable. The threat of the Germani, for instance, which Mommsen credited him with removing, did not exist. Above all, Caesar's statesmanly abilities – or rather potentialities – have been increasingly called into question. Whatever great feats of organization he performed as ruler, it is uncertain, if not improbable, that he knew a way out of the profound crisis that faced the Roman republic.

For it may well be an illusion that a truly outstanding individual can master any crisis at any time through the exercise of power, insight and organizational skill. First, however, we have to ask whether Roman society offered any basis for a solution. Rome was after all a republic, and liberty was the central element in the lives of its leading citizens. It faced problems not only of organization, but of reorientation and integration. Not every society is anxious for peace and order if the price of these is monarchy. The crisis could have been overcome only if there had been a disposition, at least in significant areas of society, to support a new order; only then would it have been possible to tackle the crisis from a political centre and thus gain power over existing conditions.

We may presume that Caesar had power only within these conditions. Command of an army, victory in a civil war, comprehensive powers, the love of the masses, a wide circle of friends, wealth, and the ability to fulfil the wishes of many – all these may enable a politician to accomplish a great deal and become irresistible. Yet in order to effect enduring changes in existing conditions he may need quite different forms of power. His will must be able to mesh with the needs, interests and opinions of others, not simply in order to bring about what he desires in this or that case, but to ensure that society as a whole changes and that the new order acquires a degree of self-regulation. Power and authority can help in achieving this end, but legitimacy cannot be engineered or attained by force.

Occasionally the stuff of the new legitimacy may already exist – if many desires, longings and interests are present and need only be mobilized and marshalled. If so, the problem of gaining control over conditions is primarily one for the politician to whom the task is entrusted. This happened in the case of Augustus, but it is doubtful whether it was possible in Caesar's day.

There is at any rate much to suggest that, in the conditions prevailing in the republic, Caesar became all the more powerful as an individual because Roman society held out so few possibilities for overcoming the crisis that afflicted it.

Does this then dispose of Caesar's fascination, or only of a few of the expectations that have been pinned on him, especially in the nineteenth and early twentieth centuries – expectations that have as much to do with a belief in the meaning of history as with a belief in the meaningful role of great men – to say nothing of expectations regarding the extraordinary ability of human beings to master problems by political means, even if this ability is concentrated in one man, a political saviour or messiah, and finally expectations based on peculiarly modern – and gradually obsolescent – demands addressed to the ordered functioning of political systems?

It may be that Caesar appears great in spite of – or even because of – his failure when confronted with the task of reconstituting Rome. Perhaps the circumstances that led to this failure also determined the special development of his personality and were responsible for his successes. These were possible because he was and remained an outsider. Caesar's brilliance and superiority, his serenity and charm, were closely linked with his detachment from the petty, stolid, ineffectual world of contemporary Rome, a detachment that he deliberately cultivated. His freedom and assurance, and the flowering of his talents, became possible only because he stood aloof from the Roman world, at first inwardly, then outwardly. This aloofness developed into opposition, an opposition so radical that it could be settled only by force of arms. Caesar may thus have bought his greatness at the price of his ultimate failure. The fact that Rome afforded no great cause to which he could harness his will may not only have diminished his chances of founding a new order, but challenged him at least to develop his personality and his world in a grandiose and imposing manner, outside established society.

After all, any serious attempt to found a new order – supposing that Caesar wished to do so – would have called for much patience and empathy, many concessions, and great forbearance, as well as a great deal of manoeuvring, calculation, persuasion, and quiet, sedulous activity. And if he had been capable of all this, would he still have been the man whose personality fascinated the European mind for so many centuries?

Can this fascination, whatever its source, still enthrall us? Can we – after Hitler – continue to speak of great men in accordance with time-honoured European tradition, above all when their spheres of activity were war and politics? Can we still be captivated by a man who launched a civil war – after a war of conquest in Gaul – for his own sake?

Yet we must also ask whether we know so much better than all the great minds who, since the Renaissance, have acknowledged Caesar's incomparable greatness, despite every possible criticism of his deeds.

'In the end', wrote Burckhardt, 'we begin to sense that the whole of the personality we find so great still affects us with its magic, across the nations and across the centuries, far beyond the limits of the mere historical record.' Can it be that the judgment of modern times was conditioned by a mysterious magic force that has only recently released us from its thrall – if it has?

While not wishing to discount the magic entirely, we would do better to think of a different explanation for the esteem in which

Caesar was held for so many centuries. Burckhardt himself speaks of a 'sentiment of the most spurious kind, a need for servility and wonder, a craving to be intoxicated by an impression of greatness and to fantasize about it.' He was thinking chiefly of contemporary impressions. Yet might there not be a need for greatness that impels us to search for models, for a yardstick for our own claims, as evidence of their rightness and the possibility of realizing them? If so, Caesar may have occupied a position that answered to the spiritual needs of modern times, so that, while subject to varying interpretations, he could never be dismissed from his role as a paragon of human greatness.

He belonged to a remote age, he was a Roman, and he counted as the first of the emperors. A good deal was known about him, so that it was possible to form a picture of him. Moreover, he was not only a conqueror, but achieved much politically. He was not bound to a particular nation, and he was untouched by current enmities. And perhaps he really was incomparable among those who count as great in the history of public affairs, by virtue of his versatility, his character, and the apparently almost classical development of his personality. And whatever he failed to achieve could be freely accredited to plans whose execution was frustrated by the conspirators. 'You should write about the death of Caesar in a fully worthy manner, grander than Voltaire's,' Napoleon told Goethe. 'It could be the greatest task of your life. The world should be shown how happy Caesar would have made it, how different everything would have been, had he been given time to bring his lofty designs to fruition.' The belief in human greatness, the dream of man's capacity to triumph over everything, could thus be institutionalized in Caesar and fortified against the many doubts that seemed justified in relation to other men. Perhaps this gave rise to a particular susceptibility to fascination.

Yet might there not be something genuinely fascinating about Caesar, about what is reliably reported by him and about him, something that can captivate the observer even today? And is this not the effect of a special greatness that cannot be vitiated even by the experiences of our own century?

Caesar's successes are imposing. No less imposing are the ways in which he achieved them and the abundance of gifts they called for. Cicero praises his intellect, his reason, his memory, his literary and scientific education, his circumspection, his resolve, and his capacity for taking pains. Three generations later Pliny the Elder found him to have been 'outstandingly gifted with mental powers'. 'I will not speak here of his energy and steadfastness or of his sublime ability to comprehend everything under the sun, but of the vitality he possessed and the fiery quickness of his mind.'

Burckhardt calls him 'a marvellously organized spirit, possessing incredible versatility, breadth, and acuity, and combining the utmost boldness and resolution with wisdom and shrewdness'. One is struck by his rich imagination, his immense technical and tactical inventiveness, his amazing ability to assess a situation rapidly and thoroughly, to see through apparent reality and find it deceptive, to perceive the reality that others ignore, to discern possibilities that normally go unremarked, and to be cautiously prepared for almost any eventuality. For he knew the power of chance and did not wish to be exposed to it. Caesar's quickness, the *celeritas Caesaris*, is famous. Remarkable too is the elasticity with which he adapted himself to everything new, his capacity to learn. In everything he did he appears thoroughly manly, at times hard and unyielding, yet at the same time there was a playfulness about him, an almost youthful profusion of potentialities. Burckhardt speaks also of his 'strength of soul, which alone enabled him to ride the storm and take pleasure in it', and of an immense concentration of the will, a ruthless single-mindedness, which he could also communicate to his soldiers. Behind this lay a self-confidence that was accustomed to success, and finally a boundless self-absorption that made it possible for him to launch the civil war for his own safety. To all these qualities should be added the magnanimity he showed to his opponents in the civil war, the much-admired Caesarian clemency. Nor, finally, should we forget the wonderful Latin he wrote, simple, lucid and elegant – not only precise and regular, but highly individual, possessing a brilliance that clearly derived from the way in which he saw and acted.

There are periods, says Musil, when one has only one choice – 'to conform with the baseness of the age (and do in Rome as the Romans do) or to become a neurotic'. In a way the late republic was such a period. Although there were some highly honourable and responsible men among the senators, their policy-making was desperate and feeble. On the whole everything was overlaid with endless self-seeking and unrestrained exploitation of position. The general picture is one of corruption and incompetence, of swimming with the tide. The historian Sallust, a man of high moral standards, blamed society for his own inability to behave as he would have wished: 'Instead of decency, self-discipline and competence, there was insolence, corruption and rapacity. Although I despised these things, being quite untainted by baseness, my insecure youth was nevertheless corrupted, in the presence of such great vices, by the desire for honours and gain and became their prisoner.' Relatively few, while not necessarily becoming neurotic, as far as we can see, felt a compulsion to be negative. And many vacillated between the two.

Caesar, by contrast, could on the one hand operate all the social levers with bravura in order to secure his own advantage, and at the same time develop an inward independence, a serene and rather disdainful aloofness. As an outsider he built up a power base that ultimately enabled him to challenge the whole of Rome. Unable to forge any real links, he stuck tenaciously to his course and continued to rely on his own resources. This accounts for the extraordinary freedom he enjoyed. He could find no *raison d'être* in the society he belonged to, and to this extent his position was subject to chance. Freedom requires a bond – what Sartre called 'the choice of a goal in the service of the past' – and this Caesar found in the old Roman ethic of achievement. It provided him with a yardstick, admittedly outside contemporary society – which was no longer the old society he belonged to. He evolved exalted standards and measured his peers against them, in order to perceive their shortcomings all the more clearly. He, however, measured up to them, isolated though he was, beset by the greatest dangers, obliged to meet the highest demands, and sustained wholly by his own energy. With a nonchalance that was no doubt aesthetically attractive but ethically dubious, he disdained to conform with the discipline of his class, which had once endorsed the old achievement ethic, and set his own personality above it. For because he was isolated in Roman society, because there was no cause to which he could commit himself or in whose name he could act, he had no choice but to build up his own position. And since the Roman empire was so immense, since Rome's ruling class was the most powerful the world has ever known, he was able to create his own world. Here he could give free rein to all the potentialities that lay within him – but at the price of no longer being able to take his place in Rome. The dynamism with which he confronted society became ever more monstrous and demonic.

The way in which Caesar played this game – risking his very existence and then raising the stakes, seeking out immense opportunities, finding them and savouring them – affords an absorbing spectacle. The costumes are historical and belong to Roman history. And the drama is set in an age that saw a relaxation of the constraints that had once held Rome together in the tension of force and counterforce. An abundance of power, which Rome had accustomed itself to producing during centuries of dull, disciplined solidarity, was now released. At the same time the city was becoming more and more receptive to the sophistication of Greek culture. All this suddenly came together in such disparate figures as Marius and Sulla, Cato and Caesar, Pompey, Crassus and Lucullus, Cicero and Brutus. Nor should we forget the brilliant escapades of the generation that grew up in this ambience, aping what they observed in the others, yet ignoring the scruples that still occasionally assailed the objects of their emulation. Nor should we forget the ladies of the Roman aristocracy, who were begining to acquire a taste for freedom, for culture, even for power, and becoming susceptible to them. Where else has there been such an imposing array of power and personal – though not institutional – brilliance?

If this period and its most significant protagonist is still able to fascinate us, this is because what we see enacted is essentially our concern too. We see it in all its seriousness. Beside and within the historical dimension the anthropological is always present.

In his *Essays on Caesar* Otto Seel remarks that Caesar's greatness – 'if one dare use this emotively charged word' – lies 'neither in the purity of a radiant genius nor in the licence of the emancipated immoralist . . . but in his problematic humanity, his possible splendour and unavoidable misery, the unhappiness he endured and the guilt he incurred, but above all in his historical activity, which both achieved and destroyed so much'.

Seel speaks of an 'interplay between the compulsively fascinating and the disturbing, between the *charisma* and the *daemonia* that must have emanated from this man, whom hardly anyone could resist –

from the simple legionary to the most august member of the nobility.' But is not fascination compounded of both these elements? Does not the impression it produces inspire both pleasure and awe? Does it not attract and repel, yet continue to enthrall?

'Not every age', says Burckhardt, 'finds its great man, and not every great talent finds its age. There may be very great men in the present age, but they have no causes.' When he goes on to say that great men – at any rate before our own age, which he finds dreary and oppressive – belong to times of crisis, we are bound to ask what kind of crisis it was in which Caesar grew up, a crisis in which there was no cause to which outsiders could attach themselves, in which society did not break down into political oppositions, but simply – or even – allowed a new reality to emerge beside the old.

3

Crisis and Outsiders

THE CRISIS OF THE LATE REPUBLIC was in many respects very curious, combining grave and at times bloody unrest with great stability, and frequent failure of the political order with a universal conviction that it was the only true one. The binding force of tradition was unanimously acknowledged; it did not necessarily determine men's actions, but it governed their thinking. And apart from the normal discrepancy between what one does and what one ought to do, there was no contradictiion in all this, for the failure of the system was not perceived as such. What was perceived were attacks on the existing order – this was how attempts at reform were generally interpreted in the Senate – and perhaps also one's own failure to match up to what needed to be done. Both were interpreted in moral terms. Hence no one was in any doubt about the validity of the inherited order.

26

Everyone, the reformers as well as their opponents, wished to preserve it, and in spite of this – indeed, because of this – they gradually destroyed its foundations. This desire to preserve the traditional order strikes us as quite remarkable, for in our day it is common for intellectuals, and at times even politicians and certain sections of society, to strive for something quite different, something new. Judged by our standards, Roman reformers seem conservative and their aims quite modest. In Rome, however, it was impossible to conceive of a different order. Hence, differences that strike us as trivial could appear enormous. The basic form of the republic was never questioned. For this reason, if for no other, the leading senators identified themselves to such an extent with the republic that they saw any plan to restrict their powers as an attack on the republic. They therefore defended it without its being attacked, and by doing so they seriously endangered it.

Some crises arise through the emergence of a new force that mounts an assault on the existing order. Others arise simply because they are apprehended as such. This is not just a question of what may be called 'crisis psychology'. Rather, if the demands made on the system are so great that those who operate it can no longer cope with them, many are tempted to infer that the system has failed. The demands may become so persistent as to exacerbate the crisis. Attention is then focused on evolving something different, something better; any disappointments are therefore bound to lead to doubts about the system, not about the demands.

What characterized the crisis in Rome, however, was the fact that the demands were limited to ensuring that tradition was properly observed. And however many citizens suffered hardship, however many attempts were made to remedy this or that grievance through legislation, no solid opposition emerged that outlasted the particular situation – no party of reform, no comprehensive programme in which the grievances came together and generated a common policy. Occasionally social and economic problems might be brought into politics, but not permanently. They did not cause a split in Roman society. Politics could thus relate only to limited individual questions of greater or lesser moment. The whole remained all-embracing, and all remained involved in it. Hence, those who concerned themselves to any great extent with particular problems were likely – indeed bound – to be outsiders. At first it was the great tribunes of the people

27

who attempted, with the support of the popular assembly, to push through reforms against the will of the Senate; foremost among them were the Gracchi, Tiberius (133) and Gaius (123/2). For a time they were able to mobilize massive support. Yet it produced no new political groupings, no opposition, that survived beyond the immediate situation.

It was the situation that determined the political positions. In consequence, very few could become outsiders. Those who did were strong, imaginative men, mostly alienated from their class by bitter experiences, who felt they owed something to themselves and were prepared to draw conclusions from their insights. There was also a number of others who for a time played at being outsiders and warmed to their role, but could not sustain it. They could not identify themselves with a particular cause. Indeed, one had to develop an exceptional degree of self-sufficiency if one wished to prosecute demands that were at variance with tradition. In this context a cause was any interest one decided to embrace. There were many causes waiting to be espoused, and no dearth of watchwords. However, there was no anti-senate cause that found a resonance in the opinions of large sections of the citizenry and offered any prospect of power or orientation. Moreover, the Senate was sufficiently strong and resolute to ensure that any outsider was embarking on an extreme course that usually ended in death – at least before Caesar's entry into politics.

All this is somewhat surprising, for the system seems to have had many shortcomings. Almost the whole of the Mediterranean world was governed through institutions that the Romans had evolved for the limited cantonal conditions of an earlier age. Moreover, the body politic had greatly increased, and many cities throughout Italy – including, since the eighties, all those south of the Po – possessed Roman law; yet all popular assemblies continued to be held in Rome. A prosperous class numbering tens of thousands had grown up, yet politics were still controlled by an aristocratic élite. Until the eighties there were three hundred senators. The structure of the body politic was changing. Economic, social and political problems arose that in the course of time weighed heavily on the city.

Originally it was the farmers who had made up the nucleus of the citizen body. It was they who had waged Rome's wars, since military service was restricted to landowners or members of landed families, and a certain minimum income was required. However, the importation of cheap grain from overseas and the partial switch of Italian agriculture to new methods of production had made the economic conditions less favourable to the farmers. When wars had to be waged in increasingly remote parts and above all lasted longer, many families ran into difficulties, as the women and children were unable to till the land. Many farms had to be mortgaged or given up. In the years before 133, a long and costly war had been fought in Spain. Tiberius Gracchus, a tribune of the people, had declared, 'The wild beasts that live in Italy have their caves. Each knows where it can take refuge. But the men who fight and die for Italy have nothing but air and light. They roam the land with their wives and children, homeless and hounded.' The decline of the peasantry posed yet another problem: where would Rome find its future soldiers? Recruitment was already proving difficult.

The consul Gaius Marius created another problem when, in the levies of 107, he took on anyone who presented himself for military service, thereby fostering an expectation among the soldiers, most of them landless country-dwellers, that they would be given plots of land when they had completed their service, which was usually of brief duration. The land he had in mind was mainly in the newly won provinces. This aroused fierce opposition in the Senate, for according to traditional aristocratic thinking such veterans were bound to feel beholden to whoever provided them with land. Marius thus gained so much power that it was feared he would no longer be bound by oligarchic equality. The Senate therefore opposed these land laws, no matter where it was proposed to settle the veterans. The soldiers consequently had to rely on their commander, and this led to serious political conflicts. Above all, the fact that the soldiers might feel a greater obligation to their commander than to the Senate provided an essential condition for possible civil wars. True, a great deal had to come together before the possibility could be exploited. But this happened as early as the eighties, long before the civil war of 49. At all events, the unsatisfactory integration of the Roman armies into the republic was the most serious problem of the age.

Moreover, during the second century the population of the city had grown to several hundred thousands. Many immigrants came from the country and the whole of the Mediterranean world; but above all many former slaves automatically became Roman citizens on being freed. Many craftsmen, clerks, tradesmen and money-changers – often highly-skilled men – came to Rome as slaves and were often freed after a time if they were successful, while many others gained their freedom through the testamentary dispositions of their masters. The lot of slaves in the ancient world was extremely varied.

There was hardly enough employment for such a large urban population. Housing was scarce, poor and expensive. There was much hardship and frequent problems with food supplies. In 123 Gaius Gracchus attempted to improve the lot of the poor by bringing in a corn law that guaranteed cheaper food. He also had large silos built for the storage of grain, so that plenty would be available even if the price increased. This law is said to have encouraged a further influx of immigrants and an even larger number of emancipations. This created a source of unrest that could lead to serious disturbances in the political life of the late republic.

Far more prosperous, influential and problematic was the large class of the knights, so called because they fulfilled the equestrian census and could therefore take to the field with their own horses. The members of this class were mainly big landowners and included the nobles of the Italian cities. Some new senators were recruited from their ranks, for in every generation a number of knights embarked on political careers. Many were engaged in business, as wealthy merchants, bankers or *publicani* (a *publicanus* being a person responsible for collecting public revenues and taking public contracts).

The *publicani* were a significant political force. For the Roman republic conducted most of its economic and financial business – the execution of public contracts, the exploitation of mines, the collection of duties, and above all the gathering of provincial taxes – not through officials, but through private agents. As such activities were extraordinarily lucrative, this group within the equestrian order was commensurately wealthy. While all the other knights could participate in politics only individually or as sections of the public at large, the publicani sometimes had shared interests and were well organized. The Senate and the magistrates no doubt showed them as much consideration as possible, but for a long time they treated them with a certain disdain. It was only in 123, when Gaius Gracchus significantly enhanced their status in relation to the Senate, that things began to change. Even then, the *publicani* never disputed the Senate's leadership, though they made various attempts to influence it and did not recoil from fomenting conflicts about it. When acting in concert, they were the strongest counterforce to the Senate, and more than once the tribunes of the people sought an alliance with them. Hence, on a number of occasions, they helped to impede and to some extent weaken senatorial leadership. True, they usually supported the Senate in difficult situations, but in doing so they were not really acting inconsistently, for at such times they and the Senate had a common interest, and they usually exacted concessions in return. They favoured the preservation of the political order, but were generally opposed to its being administered too vigorously. It was its very weakness that endeared it to them.

It is true that nearly all the economic and social problems of the late republic – the period that began in 133 with the tribunate of Tiberius Gracchus – could be settled within the framework of the existing order, provided that civil war did not erupt, but in many ways they helped to erode it. The aristocratic solidarity on which it rested began to loosen. Fierce quarrels arose, in the course of which important institutions lost some of their authority. This led to the break-up of the probably unique combination of variety and cohesion, elasticity and firmness, mobility and solidity, that had characterized the classical republic. The practice of government and the formation of public opinion became more rigid. As a consequence, new problems arose – in the field of foreign policy for instance – and greatly increased the burden on the aristocracy and the republic as a whole.

The crisis of the republican order drew sustenance from many roots, but it was in the political sphere that it developed its dynamism. It was thus in essence a crisis of the order itself and the aristocracy on which it rested.

The Roman constitution had not been laid down once and for all, but was an organic growth. This meant above all that there had never been a divorce between the social and the political constitution of the republic. The middle and lower classes, for instance, had never detached themselves from the whole to the extent of creating a purely political order and setting it against the social order. It is true that the broad mass of the population is always potentially more powerful than the aristocracy, but this potential can be realized only if it is embodied in political institutions, so that it can be brought to bear permanently and not just at moments of rebellion. It is therefore necessary to organize something in the political field that can be set against the social conditions. This had happened in Greece when democracy was created in an aristocratic society. It did not happen in Rome. Nor had a monarchy built up a state apparatus separate from the social order, as happened in modern times, when the state, as it were, transcended society.

In Rome the political order had evolved out of the social, and despite a number of changes the two remained essentially congruent. In its internal structure the Roman republic was no more than the sum of the organs, conventions, precedents and laws that had shaped society politically and made it capable of political action, after which it was partly modified by further precedents and laws and by the development of new principles, on the basis of a certain shift in power relationships and a change in political morality. Much was left open, yet the whole was remarkably cohesive.

For the overall conditions were very stable. There was a broad consensus as to what was proper and what was not. This consensus made for a powerful public opinion. The commonwealth was kept on an even course with a minimum of prohibitions. The guiding principles were generally supplied by tradition, by the custom of the ancestors (*mos maiorum*), but they were not inflexible. Because men had confidence in the inherited order and their own capacity to deal with any emergencies, it was not necessary for competences to be strictly circumscribed. In this way many things that, taken in the abstract, would have been mutually contradictory, were in fact compatible.

This applied above all to the institutions that Rome had preserved from the class conflicts. At that time the plebs, the large section of the population that was at a disadvantage *vis-à-vis* the patricians as regards civil liberties, political rights and economic conditions, had created a few instruments with which to defend its interests: the tribunes of the people, who were supposed to act as perpetual custodians of popular interests, and the plebeian assemblies, which asserted their will by means of resolutions. The tribunes gradually won a number of rights, above all the right of veto. Finally, in 287, the patricians conceded that resolutions of the plebs should have the force of law. There was no restriction on the right of legislation or the right of veto. It clearly occurred to no one that they could have been grossly abused. And if this had been conceivable, the solution could in any case have been sought only in concerted opposition. For the Romans lived to a quite extraordinary degree in the present.

Politics were conducted to a large extent in public, admittedly among the nobles, but within sight of all who were present at the forum or could be induced to attend. They were thus subject to various controls, and the scope for secret intrigues was thereby reduced. Many people took an interest in politics, were involved in them, and had time for them. The spatial presence was matched by the temporal present: men's minds were seldom distracted from the present to the past or the future; the Romans lived in a broad band of present time, which they apprehended as largely unchanging. They recognized few differences in the temporal dimension. In the old days many things had been better, and there was a danger that in future many might become worse. Of this they were certain. In spite of this – or rather because of this – the guiding principles were always the same. No one was relativized in temporal terms: the old could not be regarded as superannuated, nor the young as modern. The young were at most frivolous, the old at least authoritative and powerful. There was no question of moving from a past that had been different into a future that would be different again. With respect to the order there was no 'still' and no 'already', only an unchanging present, which had to be preserved and perhaps consolidated. This was connected, as both cause and effect, with the psychological fixation of the whole of society on the present. Men differed as little in their views on the common order as they did in their temporal orientation. This consensus determined their attitudes and their actions. There was thus a common understanding of the intention behind every institution. The less precisely its limits were formulated, the more clearly they were drawn in the minds of the citizens, at least during the classical period. Had a Greek asked a Roman at that time on what matters the popular assembly might decide, the Roman would prob-ably have replied naïvely, 'On all matters.' If the Greek, surprised by this answer, had then asked whether it could also remove magistrates from office, the Roman would presumably have been equally sur-prised and said, 'Of course not.' He would scarcely have understood

33

how such questions could have occurred to the Greek. And if anyone did not automatically know what was proper and what not, the majority would unite against him, and its success would constitute a new precedent. In this way it was gradually established that there were certain limits within which the tribunes of the people could properly exercise their rights of legislation and veto; these limits were elastic, but never breached.

Given the cohesion of the Roman aristocracy, the notion of abstract regulations, of relying on fixed laws in dealing with crucial questions, can hardly have recommended itself. For there were no conflicting claims with respect to the order; nor did the Romans live in a specialized society whose structure was so opaque as to require abstract regulations. Hence, a formal constitution was impossible. Matters were viewed in concrete terms, from the perspective of the legally constituted citizen body, not by reference to a system of laws that could be imposed upon it.

In all this an important role was played, consciously or unconsciously, by the fact that all power in Rome was concentrated in one place: in the Senate, and especially in the hands of its leaders, the *principes*. The Senate was the supreme authority. In it the general judgment of the citizens was distilled and could be effectively formulated and represented.

The Senate had responsibility for the commonwealth. It determined foreign policy, received embassies, concluded alliances, and made decisions on war and peace – though the popular assembly had to be formally consulted. It ordered military levies, appointed military commanders and provincial governors, and issued directives for the waging of war. It settled disputes between cities and ruled on all important and many unimportant matters of policy and administration.

The magistrates acted for the most part on instructions from the Senate, though they were repeatedly tempted to enlarge their scope for action. At least the senior magistrates – the two consuls, together with the praetors, of whom there were at first six, then eight, in the late republic – were in principle free to act on their own initiative. In practice, however, they regularly agreed their policy with the Senate, if necessary by way of compromise.

Admittedly the Senate could not always reach unanimity. Since the major disputes – which might affect the traditional rules – arose mainly from extraordinary claims by single nobles or whole families, many senators were involved in individual suits by virtue of kinship or friendship. Moreover, it could easily happen that the nobles in question represented the people in the form of the electoral or legislative assembly, or even, in certain circumstances, the special interests of wider circles. It would then be difficult to resist them. After all, the popular assembly could decide on any proposal put by a magistrate. And the magistrates might make generous use of their freedom of action in order to help their kith and kin.

To deal with such conflicts a highly practical policy was evolved in Rome. As a rule the Senate first let the various initiatives take shape. These naturally came up against opposition, sometimes against the tribunician veto. If necessary the leaders of the Senate seem to have persuaded the tribunes at least to threaten to use their veto. For from the third century onwards the tribunes only rarely ranged themselves against the Senate and the magistrates in support of plebeian interests. On the whole they acted within the framework of the prevailing aristocratic groupings.

At all events, the upshot in difficult cases was that the opposing parties agreed to leave matters to the Senate. It could then arrive at a rough estimate of the power relationships by observing the interplay of the opposing forces. In accordance with time-honoured custom it did not allow matters to come to a head. This would have conflicted with its strong sense of reality, its commitment to the practical and the possible. If the promoters of a particular initiative were strong, the Senate majority would be inclined to compromise and make concessions, thereby ensuring that it retained the power of decision and that the dispute did not generate precedents that would increase the power of the magistrates or the popular assembly. At times the Senate sought not only a practical compromise, but an agreement that no similar claims would be raised in future. To this extent the majority was at least able to reach agreement. Flexibility over the matter in hand and the final achievement of solidarity in 'constitutional policy' ensured that the Senate's authority was not over-stretched and that it was always preserved and handed down.

In this context one may thus speak of 'state wisdom' or an 'instinct for government' on the part of the senatorial order. At any rate we observe an extraordinary capacity for reaching agreement when the maintenance of order was at stake. It is clear that, in the face of

35

conflicts and irregularities, a series of successful decisions had created clear rules and powerful positions that enabled the majority to go on acting cohesively. In particular, the leading senators (the former consuls or *consulares*) came to assume a special responsibility for the interests of the whole – commonwealth and nobility. This was the basis of their authority. And it is likely that at least a majority of them regularly addressed themselves to this task. Thus many matters could be left to the free interplay of forces, whether these were individuals or families, and at the same time there was the assurance that they would be confined within narrow limits. In important questions, most of the leading politicians refrained from acting in a partisan fashion.

Behind this, however, stood the extraordinary power of the Senate, which ruled over a growing number of territories and ultimately the whole world. The world was subservient to the judgment of the fathers (*patres*), as the senators were called. Their judgment was not necessarily powerful because it was right, but it was right because it was powerful. One reason for this was that the Roman oligarchy did not insist that everything should run according to plan, only that it should run relatively smoothly. Its régime was thus proof against disappointments, and secure. Accordingly, the Senate remained in control of events in the city. Magistrates and popular assemblies found this all the easier to accept as their occasional claims and grievances received due consideration. Moreover, it was rare, since the class conflicts, for the people to assert itself politically against the nobility. The citizens had many ties with individual nobles and noble families, and these usually governed their actions. Only minor matters were at issue, and the citizens grouped themselves according to their allegiances. This was of course possible only as long as the points of controversy remained limited.

The Roman republic has always been admired. The wisdom of its Senate and its internal order have earned the Romans the reputation of being politically a specially gifted people. Yet every constitution has a certain capacity; none is capable of dealing with oppositions of every kind and magnitude. The Roman constitution relied on the fact that territorial conquests had for a long time made it possible to satisfy numerous interests – the farmers' interest in land, for instance – and so distract attention from politics. Whatever conflicts remained might at times be fierce, but they concerned only a minority.

However, the great problems that arose in the late republic put the system under strain. It had been able to function only because controversy was limited. This first became manifest in 133, when the plight of the impoverished and dispossessed farmers prompted Tiberius Gracchus to petition for a land law. It had very strong popular support, but the Senate majority was implacably opposed to it. Another tribune of the people entered his veto. This was clearly at variance with the nature of his office and the right of veto: one might intercede against many things, but no one had ever done so against a powerful plebeian interest. And land laws were traditionally among the most important subjects of tribunician legislation. By treating his right of veto as absolute, the tribune flouted the principles of the inherited order. Tiberius Gracchus then went a step further and demanded that the tribune be removed from office. This violated not only the right of veto, but the vital principle that magistrates could not be removed from office. After further irregularities, Gracchus concluded that he could protect himself from the threats of his opponents only by applying for a second term as tribune, whereupon the Senate majority judged that the system was in jeopardy. The tribune of the people was killed in an act of lynch justice. That a serving magistrate should immediately apply for a fresh term was deemed illicit. After his year in office it had to be possible to call him to account. On the other hand, the tribunes of the people had long been sacrosanct, the plebs having sworn that whoever harmed one of them was guilty of a capital offence. The Senate had now violated this tradition. As soon as the solidarity of the nobility no longer sufficed to bridge all the oppositions – or the oppositions were too strong to be absorbed within this solidarity – the inherited institutions were open to any abuse; they began to be dismantled.

Ten years later, Tiberius' brother Gaius had himself elected tribune. He had a great programme of reform, the most comprehensive ever drawn up in the Roman republic. In addition to further land legislation, the first corn law, and a number of measures designed to consolidate civil liberties and make Senate policy more objective, he attempted to raise the status of the knights and increase their participation in politics. He introduced a bill under which they, not the

37

senators, should serve in the jury courts that tried cases involving *pecuniae repetundae*, moneys extorted from the provinces. Gracchus wished to end the blatant exploitation of provincial cities by their senatorial governors. Yet the effect of his law was much more far-reaching: the equestrian order was to exercise control over the senatorial. In a sense it was to take over a function that had once belonged to popular jurisdiction but had long been removed from it. This put an end to the undisputed leadership and responsibility of the Senate; the knightly class was politicized and the ground laid for many future disputes.

No one entered a veto against Gaius Gracchus. He had powerful forces on his side, and by now everyone had probably learnt caution. Nor was he barred from a second term as tribune, which he seems to have obtained by adroit exploitation of an old dispensation. In his second year, however, he was faced with a rival who brought in even more popular laws. Moreover, the Senate drew the knights on to its side by conceding that they should take over the jury courts. It was thus pursuing its 'constitutional policy' – making practical concessions rather than allowing dangerous precedents. For Gaius Gracchus, by proposing substantial reforms without reference to the Senate, indeed in defiance of it, had broken an unwritten law. This must on no account happen again. The knights found that in the long term the Senate was more powerful than they; they probably also thought it the proper body to govern the republic. Finally, Gracchus made a number of mistakes; he failed to be elected tribune, and the following year a proposal was made to repeal one of his laws. The reformers were set upon disrupting the popular assembly that was to decide the issue. The atmosphere was extremely tense. Everyone was nervous, excited, and suspicious; fear of violence was compounded by a desire to protect oneself against it and if possible prevent it. Each side was increasingly dominated by its more radical supporters. It was the hour of the zealots and hotheads, but also of the resolute spirits who wanted their cause to triumph. It required little to provoke violence, and it seems as though Gracchus' supporters were the first to resort to it. Then the Senate struck.

After the experiences of 133, the senators had prepared the ground for just such a situation. They first passed the 'Senate resolution on the defence of the republic', which came to be known as the *senatus consultum ultimum*. The consuls were to ensure that no harm came to the *res publica*. This meant the use of unrestrained police power, if necessary without regard for civil liberties. This resolution later became the extreme instrument of Senate policy, its *ultima ratio*, as it were. True, it was effective only if good Roman society could be counted on to support it. It was based on the age-old legitimacy of the Senate's leadership, but was at the same time an expression of its blatant partisanship: the senatorial order could not tolerate any individual who became too powerful.

A contingent of Cretan archers, the sharp-shooters of the ancient world, 'happened' to be on hand. But the consul relied chiefly on the knights. As the consul's forces closed in, Gracchus wanted to take his own life, while inveighing, it is said, against the ungrateful people of Rome. Most of his remaining supporters, on being promised impunity, had defected to the other side. Friends persuaded him to flee. When his opponents caught up with him he had himself killed by a loyal slave. According to another account he was killed by his pursuers. Someone cut off his head, intending to take it to the consul, but a friend of the consul wrenched it from him. When he delivered it, scales were sent for, as it had been announced that Gracchus' head was to be weighed in gold. It is said to have weighed seventeen and a half pounds after the brain had been removed and the brain-pan filled with lead. The bodies of Gracchus and his supporters, said to have numbered three thousand, were cast into the Tiber.

After this second attempt at substantial reform had been put down with enormous bloodshed, much changed in Rome, thanks not only to the brutality of the consul's actions, but to the Senate's successful stand in defending him against all attacks. Its opponents prosecuted him in the popular court for violating the law on the liberty of the citizens, according to which no magistrate might kill a citizen without the prior verdict of a court. The court acquitted the consul, upholding the claim that he was in duty bound to protect the republic from harm and might if necessary disregard this law. No less significant were the proven unreliability of the knights and the weakness of the broad mass of the citizens. There was clearly no force that could be relied upon to check the power of the Senate or in whose favour the republic could be effectively reformed. Hence, no one was willing to repeat Gracchus' bold and imaginative enterprise. There were only two more attempts at extensive reform, and both were in favour of the Senate. The very thoroughness and consistency of the Gracchi's insights demonstrated the futility of pursuing far-reaching modifications of the inherited order, and the fickleness of the knights discouraged any potential successors.

However, the Gracchian reforms did lead to a relaxation of class discipline and a greater diversity in political life. A significant role came to be played by the so-called *populares*. This term designated the practitioners of a political method known as *populariter agere* ('acting in a popular manner').

Individual politicians, usually tribunes of the people, agitated against the Senate – and perhaps against the magistrates – with the aim of pushing through the *comitia* bills that had no chance of success in the Senate. This was possible because the popular assembly could resolve upon anything. Laws counted as the highest expression of the communal will, and against them the authority of the Senate was nugatory. Before 133 it had occurred to no one to lay large numbers of anti-Senate petitions before the people, and of the few that were introduced most had come to nothing. Hence, the potential for competition between the Senate and the popular assembly had scarcely been realized. After the Gracchi, however, fairly regular attempts were made to use the assembly as a tool against the Senate.

One device was to enter specific and repeated complaints against the supposedly arbitrary, arrogant and self-serving conduct of the fathers, who were portrayed as a small clique that resorted to numerous machinations and treated the commonwealth more as booty than as a charge, restricting the rights of the people and even threatening to abolish them. The *populares* claimed that they wished to restore ancient Roman liberty. They enjoined the citizens to overcome their lethargy and embrace the cause of the republic – in other words to do what the *popularis* in question wanted. A tradition grew up in which the Gracchi were extolled as martyrs and other forerunners invoked as models. A new political trend seems to have developed over the years. The method also involved various proced-ures that could be employed to mobilize a mass of supporters and push through laws against powerful resistance. Certain groups emer-ged who were instantly available – for a consideration – to act as claqueurs and voting cattle in support of 'popular politics'. Apart from this core of supporters there must have been a wider circle –

how wide we cannot tell – that was potentially susceptible to popular agitation, but could for the most part be mobilized only with difficulty and on special occasions.

However, enough members of the urban populace were obviously inclined to favour opposition to the Senate. They were not necessarily opposed to its rule; it is indeed unlikely that they were – not even the tribunes of the people opposed it. Yet they might be incensed by the manner in which it was exercised, and by the failure of the present senators to live up to the example of their predecessors. They might at times be offended by the lordly bearing of the leading nobles, just as at other times they were impressed by it – as they fundamentally expected to be. The atmosphere of the city was heavy with politics. The mass of the population was oppressed by poverty, though periodically diverted by games and public largesse. In the popular assemblies its members were addressed as the masters of the world, for the rhetoric of the time equated the small crowd that gathered there – mostly to no purpose – with the Roman people. Rome's labourers, traders and artisans were often unable to resist the call to take an active part in politics, to put things to rights, to pass laws that seemed long overdue and to assume a responsibility that was far beyond their capacity. They valued their liberties, the right to physical inviolability, the secret ballot and much else – rights that the nobles did not always fully respect. Occasionally the leading senators might flatter them – if they clearly needed them. The *populares*, on the other hand, wooed them constantly. They, being in opposition, had far less power. Only they could be expected to represent the people's interests. In case of doubt, then, there was a strong disposition among the plebs to support those who played the popular role. However little this produced in the way of a consistent policy, the senators could not buy the urban populace – or only at great expense, and not for a cause that conflicted with the usual claims of popular politics. To this extent there was indeed an opposition between Senate and people.

Yet popular action produced no political camp in which ambitious nobles could find a home. At best it afforded them a staging post on their political journey.

One politician after another played this role. Yet however similar the lines they spoke, they did not constitute a group with respect to

41

their political aims. As *populares* they had no common cause –
except insofar as they were all obliged to oil the popular apparatus in
order to make use of it.

This strikes us as very curious, accustomed as we are to modern
political parties. Was there no potential here for successful opposition
to the Senate? The urban population numbered many tens of
thousands. Even if many of them lived quite modestly or even in dire
poverty, could not their support have been marshalled in order to
push through any popular decision?

The truth is that popular politics won few major successes, and
even these were due to the interest and involvement of powerful
minorities outside the urban populace. Scarcely anything was undertaken to benefit the city-dwellers. At most there were occasional
attempts, when important measures were planned, to win its support
through laws reducing the price of grain. It seems an astonishing
situation, and astonishing arguments have been proposed to account
for it – for instance, that the urban poor were induced by donations of
various kinds to become clients of the senators. Yet clearly this
clientèle was not very effective politically on any particular occasion.
And what apparatus would have been available to the senators, had
they wished to control the voting behaviour of their clients among so
many thousands or even tens of thousands? The true reason for the
weakness of the urban populace was, in the first place, that their vote
in the popular assembly had little weight. Most of them were
assigned to four out of thirty-five tribes, and the total result depended
not on the individual votes, but on the votes of the tribes. Moreover,
the power relationships within the citizen body were determined
chiefly by the prosperous sections of the population. Since the right
of suffrage was graded according to census, it was on these sections
that the subsequent career of the tribunes of the people depended.
Some of them may have approved this or that popular action, but
they would never have tolerated an upgrading of the urban populace
– that is to say, a policy designed to serve its long-term interests, and
in particular its economic interests.

Finally, how was one to rely, save in specific cases, on an unpredictable mass? How would it have been possible, on such a basis and
with no bureaucratic apparatus, to govern an empire that extended
over the whole of the Mediterranean world? Admittedly, the different parts of the empire were largely self-governing, but countless

decisions had to be made at the centre. This would have been feasible only through a demagogic or plebiscitary tyranny. All in all, the notion of politics based on the people rather than on the Senate was presumably quite inconceivable. Hence, the poverty of the urban masses was not the object of popular politics and is unlikely to have motivated them. On the whole they could serve only as a means to various ends, and these ends were not those of the citizens at large.

We can understand such politics only if we bear in mind that party groupings in the Roman republic depended on particular issues and shifted accordingly. The Romans were unacquainted with the exigencies of modern times, which have produced parties that assert themselves cohesively in a variety of fields, so that almost all political disputes are fought out between them. In Rome no such crystallization of interests was possible. Political life was determined by the fact that in general only quite isolated interests figured on the agenda – whether at election times, when there were different candidates to choose from every year, in Senate negotiations or in court hearings; only rarely did large groups come together in support of common interests. This was due to the limited scope of contemporary politics. There were no taxes, and therefore no disputes about them; the economy was not a matter for politics; social problems seldom came up for debate; and neither education nor religion were political concerns. Even foreign and military affairs only rarely posed significant problems in the late republic. Yet because the attitudes of the senators on the countless questions that arose were determined by personal ties, they continually realigned themselves according to the matter in hand. Of crucial importance were the power and influence they commanded in the Senate, and these in turn depended in large measure on their attaining high positions in the magistrature.

Since the urban plebs had little influence on elections to the highest offices, and since large groups – of knights or veterans, for instance – rarely asserted themselves with the help of the popular assembly, such help was as a rule just one factor among many. This or that individual might canvass popular support, but this support was seldom crucial. Opposition between Senate and people constantly recurred, but tended to remain on the periphery of politics. Certain politicians might continue to support popular agitation, but they were not among the most ambitious, as such activity brought

43

insufficient influence. It was only in Caesar's day that things changed somewhat.

In a crisis, then, in which much has gone wrong, but the victims of hardship are too weak to organize themselves politically, when the discontented have no power and no ideas regarding the political order and all those who have so much as a prospect of power are contented – in such a crisis outsiders cannot bank on the security of a consistent opposition. They may of course present themselves and even win power, but their scope for achievement is restricted, and only a few can break through the restrictions.

Tiberius Gracchus, born into one of Rome's most illustrious families, was alienated from the Senate by a chance event. As a young magistrate he had served in Rome's Spanish army when it was ambushed and encircled. The Spaniards were willing to grant life and liberty to the thirty thousand soldiers in return for independence from Rome. Yet they refused to deal with the consul, for on a previous occasion they had released a Roman army after negotiating a similar treaty with a consul, and the Senate had then refused to ratify it. They trusted Tiberius Gracchus, however, because his father had once concluded a just treaty with them and seen that it was honoured. This had led to a client relationship with the Gracchi. Tiberius then concluded a peace treaty on which he, the consul and all the senior officers swore an oath. In this way the army was saved. But once again there was no majority in the Senate, and the treaty became null and void.

This deeply offended Tiberius' honour, and probably his sense of decency too. He was a proud man with high standards, about thirty years of age, and having once come to doubt the Senate's wisdom and responsibility, he was ready to take a serious stand on a particular issue over which it had procrastinated for some time – the plight of the landless peasant soldiers and the recruitment problems that had resulted from it in recent years. He therefore presented his land law, which he was resolved to carry against all opposition. He thus developed a novel kind of political claim – the conviction, quite unfamiliar in Rome, that an individual had the right to challenge the Senate. In the conditions obtaining in Rome at the time, this con-sciousness of being right when the others were wrong appeared positively sinister, especially as it stemmed not from personal achievement, but from criticism of the Senate, from the pride and sense of superiority of one man who, unlike the rest of his class, knew himself to be truly answerable for the republic. Senatorial resistance could only strengthen his conviction that the problem could be solved only by him, and that it must be solved now – even if this meant breaking the law. The fact that Tiberius' intentions were in fact conservative and that he aimed to improve the condition of the peasantry, the traditional basis of the Roman army and the Roman citizenry, was quite irrelevant. He had long since become an outsider, and, lest such action against the law and senatorial responsibility should be emulated, he had to be eliminated.

Since the attack on Tiberius Gracchus had originated in the Senate and been approved, at least subsequently, by many leading senators, his younger brother Gaius became a bitter opponent of the Senate. A highly gifted and energetic man of about twenty, he was imbued with a passionate desire to avenge his brother. He then conceived an even stronger desire to win power and continue his brother's work. He addressed his extraordinary and varied talents to the task of reforming the republic. Not content, as his brother had been, to concentrate on a single issue, he set his sights on the whole, above all on political reform. Unlike his brother, he was an innovator, though of course within the limits that Roman conditions imposed on all change. No one ever reflected so independently and carefully on the state of the commonwealth. He wished to build up a counterforce to the Senate, based on the equestrian order. However, his legislation went far beyond this; hardly any area of the commonwealth was untouched by it. Yet his death put an end to any hope of reforming the republic.

The decade between 111 and 100 saw the emergence of a political current opposed to the Senate; this rested primarily on the ambition of the knights to occupy a position of greater esteem in the republic. During this period a series of the ablest sons of the nobility became tribunes of the people and set themselves up in opposition to the Senate, but only for a time; they all went on to carve out successful careers for themselves. It seems that the bounds of political propriety had widened. The fathers did not object to their sons' seizing the new political opportunities.

Only one man came close to assuming the role of an outsider. This was Gaius Marius. He did not belong to a senatorial family, but was a *homo novus*. He had a good military record and, profiting from the current political trend, was uniquely successful in serving several terms as consul.

Marius was a man of simple nature, rough and straightforward; he was also a brave officer who had never spared himself and liked to show off his scars. His self-confidence, however, was at odds with a certain touchiness and irritability in his dealings with over-bearing nobles and their effete – or at least rather precious – sons. Like other successful *parvenus* no doubt, he was fond of boasting that he embodied the old Roman virtues far better than they. He had the added merit of being the only significant and successful general of his day. The current political mood allowed him to vent his resentments quite freely. He relished the weakness of the senators, thought nothing of doing what he deemed right, and pursued his career with a single-mindedness that was scarcely affected by the assimilating force of his class. He thus came to occupy a special position. In any case, he reacted defiantly to insults and disappointments by seeking refuge, as it were, with his soldiers, sharing everything with them and devoting himself to their welfare. He demanded much of them, but differed markedly from the other senior officers in that he felt particularly close to his men. Hence, having recruited landless soldiers, he raised a claim, on their behalf, that they should be granted land.

In 100 this claim was embodied in a land law by the tribune Lucius Appuleius Saturninus. The dispute that ensued became so fierce that Lucius was murdered. The senatorial nobility now regrouped, closed ranks, and henceforth followed a more decisive and consistent policy; in the first place, popular actions were no longer tolerated.

The last great attempt at reform before the civil war of the eighties was undertaken by Marcus Livius Drusus, who had himself elected tribune for 91. Drusus, like Gaius Gracchus, set out with comprehensive plans, though his aim was to strengthen senatorial rule. For instance, he removed the jury courts from the jurisdiction of the knights; this was amply justified, for in 92 they had acted with blatant partisanship by condemning a former consul because he had tried to prevent their gross exploitation of the province of Asia. Drusus also enlarged the Senate. To please the populace he brought in a corn law and a law on the founding of colonies. Above all, however, he wished to solve a problem that had become a burning issue – that of satisfying, wholly or partly, the demand of Rome's Italian allies for admission to Roman citizenship.

As Rome gradually extended its rule throughout Italy, it had granted citizenship, either at once or after a lapse of time, to various cities. It had also sent out some of its citizens to found colonies, mostly at strategically important points. The result was that, apart from Sicily, more than a third of the Italian cities to the south of the Po were now Roman. The rest were formally independent, but allied to Rome, and although self-governing, they were not necessarily immune to arbitrary interference by the Roman magistrates; they were also obliged to supply military contingents. Side by side with the Romans, they had conquered a vast empire for Rome. Certain families or towns might be distinguished by the conferment of Roman citizenship, but this had become rare in recent decades. True, many towns found the distinction a mixed blessing, as it robbed them of their independence and some of the individuality that they or their rulers cherished. Yet a growing number of individuals and towns were exercised less by the reduction of their autonomy than by the fact that they were at a disadvantage in relation to the Romans. Problems over military levies and other matters combined to foster a widespread desire for Roman citizenship. Rome was not only unresponsive, but in 95 began to take legal measures to revoke the citizenship of those who had obtained it by devious means. At this point many seem to have lost patience and hope; the desire for citizenship grew into a powerful demand that quickly spread among parts of the alliance.

There is no doubt that this process and the events leading up to it were closely linked with the social and political changes that had taken hold in Rome roughly since the time of the Gracchi – the reform of recruitment procedures, the corn laws (which made the city attractive), the ascendancy of the knights, the enfeeblement of the Senate, and probably the increasingly arbitrary conduct of some of its members. Gaius Gracchus too had wanted to introduce a law on the allies. These problems, which contributed so significantly to the decline of the republic, were thus not tangential or fortuitous, but essentially resulted from the crisis.

There were strong objections to Livius Drusus' bill. Had it succeeded, it would have brought him unusually wide support. For, according to contemporary thinking, the new citizens would have been bound to support him politically. Although he had undertaken to solve all the questions at issue in accordance with the Senate's wishes and had many leading senators on his side, the majority of the house refused to support his bill. They even went so far as to repeal the laws that had already been passed, on the ground of a formal error. Behind this lay a combination of oligarchic jealousy, growing resistance, and the efforts of various senators opposed to Drusus. However, their distaste and apprehensions seem to have been aggravated by the way in which he conducted himself.

Marcus Livius Drusus was an exceedingly self-confident, self-willed and ambitious aristocrat, rigorous towards himself and others, gifted with great wisdom and foresight, and accustomed to hard work; even as a boy he had never allowed himself a holiday. When he had a house built on the Palatine Hill, the architect suggested a design that would have prevented anyone from looking in, but Drusus instructed him to build the house in such a way that all could see what he was doing. He had now recognized that the Senate régime needed strengthening. He probably realized how urgent it was to admit the allies to Roman citizenship, knowing that they were determined to go to war if denied it. There was no time to lose. It all depended on him; he had to settle the matter, and quickly. He could not consider everyone's feelings. The task was so urgent that he must have been irked by the sensibilities, demands and reservations of the senators. He gave them to understand that he was too busy to listen to all their representations. On one occasion, when the Senate summoned him to a meeting, he replied that if the senators wanted anything from him they should come to him. In his proud and forceful way he probably overstepped the mark in other respects too. On his death-bed he asked his friends when the Roman republic was likely to produce another citizen like himself.

In all this there was an inescapable dialectic. Truly comprehensive reforms, whether for or against the Senate, were attempted only by someone who was unusually self-willed, courageous, resolute and, of course, intent upon pursuing power and the opportunity for effective action on the grand scale – in other words, someone who towered above the rest. And if he was young, and therefore not fully absorbed into the world of the fathers, and something of an outsider to boot, he could not fail, sooner or later, to incur the distrust and hostility of the Senate. No cause, no programme carried any weight in the face of the progressive alienation between a resolute, ambitious individual and a fearful, oversensitive oligarchy, wedded to equality, complaisance and inertia. Thus any reform agreeable to the Senate became almost impossible – unless someone was strong enough to carry it against the majority. Hence even the 'Gracchus of the aristocracy', as Mommsen called him, finally stood alone.

The struggle for the law on the allies was extremely fierce. Many Italians had come to Rome. There were big demonstrations, and much intimidation and fear. Rumours of murder plots circulated. It was therefore probably not just specific plans, but the overheated atmosphere, that cost Drusus his life. In the midst of the throng of supporters who habitually escorted him home, a dagger was thrust into his side; he died a few hours later. Rome's allies immediately went to war.

Once more a great attempt at reform had ended with the death of the tribune of the people. On each of these occasions, senators and knights had finally joined forces against the reformers, though the forces were in each case differently weighted. The lethargy, the dead weight of inertia that Rome's 'good society' – the 'good', as they called themselves – brought to bear in support of the inherited order, was immense. Various reforming laws might be passed and even survive the death of the reformer. But it was felt intolerable that one man should accomplish anything great, even in the Senate's own interest, and thereby display the kind of independence that went with being to some extent an outsider. It was seen as a threat to the political order, to which all must be bound by mutual ties.

Marius was the only significant outsider to survive. But then he was not a tribune of the people: he was a war-hero who could boast the kind of achievements that had earned men the highest fame throughout Roman history.

The deadly nature of the outsider's role did not prevent individuals from repeatedly addressing themselves to various big problems and trying to solve them. As early as 88 important plans for reform were again initiated, both for and against the Senate, by Sulpicius Rufus, a young tribune of the people, and Sulla, who was consul. The tribune paid with his life when the consul – for the first time in Roman history – led an army against Rome. Whatever attempts were made at reform, however, the political problem facing Rome was neither recognized nor solved. The essence of this problem was that the Senate could no longer master the difficulties of governing a world empire and opposed anyone who tried to do so, because this brought him excessive power, and because political independence was both a prerequisite and a product of all such attempts.

Nevertheless, the most urgent practical tasks were performed – in spite of the Senate. And since the eighties they were performed mostly by a new kind of outsider – successful military leaders in the tradition of Marius. This gave rise to the kind of situation that is always found when a society is on the wrong track, as it were: some problems are solved, but others pile up, and with every solution the real problem grows. The real problem lies in the track itself. In the late republic the individual outsiders became increasingly powerful and the conflicts increasingly fierce; in consequence the political order was progressively eroded.

On the whole, the late republic produced few new ideas regarding the political order. If it is correct to say that the purpose of political ideas is to serve as levers that can be operated in the real world, then there have to be points of purchase at which their operation can begin. Since the death of Gaius Gracchus there was little hope of this. Hence whatever was new – if it was of any significance at all – took the form not of ideas, but of human abilities – a new independence, new political positions, new ways of accumulating power, and a new isolation. Traditional Roman society was still strong enough to ensure that these possibilities were hardly ever realized, but the history of Caesar and his age shows that they already existed.

4

Birth and Family

GAIUS JULIUS CAESAR was born on the third day before the ides of Quintilis, when Gaius Marius and Lucius Valerius Flaccus were consuls (Gaius Marius for the sixth time). According to our calendar the date was 13 July 100 BC. But at that time years were still named after the consuls in office, and the fifth month of the year (Quintilis) was not yet named after the great son of the Julian house. It is not certain that the date corresponded exactly to 13 July in our solar year. For Rome still had a lunar year, until Caesar himself reformed the calendar. He may have been born in 102 BC. However, that is a matter for astrologers – or ancient historians. More important than the date are his family and the circumstances of his upbringing.

His father too was called Gaius Julius Caesar, and in 100 he had probably just completed a term as quaestor, the lowest rank in the hierarchy of Roman officials. His mother was Aurelia, the daughter of Lucius Aurelius Cotta, who had been consul in 119 BC.

The Julian family belonged to the patriciate, Rome's original nobility, and so to the small circle of the most illustrious families who in early times made up the whole of the citizenry – who in other words had once *been* Rome. They were still very conscious of this, and it was still very much part of the general consciousness. In a few institutions the pre-eminence of the patriciate was almost palpable. Of these probably the most interesting is that of the *interrex*.

This was a magistrate appointed during an *interregnum*, when for some reason there were no consuls, either because no election had been held or because both consuls had died. At such times there was a break in the continuity, so to speak. Only a consul could 'create' (*creare*) the new consuls. For they must be not only elected, but in some way assured of divine aid. This assurance was obtainable only through the agency of the consul under whose auspices the election took place. If the continuity was broken, the auspices – that is to say, the right to treat with the gods on behalf of the community – had to 'return to the fathers' (the patricians). The fathers then had to appoint *interreges* (interim kings) from within their own ranks, each for a term of five days. Only the second or a later *interrex* could 'create' new consuls.

It was as though normally the commonwealth could be left to run itself, but in an emergency only those families that had once constituted it had the power to regenerate it, by virtue of their closer links with the gods. This belief may have been doubted or even regarded with mild amusement. The reasons for it may have been forgotten. The Roman upper class had to a large extent moved away from the old religion, even if ostensibly it still adhered to it. Enlightened Greek philosophy, 'philosophical theology' (which was distinguished from mythical and political theology) had found a wide resonance in the nobility. However, the most important argument against the total rejection of traditional belief – that one cannot be sure whether there is not something in it after all – counselled caution. Moreover, everything in Rome was interconnected. Much was under threat, and the greater the threat, the less justification there was for interfering with tradition: tradition had to be maintained and observed. This conviction was so strong and deep-rooted that one wonders whether the senators could still distinguish clearly between the truth and the usefulness of religious concepts. They could not openly – or privately – allow too much to be called into question. It was therefore accepted that the patricians had special relations with the gods. Patrician pride remained intact. This was after all an aristocratic society, and the gradation of ranks within the aristocracy had to be respected. There was therefore a residue of special esteem for Rome's first families.

By 100 BC, however, a family's rank no longer depended solely on its age and patrician descent. No less important were the prestige and authority that derived from the political and military services it had rendered to the commonwealth. For apart from the patricians, a number of plebeian families had long since become great and powerful.

The upper stratum of the senatorial nobility, the *nobilitas*, comprised not only the patrician families, but also those plebeian families that had produced a consul. The term *plebs* must not be misunderstood. It embraced all who were not patricians – rich and poor, great landowners and peasant farmers, craftsmen and labourers. They had joined forces in the class conflicts, some in the cause of equal political rights, others mainly to press their economic demands. Since then a number of leading plebeian families had found their way into the magistrature.

It was on this basis that the new *nobilitas* had taken shape. Political prestige and public support were inherited. So were political ambitions. Politics and war were the only spheres in which one could establish one's noble credentials. As a rule, anyone belonging to the senatorial nobility became a politician. Conversely, few others did, except for a small number of social climbers, who were usually soon assimilated. Thus political rank (the holding of a magistracy), distinguished lineage, influence and wealth were usually concentrated in the same families.

Yet there were exceptions to the rule. Not every family succeeded in maintaining the high rank it had acquired. And the Julians were among the unsuccessful. Privileged though they were as patricians, they had long been in the second or third rank politically. In the two previous centuries they had produced only two consuls (in 267 and 157 BC). Otherwise, members of the family had reached at best the second-highest office, that of *praetor*. They cannot therefore have had a great fortune.

Nonetheless, they probably retained a more or less unassuming pride. What such an ancestry had to offer is best illustrated by a quotation from the speech that Caesar made at the age of thirty on the death of his father's sister: 'Her mother's family is descended from the kings, her father's related to the immortals. For the Marcii Reges, whose name her mother bore, are descended from Ancus Marcius; and the Julii, to whose race our family belongs, are descended from Venus. Her lineage thus enjoys both the venerability of the kings, who are supreme among men, and the divinity of the gods, to whose authority even kings are subject.' The descent from Venus was traced through a postulated mythical hero who, among other attributes, was given the name Julus and reputed to have founded Alba Longa, the old suburb of the Latini, the tribe to which Rome belonged. Various patrician families, including the Julii, were said to have come from there in early times and settled in Rome. According to one myth, Julius was the son of Aeneas, whose mother was Venus. Having come to the West after the destruction of Troy, he was held to be an ancestor of the mother of Romulus and Remus, the founders of Rome, supposedly fathered by Mars. Aeneas supplied the Romans' link with the culturally superior East; he also afforded a model of *pietas* (respect for gods and ancestors), having carried his father Anchises on his shoulders from the flames of Troy.

Modern scholars are sceptical about Caesar's claims. However, whether or not Venus was the ancestress of the Julian family, and whether the cognomen Rex is not more likely to derive from a 'sacrificial king' of the third century, such stories were invented and believed at the time. And for the young Caesar they had to compensate for the fact that, unlike many of his peers, he could boast no illustrious republican ancestors. They may at least have fired his youthful fancy and even nourished a secret pride, and if he later came to question his divine descent, the idea was too attractive not to have had at least a kernel of truth. Although it cannot be said to have pointed him along the path leading to monarchy, he may well have imagined himself to be in some way special and exceptionally favoured.

In the 80s, Sulla, Rome's most successful general and civil war leader, regarded his *fortuna* as a special favour conferred by Venus, and on his coins – and on a large victory monument – he combined Eros with the palm of victory. At least by this date, then, the goddess of love was closely associated with victory and fortune. Success and fulfilment were expected of her. And who was more entitled to expect such gifts than her late descendant – who subsequently wore an image of the armed Venus on his ring and used her name as a rallying cry in difficult military situations?

His mother's family, by contrast, the Aurelii Cottae, were of plebeian origin and, as far as we can see, had moved into the higher nobility only a century and a half earlier. As such they were no match for the Julii. Yet during this time they had given Rome four consuls, the last being Caesar's grandfather. Other branches of her family had produced four more consuls. His mother's relatives were to be of great service to Caesar in his political career.

Especially important to Caesar was the connection he acquired through his father's sister, who has already been mentioned. She had married Gaius Marius, a *homo novus*. This was not exactly the best thing that could happen to the daughter of a patrician family. Yet Marius succeeded in being elected consul and, being a brave soldier and widely esteemed, he obtained a second term (which was contrary to the law) and was charged with the conduct of the war against the Cimbri and the Teutones, who were causing great alarm in Rome. Since he was able to overcome the danger, and since there was widespread dissatisfaction with the old nobility and its practices, he was elected consul five times in succession. Such a career was unique in the whole history of the republic. Caesar's family not only basked in the reflected glory that it acquired through the surprising rise of their relative by marriage, but seems at some stage to have become quite closely connected with him – a connection that had a decisive influence on Caesar's youth and subsequent career.

5

Youth in Rome

Education

EARLY PHYSICAL AND INTELLECTUAL
TRAINING · THE PRINCIPLES OF ROMAN
EDUCATION · THE WORLD OF THE
FATHERS · THE SYSTEM OF LEARNING ·
TRADITIONAL EDUCATION UNDER NEW
CONDITIONS

ALMOST NOTHING IS RECORDED of Caesar's childhood and youth. The education of young noblemen lay largely with the family. Caesar's mother, we are told, took a leading part in providing and supervising his. His father must have taught him a good deal too, even if he was less assiduous than the elder Cato, who is said to have instructed his son in reading and writing, the law, and physical exercises – 'not only in casting the javelin, close combat and riding, but in boxing, endurance of heat and cold, and swimming through the whirlpools and torrents of the river'.

Much of a boy's education – reading and writing, grammar, the rudiments of rhetoric, and of course Greek – was entrusted to a tutor. The tutor was often a slave. Lessons took place at home, or at the houses of relatives or friends. At any rate, the sons of illustrious families hardly ever attended one of the schools, which were private institutions. Later might come a thorough training in rhetoric. It was advisable also to acquire a knowledge of Roman law; the rudiments of this, the Twelve Tables, were learnt by heart. But it was also necessary to learn about legal procedures and a large number of precedents. Caesar's private tutor, at least in rhetoric, is known to have been a former slave, educated at Alexandria and versed in Greek and Latin rhetoric.

A boy would learn riding and swimming from his father or other teachers, near the city or in the country, where the family repaired during the Senate recess from early April to mid-May, or in the summer. Great value was attached to physical training, which often took the form of competitive sports. These usually took place on the Campus Martius, which lay outside the city gates, on both sides of the present Via del Corso. 'This is an admirably large field,' writes the geographer Strabo, 'on which an enormous number of chariots and horses can race unimpeded and a host of people can daily play ball games and practise discus-throwing and wrestling.' Especially popular exercises were fencing and leaping on and off horses. Caesar is said to have been a good rider as a boy. 'He had accustomed himself to folding his hands behind his back and letting the horse run at a fast trot.' For swimming one went to the Tiber, which lay a short distance away.

All these activities took place under the critical eyes of interested spectators. Though practised mainly for sport, they were directly related to the requirements of war. All Roman education had a practical orientation. The Romans had little time for Greek athletics, which were regarded as a useless pastime that made the young soft and were appreciated only as a spectator sport.

This reconstruction, based on contemporary customs, admittedly tells us little about Caesar. It may be imagined that he received a rigorous intellectual education and trained his memory; there was naturally much learning by heart. We may be certain that he also trained his body to perform with the efficiency it manifested in later years. His biographer Suetonius writes: 'He handled weapons with great skill, was an excellent rider, and had amazing endurance. On marches he sometimes led on horseback, but more often on foot, bare-headed in sun and rain. He covered great distances with incredible speed, with no baggage, in a hired carriage, travelling a hundred miles a day. Nor was his progress held up by rivers: he either swam across them or crossed them on inflated skins, so that he often reached his destination sooner than the news of his movements.' Many marches took place at night. It was by swimming that he saved his life outside Alexandria in 47 BC, at the age of fifty-two.

We learn more, however, if we bear in mind the principles that governed the education of young men in Rome at this period. These derived chiefly from tradition, from the way in which the son of a country landowner gradually adapted himself to his father's lifestyle, accompanying him on journeys, observing everything he did, and then attempting to do it himself under his father's supervision. It amounted essentially to learning by observation and imitation, and gradually growing into one's role. This kind of education was continued in the city too, above all in politics, the chief sphere of activity for members of the nobility. In addition, more and more of a young man's time was of course taken up with theoretical training, but as a rule this seems to have remained peripheral.

It was important that he should begin to share in his father's life at an early age. He would go with him, at least now and then, to visit other people's houses. He would observe his father's dealings with friends and clients and was allowed to sit in during discussions; he could also visit the forum and the lawcourts, when his father would observe and comment on public life. Some political events might even take place in the family home, when important guests from outside Rome were entertained or when embassies from the prov-inces visited their patrons to pay their humble respects and bring costly gifts.

This practical education in the family was historically the natural result of a gradual shift from the agrarian conditions of early times to the political life of an aristocracy that ruled the world. But there was more to it than that. In the early accommodation between great families and civic communities, the share of the families, their area of competence and the rank enjoyed by their members had been extra-ordinarily great. The families had placed themselves wholly at the service of the community and accepted the concomitant discipline. Service to the city brought them fulfilment and fame; it became part of the aristocratic tradition. Their cohesion was thus reinforced by a common orientation towards politics. In the meantime, however, there had been certain shifts. The matter of politics had become too diverse and the foundations of power too fragmented to allow the solidarity of the families to continue to assert itself to the same extent. The families nevertheless maintained their individual lifestyle, a jealous concern for their descendants, and a special pride. To be the son of one's house was, so to speak, more important than to belong to the youth of Rome. This produced a particular consciousness. Cato's opinion that nothing was more important than educating one's own son was shared by other members of his class, though they might not have expressed it so strongly.

After all, there was nothing in the world to compare with the Roman aristocracy. And the aristocracy ascribed its dominance and its character chiefly to certain virtues. Ennius had written, *Moribus antiquis res stat Romana virisque*: 'The Roman commonwealth rests upon ancient customs and the men [who practise them].' These customs could not be taught in any school. Teachers might be needed to convey knowledge and method. But the nature of the Roman nobility, its maxims, experiences and connections were so ingrained and so infinitely varied that they could be conveyed only by those who had such a matchless command of them. They could not be taught, but only demonstrated and inculcated by example. The sons of the nobility were not so much obliged to absorb knowledge and methods; first and foremost they had to take over a world – from their fathers.

The Roman nobility appreciated the importance of this largely practical paternal education. This is clear from an edict issued by the censors in 92 BC, banning recently opened schools run by self-styled 'Latin orators':

> We have been informed that certain persons there have in-stituted a new kind of training for the young; these persons call themselves 'Latin orators', and the young who attend their schools are said to spend whole days in idleness. Our ancestors determined what children should learn and what schools they should attend. This new fashion, which is at variance with the uses and customs of our ancestors, neither pleases us nor appears to us right. We therefore deem it fitting to make our displeasure known both to those who run the schools and to those who attend them.

Cicero later declared that such schools blunted the mind and bred insolence. The education and learning that went with Greek rhetoric, he said, had been set aside. Obviously such schools concentrated on inculcating the methods by which the pupils could represent certain interests as effectively as possible, possibly by demagogic means. This may well have been true. Yet one should not underestimate the other motives: whole days spent at school turned young noblemen into schoolboys, alienated them from practical life and forced them into idleness. Instead of being confronted as individuals with models to be emulated, they were thrown together with their own kind and with teachers. The young gentlemen were offered little that could command their respect. They may even have been encouraged to develop a critical view of the aristocratic establishment. However, what probably told most heavily against the schools was that they estranged the young from their natural environment and fostered common youthful interests, when traditional values shared with the family and the elders should have been paramount.

The practical education of young noblemen ensured that as they got to know the world of their fathers it progressively became their own. Growing into the world of the Roman nobility required an infinite series of acts of learning. Just as public life in Rome was not subject to a systematic constitution, but conducted largely according to countless examples, precedents, conventions and rules, so it was necessary to imbibe a mass of miscellaneous knowledge in order to find one's place in it. And just as political power was not organized in groups, but built up through a multiplicity of connections enjoyed by the individual senators, it was necessary to know a vast number of people if one was to be able to assert oneself.

One might if necessary employ 'research assistants' – slaves with special knowledge, for instance, who knew everyone's name (*nomenclatores*), or learned slaves who could furnish the examples from Roman history that were continually needed as arguments. Yet one also had to know a great deal oneself. One could not, for instance, appear in the Senate with a secretary. And the *nomenclatores* not only knew people, but were themselves known; it therefore followed that if one was advised by them, the personal attention one accorded to others was devalued.

Preparation for adult life did not allow the growing boy much chance to enjoy a carefree childhood and youth. Many demands were made on him, but this meant that at an early age he was taken seriously. He was bound to feel proud, especially as the world of the fathers was exceedingly impressive. He was courted, respected, esteemed; he conducted himself in a sober and superior manner, assured and authoritative, proud yet affable, full of *gravitas*, a quiet and perhaps slightly pretentious seriousness; he appeared firm and responsible, yet at the same time energetic, and occasionally perhaps even urbane.

Rome's dominion over the known world and the leading role of its aristocracy were in the very air the young man breathed. He saw not only foreign and at times exotic ambassadors arriving in Rome to present this or that petition, but governors setting out to rule distant provinces. With friends and relatives he might escort them for a short distance or go to meet them on their return. He could watch the departure of the legions or stand at the roadside as some general rode in triumph to the temple of Juppiter Optimus Maximus on the Capitol, bringing booty and prisoners from distant parts and large-scale depictions of his battles, designed to inform and impress the Roman public. All over the city were statues of great captains and politicians; captured weapons were displayed in public or in the victors' houses, and large maps showed the conquered territories or the whole of the known world. Wealth flowed into the city from all parts. Great spectacles were staged. In 95 or thereabouts Lucius Cornelius Sulla, the praetor and later dictator, had a hundred lions and a contingent of well-trained spearmen sent to him by his African host Bocchus, the king of Mauretania. He staged a great fight in the circus, which was all the more imposing for being the first in which the lions were not chained. It is unlikely that the seven-year-old Caesar was taken to see it, but he might have been, and he must have heard reports of it. It was all part of the atmosphere of empire, which made the world of the fathers so impressive. It need not matter if one's own father did not occupy a grand position in this world. One might resolve to enhance the reputation of one's family in future.

The Roman father was more powerful than most. He had the right to chastise and even kill members of his family, and for as long as he lived his sons, their wives, and their children were subject to his authority, the *patria potestas*. True, this right had survived from early times only because it was rarely abused, and any abuse would have led to communal intervention. The father could legally be allowed to retain his power because there were enough extra-legal means of checking it. A practical compromise had been reached between the community's insistence on law and order and the families' insistence that there should be no interference in their affairs. The extent of the father's power was commensurate with his obligations. This was a patriarchal society. Even if arbitrary action was severely restricted, the families still enjoyed considerable autonomy and independence, such as comes from the exercise of self-restraint. The Roman father and the assembly of the 'fathers' – the Senate – were entitled to *obsequium* ('obedience') and *pietas* ('respect', which was due also to the gods). Behind them was the almost palpable presence of their ancestors. Polybios gives a striking account of how the masks of the great ancestors hung in noble houses as a constant reminder to the living, and were worn by servants in funeral processions. These servants, accompanied by the appropriate lictors and attired in the official garb of the highest public office the deceased had held – perhaps even the triumphal purple – walked in a long, imposing cortège bearing witness to the family's unity and greatness, its political achievements and rank. Similarly, the ancestors were represented on the ivory-adorned seats of the magistrates around the speaker's platform in the forum, combining with the statues nearby to testify to the continued presence of countless past generations. The son or a close relative would then mount the platform and extol the virtues and services of the deceased, and finally the deeds of his ancestors. In this way, writes Polybios, their fame was constantly renewed, and it would be hard to invent a more splendid spectacle as a spur to the young, who would do anything to win such renown.

Whether all this emphasis on the father and the ancestors was felt only as a spur and an inspiration and not also – or rather – as a crushing burden is a question that must be deferred for a moment. First we must say something more about the special character of a Roman education.

What the young were expected to learn in Rome was hardly the kind of dry general knowledge that might be thought to have been assembled at random. Rather, nearly all of it had a practical and individual character. Since a boy's education was essentially in the hands of his father and his relatives, its schoolmasterly component was confined to the more technical aspects. These, however, were clearly indispensable.

The study of Greek rhetoric admittedly involved a degree of erudition and philosophy. But it was optional. Moreover, erudition and philosophy had the charm of novelty and afforded new insights, which were probably viewed rather superficially and not taken altogether seriously: new modes of expression and conduct, a new style that was somewhat at odds with Roman tradition. And the world one had to understand on approaching manhood was static, stable and easy to take in. The structural difficulties and the crisis that afflicted Rome at the time may have been discernible, but they were not matters of anxious concern. The only apparent problem was the state of the oligarchy, and the remedy was seen to lie in the emulation of worthy models and the preservation of time-honoured custom. After all, it was scarcely possible for the Romans to view their world from without. In many ways one was already involved in it before one developed a need to understand it. And for the most part it required no understanding, because it was taken for granted.

The learning process was spread over successive stages of the child's development, though we do not know precisely how. There was no clear dividing line between what he learned in the home and what he learned outside it. A break came at the age of about fifteen, when he donned the *toga virilis* (the male toga), having hitherto worn the *praetexta*, an upper garment bordered with purple that was otherwise proper to magistrates. Life now became serious, as it were, and from this point on he wore the plain toga, unless he preferred something even simpler. There was a domestic celebration. Then the father took the boy to the forum in order to present him to the Roman public. He offered a sacrifice to Juppiter Capitolinus. There followed a year's apprenticeship in politics; he now applied himself full-time to what had previously been only a marginal pursuit. His father would introduce him to an important politician, in whose circle he could learn the art of politics. The young were allowed to accompany their elders to the Senate and stand by the doors, which were usually open, in order to listen to debates. At some stage they were admitted by the censor to a *centuria*, one of the divisions of the

most important Roman elective assembly. This was divided according to *census*. The most important units were the *centuriae* of the
knights; sons of noble houses were regularly placed in the first six
(out of a total of eighteen), which played a special part in Roman
elections and were consequently much coveted. As the *censurae* were
irregular during the period of the civil war, we cannot say when
Caesar obtained this privilege.

After this it was customary to serve in the army for a year; one
might then join the staff of a governor. Caesar served in the east
from 80 to 78. When he was sent to Nicomedes, the king of the
Bithynians, to take over a naval squadron, he is said to have shared
the king's bed. Throughout his life this episode furnished his
opponents and his soldiers with matter for ribald jokes. For
pederasty enjoyed no esteem in Rome. It was widespread (though
not to the same extent as among the Greeks), but even more
widespread were the reproaches – or at least innuendoes – that it
attracted. During Caesar's triumph after the Gallic War, the public
sang:

> Caesar screwed the whole of Gaul, Nicomedes Caesar.
> See now, Caesar rides in triumph, after screwing Gaul.
> Nicomedes does not triumph, though he screwed our Caesar.

Unlike most sons of great families, Caesar not only remained on
the staff, but fought in the front line. At the storming of Mytilene
he distinguished himself by his valour. During the battle he rescued
a comrade and gained the rare distinction of being awarded the
'citizen's crown', a wreath of oak-leaves that he was henceforth
entitled to wear on all solemn occasions. When its wearer appeared
at the public games it was customary for all the spectators, including
the senators, to rise to their feet. After taking part in another
campaign against pirates operating from bases along the south coast
of Asia Minor, Caesar returned to Rome.

After completing his military service at the age of about twenty,
it was customary for a young man to go into politics. Caesar
distinguished himself as a prosecutor. This was a popular way to get
oneself known and win early respect and support. Thanks to the
illegal practices that were common in the provinces and in Rome
itself, there was no dearth of defendants. Yet however guilty they
were, they were rarely convicted. But this did not matter: the

prosecution might still be conducted with *bravura*. In 75 Caesar returned to Greece to study in Rhodes.

If such an education served its purpose, the young nobleman was bound, by the end of it, to have adapted himself to the life of the Roman aristocracy, with all its rules and customs, its ways of thinking and feeling. He was now one of them. And his commitment must be total. For he was now fully involved in the activities of the ruling class, which was responsible for everything, had a hand in everything, and was at the centre of everything. Rome had no bureaucracy, no specialists, but only a small number of magistrates drawn from the ranks of the nobility. In principle the Senate had to attend to everything, though some functions fell to particular persons or families by virtue of their lineage, occupation or experience. This presupposed a large measure of homogeneity within the ruling class, and there must have been considerable pressure to maintain it. There might be scope for a degree of individuality and for the special character of certain families; one man might be more suited to this task, another to that; one might embody the ideal better than another; but on the whole they all had to conform to traditional values.

One wonders whether such an education in fact served its purpose. Learning by practice did not necessarily ensure adherence to the old rigour and discipline, to the demand for self-control, endurance and thrift, for seriousness and responsibility – in a word, to the transmission of the old Roman virtues. As early as the middle of the second century BC Polybios reports that the tastes of most young people in Rome had taken a bad turn. Some were intent upon pursuing pretty boys, others upon wenching; many thought only of carousing and other expensive pleasures. They had taken all too readily to the easygoing life-style of the Greeks. Moreover, the encounter with Greek philosophy and culture had probably eroded the unquestioning acceptance of the Roman order. The new ideal of *humanitas* was somewhat at odds with the ancient Roman code of *gravitas*, rigour and dignity. Did not all this cast more than a little doubt on the success of such an education?

What is more, contemporary conditions may have exacerbated the conflict between the generations. The young made increasing demands. If one's father, who controlled the family fortune, set great store by thrift, one could land oneself in difficulties. It is true that sons were normally given a particular property to run, and it probably did not greatly matter that by law this remained part of the family estate. Yet it might not suffice for their needs. They could resort to loans – and often did so on a grand scale – but there were limits to how much they could borrow, and violent quarrels might ensue.

And if the fathers were obliged to appear extremely powerful and important, inhabiting an immensely imposing world, could this impression survive the discovery that there was a considerable discrepancy between what they upheld as Roman virtue and authority and their actual life-style? One wonders whether opposition was not bound to arise – a desire to reject paternal authority and order things better, including the commonwealth itself. Was it not inevitable that the young, however untheoretical their education had been and however much they had been pointed in the direction of what was customarily possible – and possibly customary – should find themselves at odds with their fathers and traditional values?

After all, when Caesar was young, Rome had been going through a crisis for more than a generation, and at least some of its effects were plain to see: frequent breaches of senatorial class discipline, serious disputes, defeats for the Senate, failure to address pressing problems, and then, in the eighties, civil war.

How could the young gain access to this society, find a place in it and learn to look upon it as theirs? It was ultimately for them to decide who they wanted to be, where they wanted to stand. One wonders what opportunities they had to develop their own identity.

Freedom for the Young

THE WHOLE OF Roman education was oriented towards adulthood. The Roman child was already to a large extent a miniature adult and much involved in the concerns of his elders. It was apparently impossible to keep all one's options open for long and to revel in a wealth of opportunities. The nobleman had to devote himself to politics, and the range of roles was limited.

On the other hand, the Romans regarded 'youth' as a relatively long phase in a person's life. This is of course merely a way of expressing something that was actually quite complicated. There were different ways of dividing life up. According to one criterion, a new phase began with the end of puberty, the onset of manhood and the obligation of military service. At the age of about seventeen, then, the boy became a man. To this extent we may say that he was then grown up. According to another criterion, however, he was *adulescens* from the age of about fifteen until over thirty. This obviously corresponded to the notion that these were the years in which his mind and character had to mature. Only thereafter was he really adult. This long period of adolescence – which we can only roughly translate by the term 'youth' – reflected the respect that was traditionally – and increasingly – accorded to age and maturity. One had to sow one's wild oats and gain a measure of experience before one became truly mature. Age was what counted, and there were no short cuts. One could not become a magistrate before the age of thirty, and there was a minimum age for entry into each of the higher offices. One did not become eligible for the consulship until the age of forty-three; only then were the most successful men finally mature enough to move up into the top rank of the senators.

The Roman concept of a protracted adolescence thus points not to the pretence of staying young for as long as possible, but to the recognition that the attainment of maturity was a long process.

The process did not necessarily follow a straight line. In the late republic young noblemen were allowed a degree of latitude during their 'adolescence'; they were granted – or granted themselves – a certain fool's licence. However strict their upbringing – and here too there had probably been some relaxation – it did not always produce moderation, at least not in the private sphere, and not always in the political.

Given the city's wealth, size, and population, as well as the many tasks that had to be performed, it was hardly possible to keep a strict eye on the 'adolescents'. Nor, presumably, did their fathers entirely satisfy the standards they ascribed to the ancestors. A measure of licence was inevitable, and this opened the pores through which Greek manners and refinement could be assimilated into the Roman aristocracy. The young indulged in escapades, extravagances and festivities, and countless amorous adventures. Nor were they obliged to content themselves with common wenches – or even with high-class courtesans, who were of course by no means lacking in charm – for the ladies of Roman society too had acquired a taste for freedom and variety. At least this was true twenty years later, and there is little reason to suppose that things were very much different in Caesar's youth. The lure of love, especially illicit love, seems at times to have combined with the cultural atmosphere that resulted from the recent adoption of Greek manners. We are told that Sempronia, the wife of one of the consuls of 77, 'was favoured by fortune in respect of ancestry and beauty, marriage and children; she was versed in Greek and Latin literature, played the cithara, and danced better than was needful for a respectable woman; and she commanded much else conducive to untold delight. She found anything preferable to good repute and chastity.' She was readily credited with anything disreputable, and she was clearly implicated in the Catilinarian conspiracy. 'But she was not lacking in intellectual talent; she could compose verses, and she could jest and converse, now with restraint, now softly, now impudently; in short, she possessed much wit and grace.'

Slightly later, Clodia, another lady from the highest ranks of the aristocracy and the wife of one of the consuls of 60, was likewise highly educated, interesting, generous, pleasure-loving and charming. She was the beloved of Catullus, his Lesbia. 'Let us, Lesbia, live and love and give not a fig for all the words of sullen elders.' She found nothing good in any convention, save the pleasure of transgressing it, and did not recoil from scandal. She reputedly had intimate relations with her brother Publius, who later became an anarchic tribune of the people. She was at all events far from strait-laced and loved variety; and for this she was to incur the bitter rancour of the poet she discarded.

The picture could be elaborated, but a brief postscript will suffice. It comes from Caesar's sober biographer Suetonius and relates not only, but certainly partly, to Caesar's youth: 'All agree that he greatly loved pleasure, spent much money in order to gratify his passions, and seduced very many ladies of quality.' There follow several names, finally that of the mother of Marcus Brutus, his murderer, whom, we are told, he 'loved above all others'.

Indulgence in such pleasures might provide a young man with a means of escape; it might also be a form of protest. He might derive extra pleasure from cuckolding the great and the good; at the same time he might harass them politically, as a tribune of the people for instance, and so combine a desire for reform with the spirit of rebellion.

This may have become the more necessary as the republic slid further into crisis. The more disillusioned one was by the spectacle of aristocratic society, the more deviously one had to move in order to gain access to it. And all the while one had to prepare oneself for a political career. The one did not exclude the other. In any case, although such devious approaches to adulthood infringed time-honoured convention, they probably also enriched it. Yet they gave rise to no positive new possibilities. They were at best evasive manoeuvres, not ways to a different identity. Their outcome merely ensured that certain members of the younger generation could allow themselves an easier, more relaxed, more colourful life-style.

On the whole, however, it was hardly possible to miss one's target – intellectual and spiritual absorption into Rome's aristocratic society – whether one's aim was direct or indirect. It was probably only the zone of dispersion that had increased. There were now more fields of activity to be explored and – within certain limits – a greater variety of roles to be tried out. This was probably true of Caesar. Yet on the whole the young nobleman had no option; he must become like his peers and fall in with their ranks.

A political career was inescapable: there were no competing professions, no alternative walks of life. Moreover, it was a goal so highly prized that one had to be sick, feeble or uncommonly eccentric to avoid attaining it. It was not an option that a young man could take or leave, but a destiny to which he had to yield. For this he had to thank his family, his contemporaries – with whom he vied – and his class, as well as the confident expectations of his supporters and friends, who had ties with important Roman families and relied on them for help – and indeed the whole of the citizenry.

All this presupposed a high degree of class homogeneity, which was a potent force in the moulding of character. Roles were played with varying degrees of skill. Certain differences might be determined by one's character, temperament, intellect, experience, or family tradition. Originality was not entirely excluded, but on the whole it was the ethos of the Roman nobility that prevailed and set the tone. This was the foundation on which all else rested, the basis on which characters were formed – even in the late republic, with all its human variety.

Social reality thus exercised a dominant and far from random influence on the young nobles, and this reality appeared unambiguous. It did not present different aspects from various perspectives, nor did it offer a wide range of tempting options. It prescribed a single path, required ubiquitous activity, and allowed hardly anyone to remain uninvolved.

It left little scope for the shaping of individual identity and therefore raised few problems. In republican Rome, chances and burdens, opportunities and dangers, freedom and risk, assumed far smaller proportions than in modern times. Early in life Roman nobles must have become what they were destined to be. They were not anonymous individuals surrounded by people they did not know. All their roles lay close together. In each of these roles they were expected to display personal commitment. Everywhere they were directed towards practical responsibility and achievements that in principle were bound to seem possible. They associated with their own kind in a world that encompassed – indeed consisted in – their common activity. They took their places in it, as in a shared property, without being obliged, or even able, to strive for something different. Because their reality was so largely constituted and shaped by themselves, because they *were* this reality, they could not be confronted by it. They could not see themselves as parts of an alien whole. The whole to which they belonged also belonged to them, and they belonged to it as partners. There was no sense of a higher whole, such as is familiar in modern times, overarching an enormous variety of isolated and highly specialized members, who can easily lose their footing if it fails to hold. There was thus nothing equivalent to the enormous discrepancy between our own littleness (what Jacob Burckhardt called our 'midget existence') and our affiliation to disproportionately large abstract entities – a discrepancy that is often a problematic feature of modern identity, especially if one comes from a relatively restricted life. Young Roman nobles no doubt had their problems, anxieties and conflicts, but they were obviously not exercised by the question of who they were and wanted to be.

In this society it was far from easy to become an outsider – even during the crisis, with its great problems and its fierce and sometimes bloody conflicts. True, the Roman aristocracy may on occasion have disgusted some of its younger members – who were usually not the worst – but never in its entirety, only in some of its features. Of course, if one of them assumed a critical attitude on matters of importance and decided to proceed from criticism to reform, he would suddenly find himself opposed by the Senate majority and many others. He might in certain circumstances find himself outside the establishment. But this could only happen if he had become exceptionally estranged from the Senate. Otherwise he would hardly have reached the stage of forming his own opinion on such important matters and pursuing it in defiance of the Senate majority. Only by embarking upon open conflict did he become a total outsider. This happened in very few cases: most of those who were tempted along this path drew back because they found the aristocracy sufficiently embracing, their lives sufficiently agreeable, and ordinary reality sufficiently strong. For outsiders there was no ordinary reality, no conformity in nonconformity. Before Sulla scarcely anyone had any chance of survival as an outsider. If things changed after Sulla, after the civil wars of the eighties, this was probably due not so much to a new situation as to the personality of the outsider. The most obvious case is that of Caesar.

6

The Second Decade: Experience of the Civil War and First Commitment

VARIOUS CIRCUMSTANCES determine when the characters of the young are decisively formed. Of Caesar one can say with some confidence that the civil wars of the eighties were crucial for the shaping of his personality.

At the age of fifteen or sixteen he found himself at the very centre of one of the two parties. And because his defiant loyalty forbade him to abandon it even after it was defeated, he was committed to a very definite and very special position within the Roman aristocracy.

73

Hence the civil war, which so profoundly affected the whole of his generation, was experienced by Caesar in a quite specific way. And however much he was inwardly shaken by it, he found in it a starting-point for the gradual development of his identity, a process that was at first tentative and exploratory, but essentially involved an exceptional loyalty to himself – a self that grew so powerful that it increasingly compelled this loyalty. Especially important, it seems, was the influence of Lucius Cornelius Sulla, the most forceful personality of the age.

It was Sulla who launched the civil war in 88 BC. Its origin lay in the complications of external and internal politics that had arisen in 92, especially in connection with the Social War and the internal problems that resulted from it. Admittedly it lay also, as far as Sulla and his supporters were concerned, in the new character of the Roman army and the most recent chapter in the Roman outsider tradition, the failure of Livius Drusus.

The Social War had posed great difficulties. For the first time for over a century Rome had faced an enemy whose military might had to be taken seriously, whose legions had earlier fought beside her own and closely resembled them in training, structure and discipline. The enemy armies were strong, and they were fighting on their own ground. Rome could defend herself only by attempting, in the course of the hostilities, to weaken the opposing side by offering political terms, first to those of its confederates who were still loyal, and then to those who had remained in contact up to a certain point. Yet even if more and more of the enemy's support was thereby whittled away, this did not prevent a build-up of anti-Roman sentiment on the opposite side. Among the Samnites and the Lucani, for instance, there were many who had no desire for Roman citizenship but wished to shake off Roman rule. Once in a mood to fight, they were not easily appeased.

Fairly large contingents were still under arms when a fresh catastrophe intervened. Mithridates Eupator, the king of Pontos, a small kingdom on the southern shore of the Black Sea, invaded the Roman province of Asia and was enthusiastically welcomed by the indigenous Greeks. He had the hated Romans put to death – merchants, publicani, tourists and others. Eighty thousand are said to have perished; this was certainly a gross exaggeration, but the reality was bad enough. He then showed signs of wanting to cross over into Greece, with the clear intention of exploiting Rome's current difficulties in order to establish a new Hellenistic empire in the east. He also had relations, not previously activated, with the rebellious Italians, and perhaps with the pirates who operated from the south coast of Asia Minor, threatening the lines of communication across the Mediterranean. The problem of who should be put in charge of operations against Mithridates was then complicated by internal politics.

For in 88 fierce disputes had again erupted in Rome. These centred chiefly on the allocation of new citizens to the tribes, the subdivisions of the body politic. The Senate had at first tried to provide for them by creating additional subdivisions. Their suffrage was not to carry the same weight as that of most existing citizens. Although they would have constituted about two-fifths of the citizenry, they were to have been enrolled in eight to ten tribes, whereas the existing citizens were distributed among thirty-five. And in Roman elections the result depended not on the number of individual ballots, but on the voting of the tribes.

The senators who put forward this policy probably had no need to fear that the new citizens would vote for candidates who were not favoured by the old. Indeed, they and their supporters would still have had a good chance of success even if the new citizens had been distributed equally among the tribes. However, the elections would have had to be organized differently. New considerations would have arisen, and it would probably have been necessary to promote the interests of those who supported the politically important sections of the newly admitted communities, in order to assure them of a share in the magistracies. This would have entailed a number of changes and required a degree of generosity and farsightedness that the Senate majority was unable or unwilling to entertain.

Publius Sulpicius Rufus, a tribune of the people and a young friend of Livius Drusus, took up the cause of the new citizens. He petitioned that they should be allocated equally to all the tribes. He was supported by a few senators and many knights. For the first time in Roman history the publicani took a hand in politics; they did so massively and remained firmly committed for a long time. In recent years there had been several attempts to reduce their political influence, and in particular to put an end to the extortion that they practised with impunity in the provinces. They now seem to have set about consolidating and strengthening their political position. They had set their face against admitting the allies to citizenship, not least because they feared their competition. Now that this question had been settled, however, they could espouse the interests of the new citizens, whose tribal allocation affected them little. Indeed, they saw them as natural allies in their struggle against the senatorial nobility. For that was clearly the intended outcome. They no longer wished to be merely in a position to put pressure on the Senate – a cause in which the jury courts had been useful, though they had lost these in 89 – but to win stronger representation in the Senate itself and in the magistrature. The knights and the new citizens wanted to collaborate at the elections, so that they would be in a better position to support politicians who favoured them and to impede those who did not, and not least to help the new citizens to rise to high office and membership of the senate.

This new pretension took palpable shape in a bodyguard consisting of six hundred young knights, which Sulpicius called his Anti-Senate. The same coalition later supported Cinna's civil war régime, which was said to be characterized by the rise of young men and by its 'knightly splendour'. What was planned was not an attack on the institutions of the republic, but on the ruling circles and families. It was not intended to replace them, but to shift the balance in favour of that section of the citizen body that had so far hardly been able to make itself felt electorally, but now wished, in concert with the new citizens, to bring all its influence to bear – perhaps even gain the upper hand.

The Senate majority offered stiff resistance. Sulpicius went to work with ruthlessness and violence. Resolved to do whatever lay in his power, he allied himself with the old and still popular war-hero Gaius Marius and offered him the command in the war against Mithridates. It promised to be an easy campaign that would bring much booty and gratify the ambitious old man after his many disappointments. However, the command had already been conferred on Sulla, one of the consuls, and there was a long-standing jealousy between the two men. Sulla had served in North Africa in 107–105 as *quaestor* in Marius' army during the campaign against the Numidian king Jugurtha; Marius had then been well over fifty, Sulla in his early thirties. Jugurtha had caused them great problems. For all the

Romans' skill it was hardly possible to defeat Jugurtha militarily: whenever they succeeded in a bold attack he would withdraw into the desert and gather fresh forces. Sulla had finally captured him through adroit negotiations with Jugurtha's father-in-law Bocchus, the king of Mauretania. Bocchus hated and feared his son-in-law, who had just fled to him for protection. Although he regarded him as a liability, it was uncertain whether he would go so far as to betray him. Sulla took the risk of accepting an invitation from Bocchus to visit him at his residence, with only a small escort, and spend some time with Jugurtha, each hidden from the other. After tense and exhausting negotiations, involving a good deal of intrigue, Bocchus handed his son-in-law over. Only then did the African war come to an end. Marius was given a triumph, but Sulla claimed the credit for the coup that had ensured the successful outcome. He later used a picture of the surrender of Jugurtha as a seal, and Bocchus, much to Marius' annoyance, had the scene commemorated in a monument on the Capitol, to remind the Roman people of his deserts. Now, at the age of seventy, Marius was about to wrest the important eastern command from the fifty-year-old consul. The popular assembly, at the tribune's instance, decided to confer it on him.

Sulla would have none of this. Having already assumed supreme command of his legions outside the besieged confederate town of Nola in Campania, he appeared before them and announced that Marius was about to lead other soldiers, not them, in a war that would yield much booty. The soldiers were incensed and called upon him – indeed compelled him, it seems – to lead them against Rome in order to overturn the people's decision. The nub of the matter was undoubtedly the eastern command. Yet it would be unjust to Sulla to overlook his desire to support the Senate.

Lucius Cornelius Sulla was a strong and colourful personality. His family belonged to the oldest patrician nobility and had once been prosperous. One of his forebears, a successful general who had twice served as consul, had been struck off the roll of senators in the early third century (when things were still strict) for possessing more than ten Roman pounds of silver (well over three kilogrammes). Since then, however, the family had become so poor that Sulla at first had to live on one floor of a tenement; this was quite exceptional for a member of the aristocracy. He led an unconventional life, preferring the company of actors and *demi-mondains*, devoting himself to love affairs and a variety of other pleasures, and delighting in lively and intelligent company. He was disrespectful, full of life, and given to mockery. And it was in this way, curiously, that he came into a fortune. Having fallen in love with a rich courtesan, he was able, thanks to the charm of his company and his youthful, carefree manner, to make the transition from lover to beloved: the lady appointed this rare bird of the Roman aristocracy her heir. He was exceedingly shrewd and learned much in this milieu. And because he was so much at home in it he was more than a little contemptuous of day-to-day politics. Most of the time he was too aloof to take matters seriously. And he did not thrust himself forward.

Yet having once committed himself to a firm position, he became serious and would work energetically to accomplish any task that had been entrusted to him – not always to the ultimate degree, for he might suddenly relapse into indifference, but at least to the penultimate. If a matter was really important – and he was adept at distinguishing the important from the unimportant – he would not let go. There was thus a certain playfulness about him, and a strong vein of irony, yet at the same time he was capable of summoning up immense willpower. It all depended.

Sallust writes, 'In Greek and Latin literature he was as well versed as the finest minds. He had great strength of soul, was avid for pleasure, but even more for fame. He lived in luxurious idleness, but no pleasure ever detained him from his duties . . . He was eloquent, subtle, and indulgent to his friends; and by the incredible inscrutability of his mind he could deceive others as to his plans.'

In Sulla the heritage of the Roman aristocracy combined with Greek civilization, to which Rome had begun to open itself. Indeed, it seems as though the interest and receptiveness that flowered under Greek influence intensified the Roman element in Sulla, though in his ability rather than in his temperament. It liberated his versatile talents. This mixture of cultures and the potential for personal development that it engendered may go some way to explain how the late republic came to produce so many figures of more than life-size stature. And in many respects the mixture we find in Sulla is very similar to the one we find in Caesar – similarly rich, but not so solid.

Sulla quickly mastered any subject and soon became a match for any expert. He seems to have had a remarkable capacity for concentration, combined with intelligence, energy and courage. One of his enemies remarked that two animals dwelt inside him, a fox and a lion, but of these the fox was the more dangerous.

He is said to have been capricious. Many instances are cited. Above all, he was not wedded to Roman society: he was inwardly too free and his interests too diverse. He saw things from too many angles and had too many distractions. His standards were too high, and he was too easily disappointed, too often repelled. On the whole he was generous, unless he was concentrating on something of vital importance; he was therefore inconsistent. Not being firmly embedded in the aristocracy, he was not wholly at ease with its rules and principles.

On the other hand, he was neither willing nor sufficiently detached to evolve his own rules and make himself the central point of reference for his actions, as Caesar did later. Above all, Roman society was not sufficiently weak.

This was probably connected with his faith in fortune. He felt himself to be the darling of Aphrodite. He recalled that the enterprises he had embarked upon boldly and at the right moment had brought him greater success than those that were carefully planned. Mommsen speaks of the 'belief in the absurd that inevitably arises in someone who has lost faith in a coherent order, the superstition of the lucky gambler who feels privileged by fate to throw the right number every time and everywhere'. Caesar too believed in fortune and felt he had a special affinity to Venus. But if fortune seemed to help Caesar on a course he had consciously chosen, she seems to have preserved Sulla from following such a course, at any rate for long.

As Sulla lived life to the full, his time was precious. He was therefore impulsive and impetuous, a man of quick resolve who could not bear to beat about the bush. He seems to have been impatient with others' misgivings and with the to and fro of protracted deliberations. He saw things in relatively clear and simple terms, which naturally meant that he was out of place in the oligarchy.

It is now customary to describe Sulla as conservative. This either tells us nothing or it is wrong. For on the one hand all Romans, even the reformers, were conservative. On the other hand, Sulla showed too little consideration for the Senate and its rules to be a particularly distinguished member of that body. He was simply a realist and unable to see how Rome could be ruled otherwise than within the inherited structure. This structure can hardly have inspired him, but it dictated his actions. He planned no innovations, but simply performed the tasks that he felt incumbent upon him, though admittedly in a somewhat unconventional fashion. He was probably determined to do what was necessary, but the fact that this determination had such appalling consequences cannot really be squared with his alleged conservatism.

However, because he was extremely energetic, a distinguished officer, a superb organizer, a successful commander and an inspiring leader of men, he was able to do more than all the others in the Social War to ensure that Rome overcame its military difficulties. For this reason – and no doubt also because he had just married into one of Rome's leading families, the Metelli – he was elected consul in 88 at the fairly advanced age of fifty. It may be that, in view of the situation, the upper aristocracy had particular expectations of him and that this explains his marriage. At all events Sulla, true to his nature, tried to discharge his consular duties responsibly, and this meant crossing swords with the 'popular' tribune of the people. He failed to see how anyone could conduct such an anti-Senate policy and, moreover, form a close alliance with the equestrian order to the latter's advantage. Sulla, after all, still retained his aristocratic pride. It went quite against the grain to allow himself and the Senate to be defeated. Such disturbances and pretensions must be ended. The reforming impulse of Livius Drusus may have been halted, but it was not yet over, and it was more urgently needed than ever. Armed with the *senatus consultum ultimum*, Sulla struck, as only one consul had ever done before. There was a great difference, however: Sulla employed the legions, without waiting for the Senate's instructions.

He was determined to assert his will. He wanted to perform his task properly, in the interests of Senate, and if necessary in defiance of the senators. After the Senate majority had acted so contemptibly in abandoning Drusus, Sulla was unlikely to be impressed by its considered – and justified – reservations. He did not understand how it could expect to emerge stronger this time after an important reform. Asked why he was marching on the city of his fathers with armed force, he replied, 'To free it from the tyrants.'

* * *

Thus in 88, for the first time, a Roman army marched on Rome with hostile intent. Such were the misgivings of the senior officers that all but one *quaestor* refused to take part in the march. The second consul, however, joined it. The senate sent envoys to persuade Sulla to turn back; it was obviously not acting solely under pressure from his opponents. But the envoys achieved nothing.

Sulla's troops occupied the approaches to the city at three points. Two legions pressed on in close formation from the Esquiline gate to the centre, colours flying and trumpets blaring. They were pelted and shot at from the houses. It was only when Sulla ordered his men to set fire to the buildings and discharge burning arrows that the belligerence subsided. The soldiers pressed on to the Esquiline market (close to where the church of Santa Maria Maggiore now stands); here they were confronted by Marius and Sulpicius, who had quickly assembled a troop of soldiers. The advance came to a halt. Sulla, however, brought up reserves and sent other troops across the Subura to bypass Marius. The defenders dispersed, and many hastily left the city. During the night Sulla's legions camped in the forum. The two consuls were constantly on the move to ensure that discipline was maintained. Looting was severely punished, and casualties were few.

Caesar was now twelve years old. It is not known whether he witnessed these events, but his family must have told stories about them and expressed various surmises and apprehensions.

By demonstrating his military might, Sulla persuaded the Senate to declare Marius, Sulpicius, and ten others public enemies. The Senate complied, but with reluctance, especially in the case of Marius. Sulpicius was arrested and killed; his head was delivered to Sulla, who had it placed on the speaker's platform. The others escaped with their lives. Marius fled to Africa and joined those of his veterans who were settled there. Sulla repealed the laws introduced by Sulpicius and hastily pushed through a number of reforms in favour of the Senate. The most important of these laid down that in future no legislative initiative by the tribunes of the people might be pursued without the Senate's approval.

It was particularly important to Sulla that he should be able to leave trustworthy and energetic men behind as consuls for the following year. However, a groundswell of resentment built up against his arbitrary violence and the breaching of all the barriers that had so far prevented civil disputes from spilling over into civil war. The resentment even spread to the Senate, which could not see that it and Sulla were in the same boat. As a result, his candidates failed to be elected. One of the new consuls was an opponent of his, Lucius Cornelius Cinna, the other a rather strange and indolent man who could scarcely have been less fitted to administer the temporary legacy left by Sulla to the Senate. Sulla hesitated to confirm Cinna as consul. With stronger senatorial support he would probably have called a fresh election and manipulated the electors into producing a more desirable result. This had happened before and was by no means uncommon in emergencies. To everyone's surprise, however, Sulla contented himself, in the absence of senatorial backing, with extracting from Cinna a solemn oath that he would not tamper with his laws. It is not clear whether he believed that Cinna would adhere to this oath. But it suited him to believe it, especially as he was impatient, after so much delay, to take the field against Mithridates, who was advancing in the east.

Cinna's most pressing business on becoming consul was to petition the Senate to declare Sulla an enemy and repeal his laws. The main issue was the tribal allocation of the new citizens. The Senate resisted and was supported by the second consul, who now came to the fore. What mattered was not Sulla, but the cause. Violent street-fighting broke out. Finally the Senate stripped Cinna of his consulship and even his citizenship – which was unprecedented – but failed to act consistently by pursuing the fugitive. As a result, he was able to assemble an army, as Sulla had done a year earlier. He first recruited the legion that Sulla had left behind to continue the siege of Nola. He was joined by a substantial number of new citizens, whose interests he espoused. Marius returned from Africa and raised an army in Tuscany, calling slaves to arms with a promise of emancipation. Together they marched on Rome. Marius took Ostia and cut off all seaborne supplies.

The city was defended by regular armies, which seem to have been superior in numbers and experienced in war, but incompetently led. One of the commanders, Quintus Metellus Pius, while loyal to the Senate, became apprehensive when his men exchanged greetings with friends in Cinna's camp, and would not commit his forces, preferring inactivity to the risk of desertions. The other commander was Pompeius Strabo, the father of the 'Great Pompey', the consul of

89. He had confidence in his army, but wished to avoid fighting. He tried instead to exploit the situation in order to win a special position as the saviour of the city. As the Senate clearly did not intend to accommodate him, he decided to sit it out. He also treated with Cinna. The Senate's armies moved on to the defensive. An initial defensive battle was successful, but on the whole it was a phoney war. While the soldiers of Cinna and Marius were sure of their objectives, their opponents had no clear idea about what they were to do. Finally, when Pompeius died in an epidemic, his soldiers wanted to fight, but not under the consul, who was said to lack fortune. This impression seems to have stemmed from his rigidity. Metellus, however, whom the legions wanted as their leader, refused on constitutional grounds.

The Senate was set upon using ordinary means to master an extraordinary situation. It had long been clear that this did not work in internal politics – hence the invention of the *senatus consultum ultimum*. Yet in military matters the Senate still took it for granted that the soldiers would obey orders, even though Sulla's march on Rome had shown that the bond between them and their commander could outweigh senatorial authority if they had confidence in his leadership and were convinced of his commitment to their interests and well-being. The Senate had perhaps not yet fully grasped this unpalatable truth, which was hardly compatible with the pride of the fathers, the masters of the world. What is more, the *senatus consultum ultimum* was a demonstration of power, beside which any consideration shown to the soldiers would have appeared as weakness. To what extent they were aware of the bearing this had upon the relation between a great individual and the Senate is a question that cannot be pursued here. At all events, the Senate was supposedly sovereign. The outcome was that it was defeated. This led to massive defections. The Senate invited Marius and Cinna to return to Rome, on condition that they spared the lives of all the citizens.

Upon their return the killing began. Marius' fury against his opponents knew no bounds. The rough, ambitious and vulnerable *homo novus*, having been so shabbily treated and so often disappointed, gave vent to all his resentments. Whoever had opposed him, recently or in years gone by, and many of those who had angered the uncouth upstart by treating him with disdain, were ruthlessly put to death. Among them were one of the consuls (the other committed suicide), four *consulares* and several prominent senators. As Mommsen puts it, Marius avenged every pinprick with a dagger-blow. Sulla was outlawed and declared an enemy, his house destroyed and his estates laid waste. Marius' bodyguard, consisting of runaway slaves, spread such terror that Cinna finally had them taken unawares as they slept and murdered to the last man. The Cinnans then established their rule, which lasted from 87 to 82. Marius was elected consul for 86, but died shortly after assuming office.

We are told nothing at all about Cinna's personality. Perhaps it was too colourless; at all events, after his death this period of Roman history was almost wholly erased from memory. He must have been able, to judge by his deeds. Owing to the absence of sources, we cannot tell whether he was anything more than an opportunist. His political planning tended to be reactive. Whether he thought beyond his victory is unclear. He wanted to reinstitute the law on new citizens and put it into effect. To this extent he was energetic. It is impossible to say whether he knew what to do next. What he did was roughly what had to be done in such a situation: he had himself repeatedly elected consul, tried to restore order in the war-torn economy and sought to establish tolerable relations with the Senate.

Neither he nor the Senate had anything to gain from a renewal of the civil war. The Senate sought a reconciliation with Sulla, which Cinna apparently did not oppose. There was an obvious wish to restore Sulla's rights and leave him with his command; this was not difficult after the death of Marius, and it made sense in view of his gifts as a commander. In return he was to give up all claims to vengeance for the wrongs done to him and other leading nobles. If Cinna thought he could appease Sulla in this way he seriously misjudged him. If not, his tactics were shrewd: he offered reconciliation and a peaceful settlement. Only in 85, when war seemed inevitable, did he begin to arm against Sulla. The following year he intended to lead an army across the Adriatic into Macedonia, but was killed in a mutiny.

During this period Caesar's childhood came to an end. He donned the male toga and began his apprenticeship in the forum. At the age of fifteen he lost his father. More important, however, was the fact that a year later (clearly in 84 BC) he moved to the very centre of Roman society. Cinna gave him his daughter Cornelia in marriage, and his engagement to Cossutia, who came of a prosperous knightly family, was broken off. The connection with Cinna was an astonishing distinction. The victor of the civil war could hardly marry off his daughter to just any young man, however illustrious his family. Cinna's other daughter was married to Gnaeus Domitius Ahenobarbus, a brave young nobleman of excellent stock. One wonders whether the marriage was connected with Caesar's being a nephew of Marius. Or was Burckhardt right when he remarked of the impression that the young Caesar made on Sulla: 'Something of the extraordinary nature of the person concerned usually becomes evident at an early age,' so that the great man is exposed to special 'early dangers'? But was Caesar great? Did he already possess some of the charisma, the brilliance, the many-sided superiority and assurance of success that he later radiated? Was something of his great potential already discernible?

About the same time Cinna decided that Caesar should be appointed to the distinguished office of priest of Juppiter, which had just fallen vacant. It is not clear whether Caesar was actually inaugurated. The office carried special privileges and was accessible only to patricians. On the other hand, it imposed many duties on the holder, requiring him to perform sacrifices and conduct ceremonies, and above all it was hedged about by numerous regulations.

The *flamen dialis* was deemed to possess a magic power that must be carefully guarded. At all times – at least when he was out of doors – he had to wear the *apex*, a fur cap with cheekguards tied under the chin and a special ornament on the top. He was forbidden to mount a horse, and he must not set eyes on armed troops; on holidays he must not see anyone working. His hair could be cut only by a freeman using a bronze knife, and had to be buried in a special place together with the parings from his nails. There must be no knot in his house. A table without food must never be set before him, for he must be spared even the impression of want. At one time the priest of Juppiter was not allowed to hold a magistracy, because this required him to swear an oath on the laws and exposed him to the risk of cursing himself. However, ways had been found to circumvent this regulation, and the last holder of the office had even become a consul. Yet this priestly office cannot have been an unmixed blessing, at least not for an ambitious young nobleman. It was customary to bestow it on patricians whose poor health precluded a political career.

Caesar and Cinna may have seen it differently. To a house that had long been politically insignificant, such an honour might seem highly desirable, possibly even opening the way to a consulship. And for Cinna it was clearly important that it should go to someone close to him. For honours were exceedingly important in Rome's aristocratic commonwealth, and especially to a usurper. Moreover, the priest of Juppiter had to be married to a woman of patrician stock, and Cornelia clearly fulfilled this requirement.

Caesar thus assumed his first political commitment, not to a cause – for Cinna had no cause, once the problem of the new citizens was settled – but to certain allies, and above all against the old nobility. Outwardly Cinna ruled in accordance with the forms of the old republic – save that the leading men repeatedly had themselves elected consul. Within these forms he clearly wished to create a solid power-base, but only for himself and his allies. In some respects this seems surprising. If it was really necessary to bring a new group of politicians to power, the rules governing a career in office were a serious impediment. In particular, new men had to rise slowly through the lower grades before acquiring great influence in the Senate, since the senators held their seats for life and the hierarchic structure of the house corresponded to the grades in the magistrature. It may of course have been assumed that the united policy of such a block of electors – and pressure from Cinna – would oblige candidates for the higher offices and all who supported them to show greater consideration. Yet these intentions in no way amounted to an attempt to change the form of the republic.

Although Cinna adhered faithfully to the rules, he had dealt a serious blow to legitimacy: first through his usurpation, and secondly through all the killings and suicides that followed his entry into Rome. To the Roman way of thinking the republic consisted not only in its forms and traditions, but in the continuity of the aristocracy, which was concentrated in the circle of the leading senators, the former consuls or *consulares*. They were the *principes civitatis* ('the first men of the citizenry'). They were at the centre of all responsibility. It was their task to represent the cause of the Senate and the whole of the *res publica*, and since the senators were solidly committed to this idea, the *principes* enjoyed supreme authority. Being engaged in a common task also engendered political solidarity, at least among the bulk of the *consulares*. They and their cause embodied the Roman aristocracy; in them its age-old traditions lived on, and they were very conscious of their duty to keep them alive. Their recognition was, for instance, extremely important to a candidate for a consulship, if he was to have any prospect of success. They thus constituted a source of legitimacy. Whoever destroyed them – or decimated their ranks – struck at the heart of the nobility, even if he spared the aristocratic forms, and thereby attracted an odium of which he could not easily rid himself. Many individual nobles might make their peace with Cinna, but this was not so easy for the nobility as a whole. At that time, wrote Cicero, the republic was *sine iure et sine ulla dignitate* ('without law and without any honour').

A clear distinction must be made here. Most of the senators favoured a middle course. In the main they were against Cinna, but not at all in favour of Sulla. For, despite all that had happened in the meantime, the Senate could still not condone Sulla's march on Rome. And from a resumption of civil war they could only fear the worst, especially on the part of someone as harsh and resolute as Sulla. On the other hand, he professed to champion the cause of the nobility – not of the Senate! The honour and standing of the old ruling class had to be restored. Hence, many nobles could not forbear to look hopefully to Sulla; and they could hardly be reconciled with Cinna. Considerations of Senate policy and the hopes of the old ruling class thus pointed in different directions. For the time being the former were in the ascendant, but subsequently the latter came to the fore, until finally Senate and nobles were reunited under Sulla.

Yet however skilfully the Senate, led by the surviving *principes*, exploited whatever room for manoeuvre was still left, a mood of despair and a sense of impotence prevailed. Ultimately it was the civil war leaders who decided what happened. Between his sixteenth and nineteenth year Caesar got to know this feeble Senate from the perspective of one of the parties. Being youthful and imbued with high standards, he may already have come to despise it. Unlike many of his generation he had no reason to identify himself with the fathers in a spirit of anguished defiance, or to pin his hopes on the armies in the east. His later persistence in such a critical, dismissive and contemptuous attitude may of course have been due to Sulla himself, to the way in which he had exploited his victory and behaved towards Caesar himself, threatening him and exposing him to danger.

Sulla had at first devoted himself with energy and courage – and success – to the war against Mithridates. Receiving no supplies from Rome, he ruthlessly seized all the treasures that Greece had to offer, including the finest works of art, and converted them into money. When necessary, he imposed the severest discipline on his troops. But otherwise he allowed them a seductive degree of licence. They were to fight and be loyal to him. For the rest they might do as they pleased; even mutinies could be forgotten. His generosity in this regard matched his own excesses. Of the life of the troops behind the lines Sallust writes: 'It was at this time that the army of the Roman people first became accustomed to whoring and drunkenness. It learned to admire statues, paintings and chased vessels, to steal them privately or officially, to loot the holy shrines, to defile things sacred and profane.'

As has already been said, Sulla was a superb leader of men. He had the common touch and sometimes fought in the front line; he was a comrade whom they could meet as man to man – and who impressed them all the more by being anything but one of them. He was a dare-devil who seems to have excelled at obliterating all evidence of planning. And the soldiers trusted in the fortune of their leader.

In a comparatively short time Mithridates was defeated and forced to accept terms. At first he wanted to remain in the act and profit from Sulla's unfavourable situation. When the two met in western Asia Minor and the king approached Sulla with outstretched hand, Sulla asked him coolly whether he wished to end the war on the terms offered. When Mithridates remained silent, Sulla said, 'It is for the supplicant to speak first; the victor can remain silent.' Mithridates then launched into an elaborate self-justification. Cutting him short, Sulla said that he had now seen what a master of eloquence the king was, for even after such ruthless deeds he was not at a loss for fine words. He repeated his terms; only after Mithridates had agreed to them did Sulla greet him. Mithridates had to surrender the lands he had conquered, but otherwise escaped with impunity. This was a breach of Roman custom, and the soldiers protested. But Sulla was anxious to return to Italy.

He informed the envoys from Rome that he was ready to accept the allocation of the new citizens to all the tribes, but could not waive his demand that the guilty among the Cinnans be punished. He was deeply offended personally and on behalf of the nobles who had been murdered or driven from their lands. His was an archaic personality. He left instructions that the epitaph on his gravestone should state that he excelled all his friends in doing good and all his enemies in doing ill. In small things he could be generous, but in important matters he was not to be trifled with. However, he may also have felt that reason of state forbade him to ignore the excesses of the Cinnans and behave as if nothing had happened.

Just before landing in Italy, Sulla seems to have relented. He now insisted only on the restoration of his own rights and those of others who had sought his protection. The Senate was prepared to consider this, but Papirius Carbo, the new leader of the Cinnans, frustrated the move. No one knew what Sulla would do when he returned to Italy with his army. He could certainly have demanded a triumph. Had he really waived his claim to vengeance? Could he still be reconciled with the Cinnans? There was every indication that the civil war could be ended only through the victory of one of the parties.

The Cinnans were numerically far superior. More than fifteen legions, some of high quality and ready for action, recruited largely from among the new citizens, were matched by five, which included auxiliary troops: well over a hundred thousand men against a mere forty thousand. Before crossing to Italy, Sulla was not sure that his army would not disperse, once back on Italian soil. But the soldiers swore to be loyal to him and submit to extreme discipline, so that the war would not cause unnecessary damage in Italy. They even collected money to support his enterprise, feeling that they and Sulla had a common cause. Sulla would not accept this sacrifice. After landing he was joined by various nobles. As soon as it became clear that war was unavoidable, their place was at his side – unless they preferred to perish rather than join him in marching on Rome. Two of them came with a few thousand men. One, the twenty-three-year-old Pompey, came with a regular army. On his own initiative he had raised three legions from among his large clientèle in Picenum, which included veterans who had served under his father. Pompey had realized what civil war was. And with an audacity inspired by youth rather than temperament he sought to exploit the situation for his own ends. Sulla treated him with great respect and, contrary to all convention, greeted him as *imperator*, a designation proper to a general who commands an army in his own right. The other nobles, who, as Mommsen puts it, 'wanted to be rescued for the good of the commonwealth and could not even be brought to arm their slaves', treated Sulla with some disdain, which became all the more marked, the closer he came to victory and the more their self-assurance grew. He preferred to rely on officers who did not belong to the high nobility.

After a period of uncertainty the first battle ended in a surprising success for Sulla. His soldiers began to despise the enemy. This initial success was followed by a number of others; some were won with difficulty, some easily, but all were psychologically important. In due course large numbers of the enemy defected. When the opposing legions were close to one another and contacts developed, the Sullans were regularly able to convince their opponents of their own superiority and the goodness and generosity of their leader. In this way Sulla's army trebled in size. We hear of no defections from his ranks.

Only a few contingents, among them naturally the Samnites and Lucani, remained loyal to the opponents' cause. Their old urge for independence reawakened. The only way to be rid of the wolves that had robbed Italy of its freedom, declared one of their generals, was to destroy the wood they lived in. Finally in desperation, his defeat already sealed, he led his troops against Rome. On 2 November 82, by dint of forced marches, Sulla's army arrived outside the city just in time to save it from destruction after a hard-fought battle by the (northeastern) Porta Collina. Having delivered Rome from extreme peril, Sulla was able to enter the city. The civil war had lasted a year and a half. A little earlier, Marius' son had ordered the killing of the last of the leading senators still in Rome.

Sulla at once set about liquidating his opponents. Several thousand Samnite and Lucanian prisoners were massacred on the Campus Martius, in the Flaminian circus, or in the Villa Publica, a public arsenal. Not far away, in the temple of Bellona, he had called the Senate together for its first session since his return. When the fathers seemed troubled by the screams and lamentations of the victims, he observed calmly that they should pay no heed to them, but listen carefully to him; only a few criminals were being punished on his orders. He wished to set a warning example to all sides, especially to the Senate.

On his recommendation an *interrex* was appointed. Sulla wrote to him that he thought a dictator should be appointed to pass laws and restore order to the republic – not for a fixed term, but until such time as the city, Italy, and the provinces had been fully pacified. Hitherto dictators had usually been appointed for six months, to perform limited tasks or conduct military operations. Rome now faced an internal emergency, and there was much to be said for setting up the old emergency magistrature with its unlimited powers. Sulla did not omit to mention that he believed he could be of use to the city in solving this problem too. The law that was then passed expressly stated that all his decrees should have legal force and that he might with impunity have any citizen put to death.

When the killing showed no sign of ending, the Senate appealed to him. It did not wish to plead for mercy for those whom he saw fit to kill, but could he not free from uncertainty those whom he intended to spare? Sulla took this suggestion literally and had one list after another posted publicly: whoever appeared on the lists was outlawed ('proscribed') until 1 July 81; there was a price on his head, his property was forfeit to the republic, and his descendants barred from public office.

The atmosphere of Rome was one of increasing horror; never before had it known such systematic horror. There is much to suggest that it was only the ferocity of the long civil war that convinced Sulla of the need to annihilate all the leaders of the Cinnan régime. It may be imagined that the lists included the names of others who would have welcomed the elimination of leading Sullans. However, the proscriptions were the prelude to Sulla's reforms, yet at the same time they encumbered them with a heavy burden. Forty senators, one thousand six hundred knights, and many others were proscribed. It is clear that Sulla's particular hatred was directed at the rich 'moneybags' (*saccularii*) who had succeeded, under Cinna, in acquiring a status and a degree of wealth to which they were not entitled. Rome was to be ruled once again by the aristocracy and the Senate. If there was no other way of dealing with those who had frustrated every reform since 95, they must be eliminated. And their sons must have no chance of avenging their fathers. Sulla, though normally inclined to generosity and clemency, now behaved with murderous single-mindedness, believing that order must at last be restored to Rome. And since the Senate could not achieve this – and indeed had grave misgivings about the killings – he must act on its behalf.

Caesar belonged to the circle of Sulla's victims. As Marius' nephew and Cinna's son-in-law, he was a man of some importance; he probably posed no immediate danger, but in time he was bound to become a focus of loyalty for many old supporters of Cinna and Marius. Such relationships could carry great weight. When his mother's cousins interceded for the eighteen-year-old, Sulla demanded – as he did in other cases, in Pompey's for instance – that he should part from his wife and renounce his family. Caesar refused, and as far as we know he was alone in doing so. In consequence he was deprived of his priestly office (or his claim to it), Cornelia's dowry, and his inherited fortune. He was no longer safe. He was sick and obliged to flee from one hiding place to another; on one occasion he had to pay a large sum to buy his freedom. When his relatives obtained a pardon for him, Sulla is said to have remarked that one should be on one's guard against the sloppily belted youth, who had more than one Marius inside him. This sounds like a prophecy invented after the event. But it need not have been untrue. Why shouldn't Sulla, like Cinna, have discerned something of what Caesar had in him? Why shouldn't Sulla even have sensed a certain affinity?

Suetonius reports that Caesar was of 'imposing stature – white-skinned, slim-limbed, rather too full in the face, and with dark, lively eyes'. He is said to have paid much attention to his appearance. 'He not only had his hair carefully cut and his face carefully shaved, but also removed his body-hair, and was criticized by certain people for this.' His dress was noteworthy too. He wore his belt very loose. All this seems to indicate a sensitive, youthful pride, a certain – admittedly diffident – sense that he was something out of the ordinary, and perhaps also a feeling of isolation. Yet at the same time a degree of refinement and cultivation. Things that were probably not entirely alien to Sulla. And even if he saw nothing of himself in the sloppily belted youth, he may very well have evoked it by his warning. They were both cast in similar moulds.

Be that as it may, why shouldn't Sulla have responded to the importunate requests of Caesar's influential friends by saying, as it were, 'Very well, if you insist, but watch out!' To react in this way he need not have been a prophet, but simply prompted by a momentary access of exasperated generosity. He must have seen that it was dangerous to let a man with this family background survive in Rome – unless he renounced it. And Sulla must have found it monstrous that Caesar refused. Not least because he was used to different conduct. Perhaps he wished to draw Caesar on to his side through another marriage.

Caesar was probably outraged by the demand that he should part from his wife and felt even more bound to her and her family. At all events, his defiant decision committed him, at least for the time being, to a lost cause. For the side to which he chose to adhere had little to offer him, save for certain links with the sons of those who had been proscribed – which he continued to cultivate – and with a few other families, and a certain popular attachment to Marius. This need not have been immediately clear to him. Until recently Cinna's party had been able to rely on the strong forces it commanded throughout Italy. Sulla was not popular, and even many of the senators were against him. Why should events not take a fresh turn? Yet even if Caesar envisaged such a turn, his action was due less to calculation than to a desire to remain true to himself. He was disgusted by the way in which others truckled to the dictator. He would not be thrown off the course on which he had embarked. He knew – or was starting to learn – what he owed to himself, and this was a great deal. In this way Caesar, unlike the whole of Rome, successfully resisted the insolent demands of the powerful dictator.

This was all the more significant as Sulla's power was generally regarded as absolute. He was the victor – on behalf of the nobility, it is true, but hardly with its help. The armies had decided the issue, and the nobles had had none. Sulla could continue to rely not only on the authority of his office, but on the hundred and twenty thousand veterans – his old soldiers and those who had defected to him – whom he now began to settle, probably in close-knit communities, at various key points in Italy, as well as the ten thousand slaves he had freed from their proscribed masters.

As a token of his power he defied convention by appearing in the city with twenty-four lictors. Lictors were the official servants of the magistrates and accompanied them wherever they went, clearing a path and procuring respect for them. In the city they carried the symbol of executive authority, the *fasces* (bundles of rods); in the field they carried the *fasces* with an axe. The consul had twelve, the praetor six, and the dictator in the field twenty-four. Before Sulla, however, dictators had contented themselves with twelve in the city.

After all that had happened, Sulla wanted to introduce a thorough and coherent reform. He placed little trust in the collaboration of the Senate. He was discouraged by the experiences of 91 and 88, and probably by what he knew of current senatorial society. The oligarchs could not be used to restore the oligarchy: it had to be imposed on them. Sulla had therefore no wish to collaborate with the Senate to more than a limited extent; he may occasionally have consulted the fathers, but did not allow them to decide anything. He did, however, submit his proposals to the popular assembly for ratification.

In the first place he wanted to establish a consistent senatorial régime, in the second to restrict the scope and potential for action to their former dimensions, which he had done more than anyone to enlarge, and finally to introduce a few other arrangements that seemed to be of practical value.

His chief concern was probably to bring the tribunes' right of legislation under the control of the Senate. In future they were to introduce no bills without the Senate's approval. They thus lost a right that they had enjoyed since 287 BC. They even lost the right to have resolutions of the *plebs* formulated, which they had won early in the fifth century. In order to devalue their office still further, Sulla decreed that no one who had served as a tribune should be allowed to present himself for any other office. For the rest, he respected the people's rights, including the secret ballot, but put an end to the distribution of cheap grain.

The knights lost not only the jury courts, but their special seats at the theatre, which had publicly distinguished them, probably since the days of Gaius Gracchus, as the second order of the republic, inferior only to the senators. On the other hand, Sulla doubled the membership of the Senate to six hundred. This was no doubt prompted mainly by practical considerations: three hundred senators were no longer equal to the burdens involved in ruling the empire and acting in the jury courts, the number of which Sulla had greatly increased. It also served to broaden the social basis of the Senate. Among the new senators were members of knightly families, and certainly some of the new citizens who had remained loyal to Sulla and the nobility and rendered signal services. We may wonder whether even a body of six hundred was capable of performing all the tasks of leadership incumbent upon the Senate.

Also for practical reasons, Sulla raised the number of quaestors and practors, thus ensuring that the magistrature was better able to meet the increased requirements of the empire and the penal system. Yet in order to keep the size of the magistrature within bounds, he ruled that every consul or praetor, after completing his year's service in the city, should spend a further year as a provincial governor.

One of his laws was designed to regulate the duties of the magistrates, especially on becoming governors. It seems to have included a prohibition against leading an army into Italy, even for a triumph. The military was no longer to interfere in Rome's internal politics.

Elections to the senior magistracies were moved from the end of the year to July, to coincide with tribunician elections. This enhanced the influence, in both sets of elections, of the prosperous classes in the rest of Italy; it also gave the new magistrates more time to accustom themselves to their duties and strengthened the continuity of the magistrature. Sulla brought in a law laying down the minimum ages at which candidates might present themselves for various offices and prohibiting anyone from holding a second consulship until ten years had elapsed. The penal system was overhauled completely, to such good effect that most of the relevant laws remained in force until well into the imperial age.

Using all the legal and illegal means available to him, Sulla did everything possible to restore a consistent senatorial régime that could work and govern efficiently. He could not, of course, cancel the effect of his march on Rome. The new dimensions of politics that he had created could not be restricted by legislation. The horizon of possibilities had significantly widened and henceforth far exceeded anything that could be encompassed within the traditional pattern of expectations, fulfilment of expectations, and expectation of expectations. The outlines of a new and potentially monarchic reality were now discernible.

In 80 Sulla assumed the office of consul in addition to that of dictator. At the end of 80 or the beginning of 79 he laid down the dictatorship. Caesar later commented that Sulla had failed to master the rudiments of politics. Yet his retirement was quite consistent with his policy. He had done what he could, at enormous effort. He did not wish to expose himself to the myriad problems of day-to-day politics. It may be that the many frictions and conflicts had taken their toll. Perhaps he now lacked his earlier endurance and resilience. In any case he had never been interested in trivialities. He was never absorbed in politics. Rome was surprised to see the man who had been responsible for so many killings going about the city unguarded. It was surprised too when he allowed an opponent to succeed in the elections for 78. But he wished to leave this and other things to the free interplay of forces. The Senate should now attend to everything. He returned to his private pleasures and devoted himself once more to the company of actresses and intellectuals, which he had been obliged to forgo for so long. He died at Puteoli (now Pozzuoli) in 78.

A dispute arose over his funeral. The anti-Sullan consul was against a public ceremony, but his colleague prevailed. With regal splendour Pompey brought his body to Rome on a gold-chased litter. Many trumpeters provided the music, and the cortège was joined by senators and knights, as well as many of his old soldiers, bearing their weapons. He had left instructions that he was to be buried. It was recalled, however, that after Sulla's victory the grave of Marius had been ripped open and his bones desecrated. Sulla was therefore cremated on the Campus Martius. The ladies of the nobility went into mourning for a year.

Meanwhile the aristocracy and the Senate realized that, willy-nilly, they were identified with Sulla's cause. The 'most respected and influential citizens' had taken part in his triumphal procession and worn befitting wreaths. This is said to have been the most splendid aspect of the ceremony. Accompanied by their wives and children, they had hailed Sulla as their 'saviour and father'. This was not without its irony, for the situation he had saved them from was essentially of his own making. Soon after his return, the Senate resolved to erect a gilded equestrian statue of L. Cornelius Sulla Felix Imperator in front of the speaker's platform in the forum. Never before had anyone been so honoured in his lifetime, and it was many years before it happened again. Moreover, Sulla had been the first man, since the age of the kings, to extend the hallowed boundaries of the city. It was traditionally held that this might be done only by one who had made great conquests, and Sulla had not. But he clearly wished to demonstrate that he had re-founded Rome, and probably also that the newly marked city boundary would in future prove more effective than the one he had crossed in 88. In this way he was re-enacting the ritual founding of the city by Romulus.

It was especially ironic that the senatorial régime owed its restoration to such an ambitious and self-sufficient outsider, the kind of man it had always strenuously opposed. It was ironic too that this régime, which was generally considered legitimate, should owe its continuance to one of the leaders in a civil war. What is more, the honours accorded to Sulla as the saviour of the republic conferred a status incompatible with republican equality. Yet after all the anger generated by the power and pretensions of the nobles had broken out in the upstart Marius and all those who wished to replace the old ruling class, it was necessary to rehabilitate the nobility. This was a return to normality, and generally accepted as such, but it was also an act of partisan violence. This meant that henceforth the nobles and those senators who supported them – virtually the entire Senate – were Sullans. They were obliged, as Cicero says, not only to retain all the dictator's institutions, but to defend them with public authority, lest greater calamities should ensue.

Hardly anyone was as conscious as Caesar of the partiality of the senatorial régime. This experience probably accounts for the lack of understanding he was later to show for the objective nature of institutions.

Hardly anyone distanced himself so much from the Senate at this time. From his position as an outsider, into which he had allowed Sulla to push him, he had experienced everything: civil war and murder, the power – and, above all, the impotence – of the Senate, arbitrary action and extreme partisanship. Three times he had seen Roman legions willingly march on Rome under their leaders. Clearly anything was possible. Order seemed finally to have returned. But could it last? Could the nobility, which had been rehabilitated without lifting a finger in its own cause, continue to lead the republic as if nothing had happened? And even if so, could it carry conviction with Caesar?

Hardly anyone at this time was so determined to assert himself in opposition. It was in opposition that he evolved his standards and his criticisms. Sulla's successors were hardly able to satisfy them. And so he raised his standards, in order to see how far short of them the Sullans fell. Moreover, he repeatedly vented his fury against them, having been unable to vent it against Sulla. This determined his future course.

One wonders whether Caesar was casting around for a model and, if so, where he found it. Given the nature of his commitment, he could probably find it only in outsiders. Yet could it have been Marius, who was after all little more than a brave soldier and an able commander – and perhaps a kind uncle? Or the unsuccessful Cinna? Or the Gracchi, whom Caesar probably admired, but could hardly emulate in an age of civil war?

Only one man had acquitted himself brilliantly, and that was Sulla. True, he filled Caesar with loathing, but one wonders whether this was the only feeling he aroused. At some level of consciousness Caesar must also have been deeply impressed by this terrible and fascinating figure. Sulla's influence was not so much formative as liberating. It opened new prospects. Caesar may even have been able to sympathize with Sulla's forceful, high-handed and inconsiderate treatment of the Senate. He had measured himself against him once, and he could apply the same measure in future. A terrible, detestable model, and probably all the more detestable for being secretly admired: a nobleman whose intentions were conservative, but whose views set him apart from senatorial society, who became a statesman out of duty, but was by inclination a *bon vivant*, intent upon living life to the full; modern in his acceptance of many practical necessities and in his receptivity to all the possibilities of the age, yet at the same time archaic – in a novel way – in his political claims and his insistence on *dignitas*; a personality in which the threads that had once composed the fabric of Roman society pulled in opposite directions. It may be that no one was as susceptible to the whole of this personal reality as Gaius Julius Caesar, Sulla's junior by thirty-eight years.

The First Test
The Experience of Rome in the Decade after the Restoration (78–70 BC)

THE FARCE THAT FOLLOWED THE CIVIL
WAR · APPEARANCE IN THE FORUM ·
VOLUNTARY EXILE IN THE EAST ·
APPOINTMENT AS PONTIFEX · ROME
IN THE SEVENTIES · THE SPARTACUS
UPRISING · CONFORMITY PREFERRED
TO ACHIEVEMENT · THE RETURN
OF POMPEY · POMPEY'S CHARACTER

POLITICS ARE MORE THAN SIMPLY POLITICS if they are the central element in the life of a class. This may have no bearing on the way in which they are conducted, but it crucially affects the life of the individual. Failure in politics then means total failure, since political rank is not just one goal among others.

The unavoidability of politics restricts not only the individual's opportunities, but his expectations. The question of how to order his life – which of his potentialities to realize – does not arise. In such a world he cannot choose his life, but only lead it, following a pre-ordained path. What he expects from it is what is expected from him.

Yet he can make his absorption into the political world easier or harder for himself. And if the limits of the allotted field become blurred and he breaks through them at some point, wide vistas may seem to open. While going through the required motions, he is not obliged to think the required thoughts. He may set himself a number of goals and pursue them with such energy that his actions are directed more to the goals themselves than to the approbation of others. The world then expands. He is no longer tightly locked into his environment; the resonance within him increases, and he concentrates on more distant objectives. Certainty of the future lifts him beyond the present, creating a greater detachment from the age, enabling him to see himself differently, in a longer perspective. Personal independence is probably always a certain investment in time. This seems to have been true of the young Caesar.

When Sulla died, Caesar was in the east. Since 80 BC he had served in the army, first in the province of Asia and later in Cilicia. At that time he had been with Nicomedes, king of Bithynia, and distinguished himself as a soldier. On hearing of Sulla's death he returned to Italy – in haste, Suetonius tells us, and 'in the hope of new troubles, set in train by Marcus Lepidus'. Although Lepidus made him tempting offers, Caesar did not join him, being distrustful of his *ingenium* – his character, courage, insight and ability – 'and also of the situation, which he found to be worse than he expected'.

M. Aemilius Lepidus was the consul for 78 who had attacked Sulla's work and then sought to prevent his public funeral on the Campus Martius. He appears to have been a particularly ambitious opportunist, a weak and vain man whom the civil war had detached from the aristocratic code of conduct; this detachment, however, brought out no great talent, but merely fostered a narrow cunning, of which he was quite proud. He had enriched himself through Sulla's proscriptions and extorted a great deal of money as governor of Sicily. With this he is said to have built the finest house of the period.

We are told that he was the first to import Numidian marble for the thresholds, a costly material for a trifling purpose – merely for ostentation – and this is said to have been much criticized in Rome. Lepidus also lavishly restored the great Basilica Aemilia, which his grandfather had built in the forum.

He now perceived a chance of rallying the abandoned supporters of the defeated régime and winning a powerful position in Rome as their leader.

After Sulla's death he worked for the repeal of his laws – we do not know how strenuously or how openly. In particular he sought to restore the authority of the tribunes of the people, the distribution of cheap grain to the urban populace, and the rights of the sons of those who had been proscribed; he also wished to see the land that Sulla had confiscated for the settlement of his veterans returned to its former owners, most of whom were new citizens and had supported Cinna.

There were enough victims of Sulla in Italy, people whom he had defeated, deprived of their rights and impoverished. The Cinnans had had a wide following. Revolution and rebellion were still fresh. Lepidus could thus hope that Sulla's death would be a signal for an uprising. Sulla's opponents still had strong forces in Spain, and the local Sullan commander was no match for them. The discontented were gathering in both Rome and Etruria (present-day Tuscany). In Faesulae (Fiesole) Sulla's veterans had already been expelled by the former landowners.

Yet there was no new civil war in the making – only a farcical sequel to the old one. Both Lepidus and the Senate majority set about enacting it. It was one of the saddest chapters in Roman history.

In the Senate there had already been various complaints about Lepidus, in particular because he had levied and armed troops of soldiers on his own initiative in order to lend force to his policies. A few determined Sullans called for him to be put in his place. But at first there was some reluctance to see anything sinister in the consul's private levying of troops and his public call for the overthrow of the Sullan régime. His friends, together with all those who were unable to see where all this would lead to – the Senate majority, in other words – did not argue politically, but in class terms; they lauded the great deeds of his patrician family, the Aemilians, and refused to believe that he might be its black sheep. This was in keeping with the restoration spirit, after Sulla had re-established the nobility. It was pointed out that Rome's greatness had always thrived on forgiveness and reconciliation. The motion to oppose Lepidus gained only minority support. And for a time he seems to have tempered his demands.

When it came to putting down the rebellion in Tuscany, the Senate was again not united or strong enough to take the necessary measures against Lepidus or to outmanoeuvre him. What could they know? Perhaps they were wrong to suspect him – though no one who had eyes to see could have been in any doubt. Or at least things were not as bad as they seemed. The two consuls were therefore ordered to levy troops and proceed against the rebels. One of them, Quintus Lutatius Catulus, a firm champion of Sullan policies, soon became the leading personality in the Senate. The senators were untroubled by the fact that his colleague was so sympathetic to the rebellion and even suspected of having close links with the rebels. Perhaps they thought that, having first made the mistake of putting the fox in charge of the hen-house, they could remedy it by another – that of yoking the supporter and the opponent of the Sullan régime in a team. At all events, they made them swear a solemn oath that they would not take up arms against each other; any breach of this oath would call down special curses. Now that all suspicion had been allayed by this ingenuous device, the heirs to Rome's ancient political wisdom withdrew to the wings, as it were, and waited to see what happened. They had repressed what they knew – which was a good deal – in order to do what they could – which was nothing.

Lepidus' conduct was at first equivocal. Obviously he too wished to play the waiting game. At first he carried out the Senate's instructions and levied troops. How he used them was his affair. It may have seemed curious that he did not attack the rebels openly. But who knew what he had in mind? And for the time being Catulus could apparently do nothing either, as he was careful not to be caught between his fellow consul and the rebels. What was more striking was that Lepidus did not come to Rome to conduct the elections. Yet perhaps his duties prevented him. The elections were postponed for more than six months, until 77. And the senators were unwilling to put pressure on the consul.

By the beginning of 77 Lepidus had virtual control of Tuscany. He commanded a regular army, had garrisons in many towns, extorted money from the inhabitants, and took possession of the Po plain. When his consulate ended he was, strictly speaking only the governor of his own province of Gallia Transalpina (Provence). But he did not take such a strict view. Nor did the senators, who were probably still waiting to see whether he would in the end prove the stronger.

Everything was set for the rebellion, but Lepidus was clearly as reluctant to strike as the Senate was to recognize that it had for a long time been confronted with open insubordination. The one seems to have stared at the other as the rabbit stares at the snake – except that there was no snake.

Lepidus lacked the support he needed. The power he hoped for had apparently not materialized. Had he acted more decisively he could presumably have won many supporters. Yet he could not bring himself to do so. For the time being he was strong because the Senate was afraid, but he was unable to exploit its fear because he too was afraid – or at any rate waited until he had enough support for his rebellion. He therefore vacillated. As Sallust puts it, 'he feared peace, but he also hated war.' He probably thought that all avenues were still open to him, and so he once again tried the legal avenue: he asked for his consulate to be renewed, as though this were the object of all the preparations, as though he would then be satisfied and act in accordance with the Senate's wishes. This at last produced strong resistance. The Senate began to recognize his insubordination.

The advocates of a resolute policy now had a chance. The old *consularis* Lucius Marcius Philippus delivered a powerful speech, which Sallust records in his history (how faithfully we cannot say). One wondered, he said, whether the senators were moved by fear, cowardice or lack of understanding. They wished for peace. But they could not have it. They must resort to war, however much they hated it, because Lepidus desired it, 'unless it is proposed to offer peace in return for war'. Those who had until now spoken in favour of negotiation, of preserving peace and concord (which had long since been destroyed) had been proved wrong and become playthings of events; this was understandable 'since they wanted to win back peace through the very fear by which they had lost it when they had it' (*quippe metu pacem repententes, quo habitam amiserant*). 'By Hercules, the more eagerly you have sought peace, the more bitter will be the war when he [Lepidus] realizes that he is supported more by your fear

103

than by your reasonableness and goodness.' Philippus conjured the assembled fathers 'to be vigilant and not to allow unbridled crime to spread like rabies to those as yet untouched by it. When prizes go to the wicked it is hard for any man to be good without reward. Or do you wish to wait until he leads his army here again and then enters the city with fire and sword?' Nothing further is known of this first march on Rome to which Philippus alludes. It may have taken place in connection with the private levies that Lepidus conducted before being charged with opposing the rebels in Etruria.

Philippus' speech culminated in a motion for the *senatus consultum ultimum*, 'because M. Lepidus has led privately armed troops against this city, in alliance with the most degraded elements and the commonwealth's enemies and in defiance of the decision of this house'. This was to characterize the cause of Lepidus with excessive precision. For, as Philippus had said in the same speech, Lepidus was delaying the actual rebellion and had not made up his mind, though he may have been approaching. And he also had an army of which the Senate had given him command. However, some way had to be found to deal with this pusillanimous rebel. Philippus tried to do this by saying that although the war had not yet begun, the peace was already lost. He was certainly right in believing that the endless waiting and shadow-boxing was bound to keep the city in turmoil and increasingly paralyse the Senate's position, while the anxiety to preserve the peace increased the danger of war. Things had obviously gone far enough; Philippus was therefore able to convince the Senate.

At last the Senate bestirred itself to oppose Lepidus and succeeded in forcing him to fight. Catulus armed against him, and many Sullan veterans were ready to join him. Pompey was instructed to win back northern Italy with his own levy of veterans and clients. Lepidus marched on Rome, but was defeated on or near the Campus Martius. The Senate's contingents joined in driving him to the Etrurian coast. From there he fled to Sardinia, where he died soon afterwards – not out of despair at his situation, Plutarch assures us, but 'because a letter fell into his hands which persuaded him that his wife was deceiving him'. Even if this was not true, it was a nice invention.

He had clearly failed to see that most people were tired of civil war; even the defeated and the victims were sick of the atrocities and wanted peace. Sulla had deterred them all. Only desperadoes and implacable opponents of the Senate had joined in the rebellion.

Philippus claimed that such people had agitated against the Senate since 100. And presumably Lepidus did not realize that the political force on which Sulpicius and Cinna were able to rely had grown up in a particular situation and probably among a particular generation, that it could not simply be remobilized among the successors of the politically disaffected knights and the new citizens.

Altogether there was something unreal about the situation. Everyone waited for everyone else; no one really wanted to act. Something was seriously wrong if Rome could be kept in suspense for about a year by a man like Lepidus, who was clearly unwilling to act unless the opportunity fell into his lap and whose fear prevented him from seeing how strong he was because others were afraid. Once the farce was over, of course, things became clearer. With the greatest difficulty the Sullan constitution had managed to pass the first and fairly easy test. For the time being it still had to be reckoned with.

In view of the weakness of its defenders there was much to suggest that anything was possible. But it only seemed so. In fact there was a powerful force that was not so easy to discern – a great desire for tranquillity, and hence for restoration.

If Caesar immediately saw through Lepidus, this says much for his knowledge of human nature, perhaps for his instinct, or for the standards he required in others. At all events, he did not throw himself recklessly into the struggle against the Sullan régime, despite the hopes and wishes he may have at first entertained and – not least – despite the patent weakness of the Senate and the widespread discontent. He may also have been aware of the mood among the Cinnans. In any case, on his return from the east he probably travelled widely in southern Italy, staying with various hosts and meeting many people. He probably appreciated the general situation and the longing for calm; he may also have seen that Lepidus was short of soldiers, while his opponents had many. At all events, the twenty-two-year-old Caesar showed more than a little judgment.

During these years Cornelia gave birth to Julia, the daughter whose charm, adroitness and loyalty were later to be of such help to him.

In 77 Caesar prosecuted Gnaeus Cornelius Dolabella, the consul of 81, who had just returned in triumph from the province of Macedonia, before the extortion court. He was an old Sullan, a loyal follower of the dictator. Caesar accused him of having illegally exploited the province's inhabitants.

The prosecution of important men was a popular means by which young noblemen could get themselves noticed. Caesar had many distinguished forerunners in the field, and the practice still flourished. Through it some of Rome's most celebrated orators came to public notice in their early twenties. Cicero, in an imaginary conversation, makes the great Lucius Crassus say, 'At the earliest possible age I appeared in criminal trials, and at twenty-one I prosecuted the most eloquent orator of the day. The forum was my school.' It was mainly through practice, through engaging in forensic disputation, that one learned how to speak and perform in public. 'In this way young men quickly acquired much experience, intrepid verbal dexterity, and a high degree of judgment, since they schooled themselves publicly in verbal exchanges in which one could not say anything foolish or contradictory without its being rejected by the jury, attacked by one's adversary, or finally condemned even by one's supporters.' They got to know the other orators, and 'they also learned from a large and extremely varied public, since they could easily perceive what met with everyone's approval or disapproval' (Tacitus).

Hearings took place in the forum. Here there were two permanent tribunals (platforms for praetor and jury), and others could be set up if necessary. Other hearings took place indoors, in the public rooms of the great basilicas. Anyone who performed well was assured of an attentive audience.

Many political battles were fought out in court, and politics might occupy much of the pleading. Even if the matter itself was not interesting, the participants often were. The defendant would call upon all his friends to support him as defenders, as character witnesses, or in other ways. One of the supreme duties of friendship was to respond to this call and at times to provide the defendant with a successful advocate. Hence the most important orators, the most notable personalities among the Roman nobility, could time and again be seen doing brilliant forensic battle. Trials were thus notable events and attracted large crowds of spectators; some were attracted by a particular cause, while others came to learn, to look around, and to enjoy the intellectual cut and thrust. Cicero was an interested spectator at Dolabella's trial. As in Hyde Park, the crowd might fluctuate – 'a great audience that was for ever changing, made up of opponents and friends, so that nothing that was well or badly said went unnoticed'. Of Brutus, Caesar's murderer, it was said that he dried up when his inspiration failed him. What people saw and experienced in the forum was widely reported. Speeches and performances were an important topic of conversation in Roman society.

Given the significance of political and other cases, and of the debates that took place on numerous affairs before the Senate, the popular assembly or the magistrates, everyone had to know who was a good advocate, prepared his brief well, and could sway his hearers. One therefore had to prove oneself and attract attention, and the sooner the better. And in doing so one immersed oneself in matters that concerned one as a politician.

Anyone who espoused a cause could act as prosecutor. Rome had no state prosecutor. As in all other spheres, public functions were distributed among the citizens, as it were, not concentrated in a 'state' apparatus. There was, however, a group of men who specialized in prosecution; these so-called *accusatores* were not highly regarded, but performed a necessary function. They were joined, from case to case, by young, ambitious nobles eager to win their spurs. If several applied to conduct a particular prosecution it was necessary to decide to whom it was to be entrusted.

It is not known how Caesar came to be chosen by the Greeks who wanted to sue Dolabella for compensation. He may have sought them out specially, even encouraged them, or been recommended to them. At all events, he must have espoused their cause with enthusiasm; and he was a good orator. Cicero later praised him for his correct, precise and elegant Latin and for his perfect delivery. In Cicero's opinion it was like seeing well-painted pictures displayed in a good light.

All the same he had no success. Roman courts – especially when the jury was made up of senators – were inclined to acquit persons of rank. The exploitation of provincials was regarded as a trifling offence, unless it was carried to extremes. Moreover, Dolabella was defended by the two most distinguished advocates of the day, Caesar's uncle Gaius Aurelius Cotta and Quintus Hortensius. Caesar's defeat was thus probably predictable; at any rate he could console himself. He certainly achieved his main object: he became known as a brilliant speaker.

The following year the Greeks asked him to take on the prosecution of Gaius Antonius, a Sullan officer who had shamelessly enriched himself during the war against Mithridates. In this case Caesar almost succeeded. However, Antonius appealed to the tribunes of the people, and they intervened in his favour. Then in 70 the censors expelled him from the Senate, but he returned in 68, and in 63 he even became consul, together with Cicero.

Soon after this Caesar went to Rhodes to study rhetoric under the celebrated Apollonios Molon. He is said to have gone there not only to pursue his studies, but to avoid certain charges arising from his prosecution of Dolabella. It was not yet common to study in Greece. But Caesar attached particular importance to rhetoric, culture and style. Apollonios Molon was highly esteemed. When he had gone from Rhodes to Rome as ambassador in 81, he was the first person to be allowed to address the Senate in Greek. Cicero had studied under him in 78/77 and owed much to his teaching. But it is questionable whether all this would have been enough to draw Caesar away for any length of time from the political career to which he was so committed. Moreover, although such charges were common, was it not inevitable, when the restoration seemed so precarious, that the leading senators, and public opinion generally, should construe Caesar's prosecution of prominent Sullans as an act of anti-Sullan policy? And had he really taken much trouble to conceal this aspect of his attack on the corrupt grandees? Another reason why Caesar may have gone into voluntary exile at this time becomes clear from the later course of his career.

On the voyage to Rhodes Caesar was captured by pirates off the island of Pharmacussa, six miles south of Miletus. When they demanded a ransom of twenty talents, he is said to have replied haughtily that they obviously did not know who their captive was; he would offer them fifty. This may be an anecdotal embellishment. What is certain, however, is that his companions and slaves visited the coastal towns nearby in order to collect the money and clearly held them responsible because of their inadequate control over local waters; a member of the Roman ruling class was worth a great deal even to the Greeks of Asia Minor. Caesar is said to have spent nearly forty days in captivity and to have conducted himself with aristocratic nonchalance, demanding quiet when he wanted to sleep, composing poems and reading them to his captors, and jocularly threatening to string them all up once he was freed. When the ransom was collected he made the pirates give hostages to the towns in order to ensure his release. Once free, he chartered a few ships at Miletus, pursued the pirates, and captured them. He confiscated their booty; it is not recorded whether he returned it to the towns. When the governor of Asia hesitated to punish the prisoners – he reckoned on taking a ransom himself – Caesar had them summarily crucified. But for old acquaintance' sake, he first had them strangled.

All this was possible, given the conditions prevailing at the time. But there is no doubt that such self-sufficiency and arrogance were exceedingly unusual: such decisive action in the name of Roman rule, or at least in its interest, in the interest of bold efficiency and as a demonstration of Roman power – and executed with such energy!

Caesar now applied himself to his studies. But he immediately interrupted them when, in 74, Mithridates re-opened the war. He crossed over to Asia Minor. The king's real attack took place in the north of the peninsula. He marched into Bithynia, which the Romans had just acquired through the will of the former king, Caesar's friend Nicomedes. At the same time he tried to persuade the province of Asia to sever its ties with Rome.

Caesar tried to halt the troops that were sent to the southern flank to accomplish this. He demanded soldiers from the local towns. With these he moved against the enemy commander and forced him to quit the province. It is not clear how significant this military achievement was. Mithridates can hardly have reckoned with much resistance. There can have been only small Roman detachments stationed there. Hence the king probably sent only a small force. At all events it was important that Caesar, by his prompt action, prevented the towns from going over to the king. The Roman merchants and tax-farmers who were settled there would have been likely to flee, remembering how many thousands of Romans had been murdered during the earlier Mithridatic war.

The way in which this twenty-six-year-old aristocrat, with few friends, no office and no instructions, took it upon himself to seize command in a difficult situation was quite unparalleled. Normally Rome fought with regular troops. If individual citizens took independent action it was in areas where they were at home or had clients. But even if Caesar had a few clients there, his intervention remains an amazing act of presumption. There was a model, however, in Pompey's initiative during Sulla's civil war. And Caesar was not lacking in self-confidence and audacity — or should it be called impudence? He was thus able to take the provincial towns by surprise and carry them with him, giving the impression that he had Rome's backing. Moreover, he assumed responsibility for Roman sovereignty, no doubt fully aware of what he was doing, and thereby demonstrated to his peers how well he, unlike them, could live up to the obligations of a Roman aristocrat. At all events, his actions evinced a high claim to responsibility for the commonwealth. He may also have had an immediate reason for taking such an extraordinary risk in an emergency situation. For it now becomes clear why he left Rome in 76.

Late in 74 or early in 73 Caesar received news that he had been co-opted into the college of the *pontifices*. He at once set off for Rome. When crossing the Adriatic, he had to take care to avoid pirates. He chose to cross in a small fishing boat and arrived safely in Italy, with only two friends and ten slaves.

One wonders where Caesar obtained the money for such a retinue and such journeys, considering that Sulla had confiscated his inheritance. Perhaps it had been restored when he was pardoned; it is likely too that his mother and his relatives helped him out when necessary. In the aristocracy there was always money to be had somewhere; something could always be engineered. When making a will, for instance, it was customary to provide not only for one's heirs, but also for friends and benefactors, as a kind of compensation in a society where there was so much that could not be paid for but still had to be requited. This did not yet apply to Caesar, but it was certainly symptomatic of the way money was treated and the way it circulated. Finally, Caesar had the booty from the pirates, and he could always borrow. In any case, he could hardly go around without servants. What would people have thought of him? The fact that he was attended by servants made it clear who he was. And who would he have been without any? How could he have achieved anything in the circumstances of the time?

There was no obvious reason for his hasty departure. There were no pressing duties awaiting him that brooked no delay. Rather, the real object of Caesar's absence had clearly been attained; this was to let the grass grow over the ill feeling caused by the Dolabella case. Caesar was now fully rehabilitated — and more than that.

★   ★   ★

The appointment as *pontifex* was a great honour and distinction. It brought a certain influence. The fifteen *pontifices* constituted the highest religious authority in Rome. They had to see that all rites were punctiliously performed and, when necessary, to rule on all religious matters. In particular they had to supervise and regulate the Roman calendar. Rome still had a lunar year – until Caesar's reform of 45 – and an extra month usually had to be intercalated after February every two years. It was the *pontifices* who decided when this should happen; there was a political aspect to their decision, as it involved the question of whether or not the annual term of certain magistrates should be extended. Apart from this, the college could exercise political influence by tendering opinions or rulings on infringements of the proper procedures and on how such infringements should be expiated. It had previously been responsible for Roman law and kept the magistrates' rolls. It enjoyed high and time-honoured authority. Since the third century the *pontifex maximus*, the city's supreme priest, had been elected from its number by a special popular assembly.

The fifteen augurs and the *pontifices* were the most important priestly bodies. Neither was there to carry out the ceremonies themselves; this fell either to certain priests such as the *flamines*, or to the magistrates, who represented the community in its dealings with the gods. Rather, they were the experts, the supervisory authorities. Seats in the two colleges were greatly coveted, being limited to thirty (out of 600 senators). And at the time of their election not all were senators.

The priests often co-opted quite young noblemen – as in Caesar's case. Care was taken to ensure that the college had a wide age-range. There was much to be learnt and handed down. It was therefore desirable for at least some members to serve for a long time. It was probably also important, unless it was politically inexpedient, to recruit particularly intelligent young members. However, the choice was not easy. This was no doubt one of the main reasons why it had become customary, when a member died, to replace him by a relative. This made the choice easier. But it was by no means the rule. There was also a wish to confer the priesthood on respected *consulares* and thereby strengthen the authority of the college. And there might of course be objections to the relatives of a deceased colleague.

The seat to which Caesar was co-opted was that of his mother's cousin, Gaius Aurelius Cotta, the consul of 75. The family connection was certainly of benefit to Caesar, as was his membership of the patriciate. At the same time it was by no means self-evident that Cinna's son-in-law, who had prosecuted prominent Sullans, should be appointed to the college, dominated as it was by the ruling aristocracy. Caesar's friends must have used their good offices on his behalf. One of the *pontifices* is mentioned among those who interceded for him with Sulla. Another had been commander in Cilicia when Caesar served there. A third, Quintus Catulus, may have been won over by his cousin Servilia, the mother of Brutus, if she was already Caesar's mistress or at least looked upon him with favour. Moreover, his exceptional conduct in Asia Minor no doubt made up for much. Finally, the establishment may have hoped to draw this promising young nobleman on to its side.

It was certainly useful to him to have been away from Rome for a while, presumably on the advice of well-wishers. This would explain the study journey to Rhodes. He may have decided to absent himself so that he would not have to choose between being the kind of person everyone was expected to be and falling out with the aristocracy. His individuality was not to be confined to his sartorial tastes or private extravagances; he wished to be seen by others as he saw himself, to be wholly himself. This may be connected with the fact that he went away to study and not for another tour of military duty. This of course allowed him to distance himself more from others and deliberately opt for a role that still had to be developed.

We do not know whether he was prepared to reconsider his position, given the opportunity that the aristocracy now held out to him, and decided to heed the warnings that he doubtless received from well-wishers by conforming to a conventional pattern of behaviour. This possibility should perhaps not be dismissed. Some time had elapsed since Sulla's death; he had made the best of his youth, shown his mettle, and been treated with generosity. However, we know something of what Rome – and in particular Roman aristocracy – was like at the time. And that cannot have made it easy for Caesar to change his ways, even if he wanted to.

The picture presented by the Senate was anything but imposing. The immense losses it had suffered in terms of biological and moral substance had become all too obvious. There were hardly any *principes*. In normal times the leadership of the house was in the hands of a group of twenty *consulares*. Only two had survived from the pre-Sullan period, and one of these, Lucius Philippus (the consul of 91) had died shortly after 75. The other (the consul of 92) lived to be eighty-nine, but we do not know what help he could still give. The consuls of 81 were Sulla's creatures and had clearly been elected because they were faithful followers and political nonentities. Of the consuls of 80, 79 and 78 only three were still alive, and one of them was in Spain. The consuls of 77 were notoriously so incompetent that when it came to sending a good general to Spain, the aged Philippus proposed that Pompey should be chosen 'in lieu of the consuls' (*pro consulibus*). A commander who was not a magistrate could hold his command as 'proconsul' (*pro consule*). This then became Pompey's title. Philippus meant that he should be sent in lieu of both consuls, who were patently not equal to the task. One of the consuls of 76, Gaius Scribonius Curio, was an energetic man who later became one of the most important *principes*. But he first had to go to Macedonia for three to four years in order to wage war on the Thracian tribes who were threatening the borders of the province. The other was severely incapacitated by pains in his limbs. One of the consuls for the following year was Caesar's uncle, the alert and ambitious Gaius Cotta, but his colleague was described as negligent and idle.

The truth was that two men, Quintus Lutatius Catulus (consul for 78) and Publius Servilius Isauricus (consul for 79) had to do what was formerly done by all the *principes*: they had to direct the business of the house and see to it that the Senate's authority was upheld and its code of conduct observed. And they had to do this after the earlier discipline had been shattered by civil wars and proscriptions; they were faced for the most part with new members who had not risen gradually to prominence under the guidance of the *principes*, but were in many cases favourites of Sulla, and with a body whose membership had been increased from three hundred to six hundred.

Senators held their seats for life and were subject to no control, as they also sat in the lawcourts. Their power and responsibility were all-embracing. If the leading senators did not succeed in discouraging unhealthy developments and setting an example, in gradually educating the new members to adopt the traditional image of the senator and setting the right tone, it was hardly possible for the assembly of the fathers to perform the functions entrusted to it. But how could it do so if the ruling caste had yet to mature? And, moreover, they were all faced with the countless tasks resulting from the legacy of the civil wars in Rome, in the empire, and on the periphery of the empire.

In Spain, Quintus Sertorius, one of the ablest Cinnan officers, had created a power base for himself. He had been governor there, but in 81 he was driven out by a successor sent by Sulla. Whereupon the Lusitanians, the inhabitants of what is now Portugal, had called upon him to lead them in their struggle for freedom. He was a master of guerilla tactics and at first scored great successes. His opponent Metellus Pius, who was consul for 80, employed regular fighting methods and proved no match for him. Disaffected elements in Rome pinned their hopes on Sertorius, and some prominent nobles begged him to march on the city. Obviously they could expect more of him than of Lepidus, whose rebellion had never materialized. The Senate therefore deemed it necessary, in 77, to send Pompey to Spain with a second army. But it was another five years before the Spanish rebellion could be put down. When the enemy camp was finally overrun, Pompey found letters that had been sent to Sertorius from Rome. He had them all burnt without reading them; he is said to have feared that otherwise worse internal conflicts would arise than those he had just ended.

During these years Sertorius had allied himself with the pirates and, more importantly, with Mithridates. For decades the pirates had profited from the feebleness of Roman rule. From their bases in Asia Minor and along the Adriatic coast they ranged over the whole of the Mediterranean, attacking and robbing not only ships, but towns and whole regions, and threatening, even blocking, the supply of goods to Rome. They were so successful that even prominent magnates made common cause with them. They gained control of whole stretches of the coast and secured them with observation towers.

The pirate ships were not only very seaworthy, but are said to have been splendidly equipped with gilded masts, purple flags and silver-mounted oars. Flutes and stringed instruments, singing and carousing could be heard on every beach; abductions and forced contributions were frequent. In the end the pirates are said to have possessed a thousand ships and attacked four hundred towns.

In 74 the Senate, under the influence of Marcus Aurelius Cotta, another of Caesar's uncles, and thanks to the machinations of

Cethegus, a powerful intriguer, had placed the entire Mediterranean littoral under the command of the praetor Marcus Antonius. He is reputed, however, to have outdone the pirates in plunder – understandably, for Cethegus certainly did not give his help *gratis*, and there had to be something in it for the praetor too. In any case he lacked any understanding of the task he was expected to perform. Had he wished for major successes he would have had to think, organize and operate on a grand scale. As it was, he failed to score even minor successes, and became involved in a war on Crete, in the course of which he died, having achieved nothing. He was posthumously accorded the ironic cognomen Creticus.

In this situation, Mithridates went to war again, probably surmising that the Romans were in no position to confront all their enemies at once. They were indeed almost overstretched. The consuls of 74 were sent to the east; one of them, the *bon vivant* Lucius Licinius Lucullus, owed his command to an intrigue that had earned him the goodwill of Cethegus' mistress. He had learnt a good deal about warfare under Sulla, and on the voyage he took a number of experts and many books with him; it is said that by the time he landed he had mastered the whole art of war, thanks to his exceptional powers of perception. He waged war for a good seven years, with skill and success as a general, though he soon failed as a leader of men. For all his intelligence he seems to have been too haughty and decadent; he lacked the common touch, and this led to mutinies and reverses, so that his conquests came to nothing.

External and internal problems were interconnected. Revenues were insufficient. Supplying the armies was expensive. The inadequate provision of grain led to unrest in Rome. The tribunes of the people exploited the situation to mount a vigorous campaign of agitation against the Senate. They demanded the restoration of their official rights. Despite opposition from the leading Sullans, one of the consuls was prepared to bring in a law enabling the tribunes once more to apply for other offices and so to rise in the political hierarchy. The Senate must have agreed. This was the beginning of a slow retreat.

Yet the tribunes of the people and those concerned with the restoration of their powers were at first too weak to effect any decisive change. And the senators procrastinated on the ground that they had to await the return of Pompey.

From the mid-seventies Rome lived in the shadow of his return. His intentions were not clear. If he wished, he could certainly lead his soldiers to Italy, despite the legal prohibition. Probably not in order to seize control of Rome – that would have been too difficult, given the enormous area ruled by the city – but to assert his will in this or that question. Ever since the days of Marius and Sulla, soldiers were more beholden to their leaders, in case of doubt, than to the Senate, and Pompey's would certainly have followed him. It was therefore inadvisable to give him a pretext for marching on Rome, either by agreeing to restore the rights of the tribunes – which might have offended his Sullan sensibilities – or by firmly refusing to do so and thereby enabling him to act in the name of the people's rights and on behalf of everyone who had ever wanted to prevail over the Senate in some matter. Perhaps he could be enrolled as an ally against the tribunes. All this was an open question.

Pompey himself informed the Romans only that if the Senate and people had not agreed before his return he would try to promote agreement. Ultimately it was not only a question of the fate of the Sullan régime, but of how to reach an accommodation with a powerful outsider. Rome thus entered a power vacuum. The Senate did not know what to do. To attentive critical observers it must have presented a sorry spectacle.

There was a feeling that the Roman order was out of control. There was a discernible lack of leadership, cohesion and certainty; nothing could be done to alter the prevailing conditions; they simply had to be accepted. No one knew what was happening. As a result, an essentially trivial event could assume such proportions as to rock the whole of Italy once more.

In 73 about seventy gladiators broke out of their barracks in Capua. During their training, which was in any case very harsh, they had been subjected to such harassment that they had joined together to plan and make good their escape. They fled to the south and at first sought a hiding place near Vesuvius. Presumably they intended to operate as a robber band.

Gladiators ('sword-fighters') were trained to perform in life-and-death contests before an audience. Such spectacles are attested in Rome from 264 BC and had once formed part of the funeral rites of prominent nobles, but the link was later dropped. From 105 BC they were sponsored by magistrates. The institution seems to have originated in Etruria or Campania and to have been associated with human sacrifice and the cult of the dead; presumably unconditional human sacrifice was modified so that the victims could decide by combat who was to die and who to survive. It then became a form of entertainment. The Roman people – or certain elements in it – took pleasure in this atrocious spectacle, just as other peoples, at other periods, have taken pleasure in public executions or sex and crime. Although there was always something going on in Rome, the bored, apathetic populace needed titillation; they delighted in the fascinating spectacle of killing, the sadistic identification with a deadly sport, even though – or because – they may have been disgusted by it. It even occupied their dreams. As masters of the world, though living in miserable conditions, they must have felt such games to be something splendid. They followed them with passionate enthusiasm. And the nobles vied in staging them, until in the late republic hundreds of fighters were pitted against one another. The games were usually held in the Forum Romanum, where wooden stands were erected for the spectators, and perhaps also in the Circus Maximus. At a later period amphitheatres were built in Rome and many other cities; the oldest known example, from about 80 BC, is at Pompeii.

Most of the gladiators were slaves imported from Thrace, Gaul or Germany, but among them were also some freemen, attracted by the money, the danger, the publicity, and the thrill of defying death. In the following century there were even women gladiators. In the reign of the Emperor Domitian contests were held by torchlight.

Gladiators had to swear a solemn oath by which they undertook, on pain of death, to subject themselves unquestioningly to the harshest training and to every fight, to 'let themselves be whipped with rods, burnt with fire and killed with iron'. They were trained to be highly skilled fighters; careful attention was paid to their health; they were given massage, baths, and a special diet devised by medical experts: so valuable were they to their masters – and to the public. As the demand grew, their numbers increased. The most important gladiatorial schools were in Campania.

At the games they made a grand entry into the arena. This was followed by a weapon test – a mock fight with blunt weapons. Then, after a flourish of trumpets, they fell upon one another in earnest.

Precise details are attested only for the imperial period, but they are unlikely to have been much different under the republic. Timid fighters were driven into combat with whips and red-hot iron bars. As at modern games, the spectators excitedly shouted encouragement, reproaches and insults at the combatants: 'Why is he holding back?' 'Why doesn't he want to die?' 'Why doesn't he go for the kill?' The promoter of the games decided on the fate of the defeated – life or death – with a thumbs-up or a thumbs-down. During the contests the spectators took sides for or against particular contenders, but especially against the more timid. Solicitous promoters provided fine mass graves for their fallen gladiators.

Such men were naturally not easy to deal with; they had to be subjected to iron discipline. They had their own fierce notions of honour. The breakout of 73 is unlikely to have been the first, and it was certainly not the last. It differed from the others only in that it had extraordinarily wide repercussions.

For the seventy men did not remain alone. Having defeated the Roman troops sent to quell them, they were joined by many others – slaves and freemen, agricultural labourers, shepherds, small farmers and – not least – numerous booty-seekers. For the proceeds of their pillaging expeditions were always divided equally.

Many of the slaves employed on large estates had a miserable existence. They worked hard under strict overseers; their rations were frugal and often inadequate; they lived together in large quarters, and were often chained up at night. Others may have been better off, but hardship, deprivation and lack of rights were widespread not only among slaves, but also among agricultural labourers, tenant farmers and smallholders. We may presume that such conditions were not peculiar to this period. The most one could look forward to was even greater hardship, thanks partly to poor harvests and the lack of supplies from abroad. Peculiar to the period was the widespread feeling that current ills could be escaped by joining the gladiators. The existing order was not taken seriously because it seemed unreal, not least on account of the civil war, and because its champions did not strictly adhere to it. It is unlikely that a different social order was envisaged. Nor did anyone consider the abolition of slavery: that would have been inconceivable. But some imagined that they could do what they wished. Conditions were so unstable as to engender unreal hopes, not because there was any reason to expect them to be fulfilled, but because there was reason to hope, if only for vengeance against society. There are also indications of the re-emergence of some – though not many – of the old Italian resentments against Rome.

It was not long before the number of rebels ran into thousands and tens of thousands. At its height the uprising is said to have attracted seventy thousand supporters; one source even speaks of a hundred and twenty thousand, but this probably included sympathizers. It was led by three men, of whom the best known and most important was Spartacus.

Few figures have attracted as much retrospective interpretation as Spartacus. Karl Marx called him the 'finest fellow that the whole of ancient history has to show, a great general (no Garibaldi), a noble character and a true representative of the ancient proletariat'. This is of course nonsense. In reality Spartacus seems to have been a robber chief on the grand scale. We have no evidence that his abilities and intentions went beyond this. His task was difficult, impossibly difficult – to lead a large, motley, undisciplined host that had no land, no military bases, no proper weapons, and no common objectives beyond plunder and booty, in such a way that it could assert itself against Rome. This meant living and surviving as an excessively large robber band – somewhere, somehow.

For a long time Spartacus performed this task with *bravura*. He was fortunate in that his force included some outstanding fighters – gladiators trained to despise death – and that a large part of his army consisted of runaway slaves who knew that if they took to pillage they could expect no pardon from their masters. It was to his advantage too that the Romans at first underrated the uprising and that their regular methods of warfare were unavailing against this particular enemy. For a long time Spartacus had more difficulties within his own ranks than he had with the Romans. He was a great tactician, but it is not clear whether he had any strategy.

The rebels first turned south, towards Metapontum, then moved northwards through the whole of the peninsular. Although one of their detachments was wiped out, the bulk of their army was victorious in numerous encounters. After they had defeated a Roman army at Mutina (Modena) in the Po plain, the whole of northern Italy lay open to them. Had they wished, they could have crossed the Alps into Gaul, Germany and the Balkans and been free.

Perhaps Spartacus had intended to lead them to freedom; they may even have wanted him to. After their recent victory, however, they were convinced that they were now strong enough to raise the stakes – a conviction based wholly on the inner logic of a victorious army that had no clear objective resulting from policy and that overrated itself on the strength of wild, short-term expectations. There is much to suggest that the rebels' appreciation of the situation was as unrealistic as their hopes. At all events, they turned back and moved south to Metapontum and from there to Bruttium.

Meanwhile, at the end of 72, the Romans had placed Marcus Licinius Crassus, who had recently been praetor, in charge of the war against Spartacus. Crassus had served under Sulla. He was the richest man in Rome and exceedingly ambitious – which is probably why the Senate gave him the command. Rome sent six to eight legions into the field. Crassus encircled Spartacus with a long trench at the southern tip of the boot of Italy. Spartacus is then said to have treated with the pirates, in the hope of breaking out to Sicily. When this came to nothing he filled in the trench at one point and broke out in the direction of Brundisium (Brindisi). Faced with this situation, Crassus asked the Senate to send Marcullus Lucullus and Pompey to his aid; the former was on his way back from Macedonia, the latter from Spain. Before this had happened, however, Crassus had defeated the rebels.

Spartacus seems to have fallen in battle; the scattered remnants of his army were tracked down. Crassus had six thousand captured slaves crucified along the Appian Way – any freemen among them were similarly punished – as a warning to all who passed along it. A troop of about five thousand slaves that had broken through to the north was wiped out by Pompey. Despite his customary clemency, he had all the prisoners massacred. But at that time runaway slaves deserved no better. In the early summer of 71, the 'Spartacus War' was over, having lasted for two years. It had never posed a serious danger to Rome. But it was symptomatic of the laxity of Roman order and yet another sign of how ineffectual senatorial leadership had become. The whole peninsula had been open to the slaves, and it had taken three large armies to defeat them.

When the seventies ended the Senate was obviously in the same parlous state as when they had begun. At all events it still had the utmost difficulty in performing its tasks, and the commanders it chose from among the magistrates or its own ranks were repeatedly found lacking in fortune. If we are not content with such formulations, we must ask what it was that the Senate and the senators lacked. Were the senators no longer men of substance? Were they different from their fathers and forefathers? Referring to an earlier period, Mommsen wrote: 'It was not so much that different men sat in the Senate, as that it was a different age.'

This seems to be the correct explanation. Between them the men of this period created different situations, different tasks, different expectations, different sets of circumstances. This was what distinguished them from their fathers and forefathers – which is not to say that their individual talents were necessarily inferior, or even different. It is a question of the specific possibilities that a given period affords for realizing and developing talents or letting them go unused; for rising to challenges or despairing and becoming paralysed in the face of them; for realizing one's full potential – in other words, what one can expect of oneself within the framework set by one's identity and the demands one has to meet – or succumbing to self-pity. It may be that a living élite cannot have the feeling of being able to do what it must. Yet part of its effectiveness probably derives from the fact that the gap between ability and obligation is not so great as to inhibit them from bracing themselves to discharge their obligations to the full and so realize their potential. What seems to have been at work at this period, however, was not an awareness, perhaps not even a feeling, but a fear that their obligations exceeded their capacities. This led to distractions, evasions, constrictions. Other talents might develop individually. The aristocracy still recognized the tasks confronting it, but on the whole it lagged behind its forefathers in responding to them; it no longer had the cohesion and force with which the latter, for all their weaknesses and occasional failures, had devoted themselves to their institutions and the life of the *res publica*.

The reason for the Senate's many failures at this time was therefore not that – Pompey apart – it had chosen the wrong men for the more important tasks. This may admittedly seem to be so when one reads accounts of the great influence wielded by the intriguer Publius Cornelius Cethegus. He was an extremely shrewd and skilful senator from the patrician nobility. He had at one time supported Marius; in 88 he was one of the twelve men who, after being proscribed by

Sulla, went over to his side; in 78 he sided with Lepidus, but his power was not diminished when Lepidus was defeated.

Cethegus knew the republic better than most, or – more precisely – he knew all the players in the political game; he knew their strengths and their weaknesses, what they could offer and what they could demand. He was a good speaker and seems to have had an attractive personality. He was an able mediator, in the Senate and at elections, and for a time he clearly succeeded in making himself indispensable. For Cethegus politics were a great bazaar. The biggest deals were done over his counter, and there was always something in them for him – money, connections, dependency. In this way, so Cicero tells us, he attained the same authority in the Senate as the *consulares*, though he never rose to this rank. The chief way to gain access to him was through Praecia, who according to Plutarch was famous 'for her beauty and her waywardness', though 'otherwise she was probably no better than a common whore. Yet because she knew how to use the men who consorted with her for the benefit of her friends . . . she gained a reputation not only for her charm, but for being a loyal friend and a woman of some energy, and won great influence.' Cethegus is said to have been exceedingly devoted, if not in thrall, to her.

It was with Cethegus' help that Antonius was put in charge of operations against the pirates. And the able and intelligent Lucullus, who was very proud and probably averse to any kind of amorous excess, had put Cethegus and Praecia under an obligation to him in order to obtain the command against Mithridates. Their choice thus fell sometimes on the right person and sometimes on the wrong. They cannot therefore be held responsible for the failures of Rome's military leaders.

The real reason for these failures seems to have been that the aristocracy was guided by the wrong principles – or rather that it encouraged the wrong kind of behaviour.

They lived in constant fear that ambitious individuals might again defy the rules of senatorial propriety and violate the fundamental equality that prevailed among the senators, or at least among the *principes*, which called for solidarity within the traditional framework and a willingness to be bound ultimately by the judgment of the Senate. No one should again behave like the Gracchi, let alone like Marius and Sulla. This sentiment had existed for some time. But it was now embraced with particular fervour, under the pressure of the widely felt need for restoration. Conformity was at a premium; it was inculcated, demanded, encouraged and rewarded.

From the earliest times Rome had set great store by preserving and handing down the customs of the fathers. And as no one knew or could even imagine that the Roman order as a whole was no longer able to respond to the exigencies of the age, the only possible explanation for present crises and emergencies was that the old customs were no longer properly practised. It was therefore necessary to be all the more punctilious in observing them.

Hence the Senate, and with it large parts of society, sheltered behind tradition. They followed the rules and sent the consuls against the slaves because they were consuls; and the consuls waged war in the conventional manner, although they had quite different opponents. It seems to have occurred to no one that pirates could not be dealt with by standard manoeuvres on sea and land, carried out at fixed points, and Lucullus took it for granted that his men should simply obey orders, no matter what was demanded of them, with the result that they finally mutinied. It was because the problems, or at any rate the demands, were so great that the Romans adhered to convention. Rigidity thus led to failure, failure to rigidity. Being clearly reluctant to contemplate reality, they were reduced to compulsive self-constraint and anxious immobility. In earlier times, they had been willing, despite their reverence for the old, to explore new avenues, to devise new measures appropriate to new situations. Respect for the old, formerly a rule, now became a binding law. Often it was no longer the rules of the ancestors that were raised to the status of dogma, but what was written about them, as it were, in the history books.

Yet because every effort went into ensuring that everyone adhered to the rules and that no one became too important, there was no longer the flexibility that was necessary if exceptional tasks were to be adequately performed. A dichotomy emerged between the defence of conventional conduct and order – on which everyone concentrated – and the solving of urgent practical problems, which were consequently neglected. What had once gone together became an alternative. What was expected was not achievement, but conformity.

Yet this meant that even conformity had to be more narrowly construed. For if such a wide gap exists between ability and obligation, one may decry the new, yet still fail to preserve the old from decadence. If duty seldom calls, it becomes difficult to perform. The Romans were thus small-minded, in that they expected no one to take exceptional political risks or attempt anything on a grand scale. Yet they were also broad-minded – perforce – in that they tolerated countless minor deviations from the customs of the fathers, at least in the more private areas of life. Corruption was rife. Politics were seen chiefly as a struggle for office and position. Little was demanded of the sons of the great. The educational rigour of earlier times became a rarity. A life of luxury was preferred.

What was not permitted in politics found an outlet in other spheres. Men vied with one another in devising new forms of indulgence, housebuilding and entertaining. Roman nobles had themselves portrayed in the manner of Hellenistic monarchs and delighted in being venerated as gods by provincial peoples.

A particularly good example of the escape into harmless freedom in a more or less private sphere is afforded by the coinage of the period. The striking of coins was entrusted to three annually appointed mintmasters, often young sons of noble houses. They were free to design the die as they liked and took more and more to commemorating their illustrious ancestors. The republic's currency thus became a medium for advertising the fame of one's family, a symbol of how the city's nobility saw the commonwealth less as a trust than as a possession. Some of the designs and legends were so cryptic as to baffle all but the initiated; their interpretation was a party game that few outside the leading families could play. Even themes from high politics were often represented in such a way as to draw attention to the fame of the mintmaster or a family to which he claimed to belong. Marcus Brutus, for instance, commemorated the founding of republican liberty by the old Brutus, Rome's first consul, whom he obviously thought of as his ancestor.

This and other peripheral phenomena, though of little interest in themselves, are significant as symptoms of the narrowing of senatorial standards within politics: the fame that the young mintmaster accredited to his forebears was no longer available to him in the present age, or at least not in opposition to the Senate. Their signet rings bore similar devices.

Metellus Pius, Sulla's fellow-consul in 80, who was later sent to Spain to lead the Roman forces against Sertorius, is said to have had his arrival in Spanish towns celebrated like that of a god, complete with altars and incense. At his banquets he wore a garment embroidered with palms, like the triumphal robe of Juppiter Capitolinus, while figures of victory, lowered from the ceiling by an ingenious system of ropes, set golden wreaths on his head. And all this in the inhospitable atmosphere of Spain, with a difficult war in prospect. The noble senators are later said to have conceived a passion for fish-breeding. Such activities distracted them from ambitions that could seldom be fulfilled in politics and had to be satisfied in other ways.

All this was presumably to be expected: the ruling class may have been overstretched, but it still enjoyed unquestioned authority and found a variety of outlets for the overwhelming pressure of its normal image of itself. Some of the leading personalities – and others too – would have liked things to be different and behaved honourably, shunning luxury. Yet if they wished to influence affairs they probably had to fall in with the general tendency of their class. Anyone who wanted to break out of these narrow confines and think independently had to remain aloof from the majority or try to distance himself from it. And clearly only a few could do this: those who tried had many reasons for doing so, but few points of purchase. Mommsen's sarcastic judgment of the senators was objectively correct: 'Their political wisdom was confined to a sincere belief in the oligarchy as the only guarantee of salvation, combined with a fervent detestation and courageous condemnation of demagogy and any force that emancipated itself.' It is unlikely that the senatorial order itself could have evolved an alternative wisdom. Only outsiders could do this. It had nothing to do with their capacity for abstract thought, but with the place from which their thought proceeded, their viewpoint – in other words, the position that the thinker occupied. And the senators can hardly have been in two minds as to whether they could – or should – capitulate to outsiders.

The most important outsider was Pompey. The problem of whether he would defy the law by leading his army into Italy after his Spanish campaign had meanwhile been resolved, as the Senate had actually invited him to do so. After his victory over the remnants of the rebel slaves he marched to the outskirts of the city. But he reassuringly declared that his sole intention was to celebrate his triumph; after that he would dismiss his army forthwith.

Senators, knights and others went out of the city and received him with honour. Negotiations took place. The outcome was that the Senate granted Pompey not only a triumph, but the right to submit himself immediately for election as consul, without having previously held office as quaestor or praetor. He would normally have had to wait another seven years, but after commanding Rome's armies for thirteen years with scarcely a break he could not be expected to start his political career from scratch. And one could hardly make him wait for his consulship. The leading senators also promised to pass a land law in favour of his veterans. This was to be the only land law of the late republic – save for that of Livius Drusus – that had the Senate's approval. However, lack of funds delayed its implementation for so long that it became a dead letter.

There are grounds for thinking that the Senate did not even find it hard to make these concessions. Pompey was an old Sullan. True, he had vexed a number of the senators, including Sulla himself. Sulla had at first deeply offended the others during the civil war by greeting the youthful leader of a private army as *imperator*. He had then sent him to Sicily and Africa to defeat his opponents there. Having accomplished this task speedily and efficiently – and with a degree of clemency – Pompey was ordered to dismiss the bulk of his army and wait, with just one legion, for the arrival of the new governor. The soldiers had been incensed, as they wanted him to lead them back to Rome. With a show of reluctance he let himself be forced into compliance. Outside Rome, Sulla greeted him with the title *Magnus*. It is said that a certain resemblance to Alexander led people to call him 'the Great', at first ironically, then flatteringly, and finally with conviction – because everyone did so. Sulla decided to follow suit, no doubt hoping to forestall Pompey's demand for a triumph, for only magistrates were allowed a triumph, and Pompey was not a magistrate: from now on the law was to be strictly observed. Pompey demurred: standing before Sulla and pointing to the sky, he declared that more respect was due to the rising than to the setting sun. Seeing that it was futile to argue the point, Sulla acquiesced in this further advancement of the young man's career. When the Senate dispatched Pompey against Lepidus and Sertorius, this was not just because it needed him, but rather because he pushed himself forward; in any case the senators were glad to have him out of the way. Since then he had served them well, and they probably hoped that by accommodating him they could win him over.

However, Pompey was approached by all those who wished to see the rights of the tribunes of the people restored – the knights, a number of senators and others. Caesar may have been among them. Soon after being elected consul, Pompey declared his support for their cause. He also advocated a reform of the courts, following a number of scandalous verdicts delivered by juries composed of senators, and the election of censors. Not even this programme was opposed by the Senate, as the abuses he complained of were too blatant to be denied. Some senators even seem to have felt that he was relieving them of a burden by removing indefensible features of the Sullan régime. The Senate then seems to have approved the law restoring the rights of the tribunes.

It can hardly have been unaware that this was bound to lead to grave conflicts in future, not least with Pompey himself, who obviously hoped for further commands. And he must have known that the Senate was bound to be apprehensive about enabling him to use the popular assembly, if he needed to, as a means to gain his ends.

Gnaeus Pompeius Magnus was fundamentally averse to conflict. He was not a man to assert himself, but preferred to please everyone. Alfred Heuss compared him with a head boy who goes around showing off his unquestionable achievements like a school report. He was vain, eager for applause, and full of respect for the inherited order, even for the Senate. He sought honour and fame, rather than power and influence. Politically he liked to stay in the background and was reluctant to become involved. He did not champion others; he avoided controversy, preferring to be seen as a kind of figurehead. On his rare appearances in the forum he was attended by a large retinue, 'whereby he lent weight and grandeur to his presence, believing that he must preserve his dignity by avoiding contact with the crowd'.

Yet it was his ambition to go on being entrusted with all important tasks. This was the basis of his position and his fame. In this way he could demonstrate his abilities. He was above all a superb organizer in both the military and the administrative sphere. His campaigns were marvels of planning. He led his troops with a superiority that must have communicated itself to them. At the same time he was a master of strategy and tactics, and, having been schooled in the civil war, knew how to deal with soldiers. He liked to command and govern in the grand manner, brilliantly representing Rome's claim to empire. Having penetrated (by ancient standards) deep into Africa, as far as the borders of the Numidian empire, he hunted lion and elephant, believing 'that even the wild beasts that inhabited Africa should learn of the power and bravery of the Romans'. Such imperial gestures were characteristic of Pompey. What had he to do with the pettifogging business of day-to-day politics, with the hustle and bustle, the vanities and the intrigues?

He thus stood on the periphery of the aristocracy, though he found no fault with the Roman order. His sole concern was to occupy a special position in it, from which he could carry out the many tasks that he understood better and more clearly than others – precisely because he was not taken up with everyday affairs and could appreciate the reality of empire, which was gradually taking on a new quality.

Even in his youth Pompey had stood aloof. His father too had sought a special role for himself; having contributed to the Senate's defeat at the hands of Cinna, he was 'much hated' by the nobility. Under Cinna, Pompey had held back. Yet he also judged that his position in the nobility would not be easy, once its rule was re-established. Taught by his father to maintain a fairly impartial stance, and having convinced himself that a man's value was commensurate with the number of soldiers he commanded, he had raised his own army – in the interest of Sulla and the nobility; this was probably important to him too. In this he differed from other nobles, to whom it never occurred to do anything so practical, so appropriate to the situation. Right from the beginning he was much more open and energetic than the rest. Yet he probably failed to see not only that this set him apart, but that he was moving into the position of an outsider, for which he was not really fitted. At first this did not trouble him; he relished the position because it was special. But then the difficulties became obvious. It was because he met with envy, jealousy and mistrust – not just because he was ambitious – that he decided to pursue the course he had embarked upon, in order to prove his worth through fresh achievements and make himself indispensable. He is unlikely to have known that the Senate majority was not interested in such achievements. To him they were a form of legitimation, but he did not overemphasize his claims based on achievement.

Moreover, he was always mindful of his beginnings, in the civil war under Sulla, and knew that the same situation might recur. He constantly made provision for this eventuality. Wherever he campaigned, he was careful to establish a clientèle and so build up reserves of power in different parts of the empire; and he certainly intended, should the need arise, to use them, as Sulla had used his, in support of the Senate.

Hence, for all his caution, Pompey was ready to reach an understanding with the Senate. He wanted nothing more than to go on acting as its premier general and trouble-shooter and to garner the respect that went with this role. He sought pride of place within the Senate, not in opposition to it.

For the time being this was still possible. He was allowed to enjoy his fame and rest on his laurels; the senators were glad that he seldom meddled in politics – of which he understood little – and behaved with such moderation. In the end, however, it became clear that Pompey's aims were incompatible with those of the Senate majority.

On the last day of 71 Pompey entered Rome in triumph; he then dismissed his soldiers. He became consul, together with Marcus Licinius Crassus. He pushed through his programme. The reform of the courts, despite the original slogans, was in fact extremely moderate. From now on the juries were to consist one third of senators, one third of knights, and one third of *aerarii*, a group whose members, according to the census, belonged to the equestrian order, but in which the tax-farmers, who could put pressure on the Senate, were poorly represented. The censors struck sixty-four unworthy senators from the Senate roll. Their review of the knights was famous. They sat at the tribunal in the forum with their heralds and comptrollers. One by one the members of the equestrian centuries passed before them with their horses. Some were senators' sons who had not yet been admitted to the Senate; the others were the most distinguished members of the knightly class. 'Then Pompey was seen coming down to the forum, resplendent in the insignia of his consular office, but leading his own horse by the bridle. When he had come close and could be seen by all, he ordered his lictors to stand aside and led his horse up to the tribunal. The populace was amazed and fell silent; the censors were filled with a mixture of embarrassment and pleasure at the sight. Then the senior censor said, "I ask you, Pompeius Magnus, whether you have taken part in all the campaigns required by the law," and Pompey answered in a loud voice, "I have taken part in them all, under my own command." Hearing this, the people applauded loudly and could hardly contain themselves for joy. The censors rose and escorted Pompey home in order to please the citizens, who accompanied them with much applause' (Plutarch).

From 16 August to 1 September Pompey gave a great festival of games, as he had promised while still in Spain. During this time, as always when public games were in progress, social life came to a halt. Crassus, not to be outdone, sacrificed a tenth of his fortune to Hercules. It was customary to offer a tithe to the god after a successful enterprise, whether private or military, but this was usually a tenth of the yield or the booty. The victorious Sulla, however, had offered a tenth of his fortune, and Crassus now matched Sulla's example. All the gifts had to be edible and were offered on the *Ara Maxima* (the 'greatest altar') in the *Forum Boarium* (the cattle market). None might be taken back home: Crassus entertained the populace at ten thousand tables set up in the marketplace, the streets and the surrounding squares. The junketings must have gone on late into the night, at successive sittings.

Moreover, Crassus supplied the citizens with corn for three months. In this way the two most ambitious Romans of the age vied with each other. It was important that the citizens should take them to their hearts.

In this year Caesar reached the age of thirty. In one essential respect Pompey was able to serve as a model and a spur: in the way he tackled practical problems and won power and prestige as a commander – and no doubt also in the way he distanced himself from the Senate. Caesar admired the achievements of Pompey, whose fame and splendour were still undiminished, and above all untarnished. Pompey's high-handed conduct may have inspired Caesar when he was in Rhodes and decided, equally high-handedly, to intervene in the Mithridatic war. He may even have been in contact with the great man when campaigning for the restoration of tribunician rights. In 73 or 72 he obtained his first elective office, that of military tribune. Each year the popular assembly appointed twenty-four military tribunes, who were charged, when necessary, with levying troops and given minor commands. In this capacity Caesar may have taken part in the war against the slaves. In 70 he supported a bill permitting the surviving supporters of Lepidus and Sertorius to return to Rome. This bill accorded with Pompey's conciliatory policy. Its sponsor may have been the same man who put through the land law for his veterans. Caesar's brother-in-law Cinna was among the beneficiaries. In these years, then, it became clear that he wished to champion at least the supporters of the Cinnan cause and attached great importance to popular politics.

We may presume that in his twenties Caesar could hardly fail to maintain his inward opposition to Rome's ruling circles, and that he was obliged to seek – and limit – his freedom by refusing to commit himself to them, even if this went against the grain. For there is much to suggest that a fairly straight line can be drawn between Caesar's beginnings under Sulla and the course he subsequently pursued. To draw such lines may be to read into a person's life a meaning that was not originally there and to confer significance on what should properly, viewed without hindsight, be ascribed to a series of contingent events.

Yet what appears superficially to be contingent may sometimes be guided by an innate disposition that triumphs over chance and, while viewing certain fortuitous events as random, can see others as significant, so that they appear in retrospect to have been preordained. As Hofmannsthal puts it, fate does not necessarily attack one as a vicious cur attacks an unsuspecting peasant child carrying a basket of eggs on his head.

With Caesar this disposition probably derived from an urge to be different from others. It no doubt involved a degree of vanity and levity, of audacity, vindictiveness and pretension, perhaps even a fanciful notion that, being descended from Venus, he enjoyed the goddess's particular attention. This and probably much else led Caesar to withdraw inwardly from the society he belonged to. Ties and attachments normally lead outwards and lock the individual into the world around him, but in Caesar's case a special pride, nourished by a determination to achieve something special, seems to have detached him from this world; this was the particular disposition that

131

emerged in the years that saw the development of Caesar's immense self-sufficiency.

Such inward detachment indicates both sensitivity and insensitivity: sensitivity to one's own claims and insensitivity to others' demands. Caesar's experiences in the civil war, together with the spectacle that the republic presented in the seventies, may have inclined him to consider the existing order provisional and its institutions superficial, and to judge its leading figures not according to their rank, but according to their nature – wearing institutional robes that on the one hand fitted them too tightly, but on the other had become too capacious. He could not appreciate the huge burdens imposed on them. He can have had no sympathy with them. His standards required them to be measured not by the yardstick of the possible, but by that of the necessary.

If this was so, then in this and the following decade he must have been somewhat restless, both secure and insecure; the reality that the republic still represented must have imposed itself on him only gradually. None of this is now clearly discernible. Yet it seems clear that for all his vacillations he was constantly inclined to scorn this reality and remain an outsider.

8

The Political Rise of the Outsider (69 to 60 BC)

THE WILFULNESS OF THE OUTSIDER ·
THE DEMANDS OF A POLITICAL CAREER ·
POMPEY'S GREAT COMMANDS · THE
ROMAN PLEBS · CRASSUS · THE
EVENTS OF 63 · ELECTION TO THE
OFFICE OF PONTIFEX MAXIMUS ·
CATILINE · CAESAR'S IDIOSYNCRASY ·
THE CATILINE CONSPIRACY · THE
SPEECH OF 5 DECEMBER 63 · CAESAR
AND CATO · POMPEY'S RETURN FROM
THE EAST · ELECTION TO THE
CONSULSHIP · THE TRIUMVIRATE

T HE ROLE OF THE OUTSIDER, however arduous, has one great advantage: it enables one to preserve one's integrity, to remain untouched and uncorrupted, to retain the candour and high expectations of youth.

Participation in the normal business of living, on the other hand, may seem to involve a degree of complicity and the attainment of maturity to imply acquiescence in a web of weaknesses, impositions, dubieties, and half measures, or at least a readiness to make concessions. There is thus a certain attraction in remaining on the outside.

On the other hand, if one approaches practical living not just with scepticism – or even contempt – but with the intention of making something of oneself, if one sets out with the pretension of being in some way special, one must not be too fastidious, but adapt oneself to what cannot be altered. Experience teaches that the more a person imagines himself to be above the dubieties, constraints and entanglements of existing society, the less compunction he has in conforming; the less respect he has for a flawed morality, the more assured he is of his own superiority; this gives rise to many opportunities.

Caesar took advantage of them all, proceeding along a preordained course and certainly making compromises, yet at the same time behaving with a certain wilfulness and a detachment from convention that oscillated between criticism and hostility.

This readiness to swim with the tide, in which blind trust and indifference combine to afford the pleasure of yielding to one's own impulses without undue circumspection – this mixture of audacity and complaisance that we call wilfulness – need not lead to total absorption in the ordinary, which can sometimes be quite devastating. If a person is capable of calculating, of respecting certain limits, of drawing back in extremities, and, above all, if he is a player in a big game and has something to stake in it, if he is sufficiently disciplined, sufficiently cool and collected, and at the same time sufficiently rigorous towards the high expectations he has of himself – then wilfulness may generate not only a heightened self-awareness, but, if success comes, a particular trust in his own fortune. Success did not come early to Caesar, but he did not cease taking risks or renounce his desire to remain outside society rather than be absorbed in it. His political career was therefore not only successful, but ultimately became a special chapter in the remarkable history of Caesar and his fortune. The office of quaestor was the lowest rung of the Roman career ladder. Caesar probably became quaestor in 70, at the age of thirty. This was the earliest age at which one could normally apply for the office. It may be, of course, that Caesar, having won the citizen's crown, was accorded the privilege of applying earlier. At all events, he subsequently became praetor and then consul two years before the prescribed age.

At about this time his aunt Julia, the widow of Marius, died. He delivered the funeral oration in the forum. He was assured of wide attention, for everyone was eager to hear what the extravagant young nephew would say about Sulla's old enemy, the victor in the Cimbrian war. The content of his speech is not recorded. We learn only that he extolled his family's descent from Venus and the Roman kings before the assembled crowd and that the image of Marius was carried in the cortège. This is said to have caused much displeasure. But there was also much applause, and not just for political reasons: people wished to honour a distinguished general, whom Cicero too had recently praised. The Sullans were naturally bound to see it as a political demonstration. However, Caesar scored a great success: he became known, and it was important that the Romans should take note of his name.

When Caesar's young wife Cornelia, Cinna's daughter, died shortly afterwards, he delivered a funeral oration in the forum for her too. This was contrary to custom, which allowed such an honour only to matrons. Caesar disregarded custom. He needed to praise her in public and let his fellow citizens share in his sorrow. Those present are said to have been touched by the passionate grief of the handsome and elegant young man. We are not told whether Caesar spoke of his father-in-law or had his image carried in the procession. He may have deemed this unduly provocative – if one supposes that Caesar calculated every step.

There are many indications that Caesar now began to attract attention in a quite different way – certainly not as a future ruler or even as a promising politician. He became known as an extravagant, bold and disrespectful young man, certainly not without arrogance, though of a rather charming kind – at least to those not directly exposed to it. He was carefree and high-spirited rather than overbearing. He led a fairly easy-going life. So, of course, did many others. But Gaius Julius Caesar probably outdid them.

On his admission to the Senate after his term as quaestor he was entitled to wear the broad purple band on his tunic. Rome had strict sartorial regulations. Just as the toga was reserved for Roman citizens, there were special forms of dress for the various classes of citizen – patricians, senators, knights. Senior magistrates wore a purple-edged toga, senators a broad band of purple on the tunic. There were special senatorial shoes, even special patrician shoes, which Caesar was entitled to wear in any case. Only senators and knights might wear a silver ring. Caesar is said to have worn a belt over his purple-banded tunic, contrary to normal usage – though it hung loose, more as an adornment; and at certain points the purple band was fringed. He clearly felt it important to project an unmistakable image of studied casualness.

Equally unmistakable was the skill and artistry of his rhetoric. Cicero praises not only the correctness and precision of his Latin, but its unpretentious purity. Caesar himself later justified this in one of his writings, saying that one should avoid an unfamiliar word as a ship avoids a reef. Such simplicity of expression clearly had a special elegance and charm. A later author writes that there was such force, sharpness and fire in Caesar that he obviously spoke in the same spirit as he later waged war. In Cicero's opinion Caesar was unmatched as an orator. He had a brilliant and far from conventional delivery, to which his voice, his gestures and his figure lent nobility and grandeur. These attributes do not properly belong to rhetoric. Rather, they characterize the whole man. Cicero's description of Caesar as a speaker thus embraces the whole of his performance.

Caesar had little money and, though living far from frugally, he used whatever means he had – which were largely borrowed – purposefully and effectively. At least he lived in cheap accommodation in the Subura, a populous district on the slopes of the Quirinal, Viminal and Esquiline hills, just above the forum. There were more taverns here than anywhere else, and life was especially noisy and bustling. As well as artisans and shopkeepers, many whores plied their trade here. The main street led to the Argiletum; from there, passing to the right of the Basilica Aemilia, one reached the forum.

Caesar often appeared in the forum. A politician needed connections, and Caesar still had to establish his. It was not particularly difficult to become a quaestor, since twenty were appointed annually. As a rule, the next step was to seek one of the ten posts available as a tribune of the people or one of the four aedileships, after which, at the age of forty, one could move up to the office of praetor. There were eight praetors, and so less than one senator in two could obtain the office. And only one praetor in four – and one senator in ten – obtained a consulship. This was the goal of every ambitious noble, and the competition was accordingly fierce. Moreover, the most prominent nobles set great store by obtaining these offices at the earliest possible date (*suo anno*), but this was not easy, especially if one had little inherited influence and no strong family backing.

In these circumstances one normally had to spend much time gathering the necessary support. Everyone had to build up his own personal constituency, for there were no parties, no large groupings offering a ready-made power-base.

Roman voting customs were determined largely, though not wholly, by a system of personal ties that had grown out of the old system of client relations. In early times the citizens were divided more or less into groups supporting different noble families. The nobles represented their clients and spoke on their behalf in the courts, before the magistrates, in the Senate, and in the popular assembly. Conversely the clients had an obligation to support their patrons, especially at elections. Then, as the citizen body grew and the old and new citizens combined to form a prosperous new class below the senatorial nobility, the old client relations increasingly gave way to 'friendships', on a basis of near-equality. Moreover, the members of this new class usually had various simultaneous allegiances. In Rome they had to deal with different magistrates every year, in the provinces with different governors almost every year. Even in the Senate they did not always have to attend upon the same masters.

It was therefore necessary to establish many connections, for one was reliant on them. The best forensic speakers, most of whom belonged to the high nobility, could not be retained for a fee like modern advocates; their services were secured through the good offices of others – or they offered them because they counted on the gratitude and influence of those they were to represent. If a slave ran away, there was no police force to turn to – only the governors in the provinces where he might be apprehended. Again one needed connections. The same was true if a town was in dispute with another and the matter came before the Senate, if a tax-farmer sought the attention of a powerful citizen – or wished to evade it – or if someone needed protection, help in collecting debts, or various other matters.

Individual citizens, associations or towns often had to appeal to powerful senators to help them obtain the support of the political authorities or the courts. This usually involved individual interests affecting only a few politicians or small sections of the citizen body. Yet in each case a ruling had to be obtained. This is why it was so important to have a variety of connections and to show one's gratitude; one's mutual obligations remained and were even bequeathed to one's heirs.

When Caesar prosecuted Marcus Juncus on behalf of a number of Bithynians after the death of Nicomedes, he justified his action by saying, 'Whether out of the duty I owe King Nicomedes for his hospitality or because of my obligations to those whose cause is involved here, I could not shirk this task, Marcus Juncus, for men's memory must not be obliterated by their death, so that it is no longer preserved by their closest friends, nor may one desert one's clients without incurring the greatest disgrace, for it is our custom to ask even our friends for help on their behalf.'

This was all quite natural and self-evident. There was no place for larger groups; there were no major interests around which they could form. As a rule, then, the various interests remained isolated and had to be represented in differing combinations, depending on who brought his connections to bear against whom.

Likewise the citizens sought to discharge their obligations as electors. They had to present themselves when their patrons and friends were candidates for office or strongly supported the candidacy of a close friend. As many citizens maintained a number of simultaneous connections, it was naturally difficult to mobilize support. No one, for instance, could be expected to make a lengthy journey to vote for the friend of a friend unless he was exceptionally beholden to the latter. Those who lived at some distance from Rome presumably had to present themselves only when their own friends stood for election and sometimes perhaps only when they sought one of the higher magistracies. Because relations based on obligation so far outlasted the immediate situation, they could nevertheless rely on the support they needed from time to time in the Senate.

The mutual ties entertained by the Romans were above all of a personal, not an objective nature. The nobles were involved everywhere – as advocates, intermediaries and decision-makers. They could thus bring their influence to bear in all matters. The precondition for this was that important subjects only occasionally appeared on the agenda. Otherwise Rome could hardly have retained the old system, whereby elections were held only for the appointment of magistrates; elections only indirectly affected the composition of the Senate and the weighting within it, as only about a thirtieth of its membership changed when the twenty newly elected quaestors were admitted and the other magistrates moved up in the hierarchy. An objective 'will of the electorate' could not take shape; what mattered were the interests of the individual citizens in promoting their particular candidates.

In one's own career one was therefore thrown upon one's own resources, at least insofar as one did not belong to any solid groupings. Admittedly one had ties with many friends and supporters, but everyone had his own combinations of ties and had to cultivate them assiduously.

When a man like Caesar entered this world, relatively new and without many inherited connections, he had to try to secure the goodwill of a particularly large number of people by speaking or acting on their behalf in the courts and the Senate. He had to hold himself in readiness for many people. And we know that Caesar did so to a considerable extent.

One route to electoral influence lay through the tribes, the thirty-five subdivisions of the citizen body. It was customary for young noblemen to show special concern for the interests of the electors within their own tribes. In such a world anything one had in common with others was naturally a ground for calling upon their good offices. Hence a certain solidarity developed among the members of the tribe, and they were proud when one of their number was elected. And the young nobles who had to build up their careers would then try to persuade their fellow members to vote for certain candidates so that these would offer their services in return when suitable opportunities arose. In this way electoral influence could be organized on a rather wider scale. The career of the young Caesar followed this pattern.

'Toiling day and night with unremitting zeal': this was how the poet and philosopher Lucretius described a man's rise to the highest honours. The younger nobles had to make great efforts and endure many hardships. Their endeavours must not, of course, degenerate into vulgar sedulity or even into the censorious efficiency of a *homo novus* such as Cicero. By no means everything could be achieved through industry, as though it was just a matter of accumulating more and more useful connections.

For there was another complex of electoral motives: what was known as *existimatio* (reputation or prestige). This embraced a number of factors that might convince others of the candidate's worth – especially those who were not directly beholden to him. Among these were the age and standing of his family, the achievements of his father and forebears, the recollection of splendid games he had sponsored, of public buildings he had erected, of the grain he had distributed, and perhaps of his military successes or signal political services. Then there was the way in which he conducted himself, his style and bearing. His treatment of clients and the support he gave to those who sought his help likewise contributed to his *existimatio*, as did the size of his retinue, the extent to which he was known, the number of hands he shook. It was carefully noted who received the support of the leading nobles.

In all this the candidate probably had to discover and maintain a suitable mean between dignity and eagerness to please; he must not seem plebeian, but he had to do much that was beneath his dignity if he wished ultimately to rise to the rank, honour and dignity of a consul and take his place among the *principes*. Special conditions applied only to the *homines novi*.

A man's *existimatio* was always important, as it compounded the sum of his connections, conferring a chance of being elected and holding out the promise of success. As every voter at consular elections had two votes, one of which might still be uncommitted, there were usually many spare votes that could secure the success of a particularly attractive or promising candidate.

It is not clear how Caesar planned his career. Did he really wish to distinguish himself by diligence and hard work? Or did he prefer not to? Or did he simply not want it to appear so? Was he quite serious about the regular canvassing of more and more useful connections, or was he to some extent a gambler who preferred to risk sudden leaps?

And if he had at least an inclination towards the role of the outsider, did he allow it to become obvious? Did it help him or hinder him? And not least: if the majority tended to row in the lee of the shore, as it were, while he steered a course farther out to sea, did his ship not need a suitable sword to stabilize it? Was an abstract understanding of his duty to the commonwealth sufficient? Could Caesar, like Cicero, find that sword in the realm of theory, in the largely theoretical desire to bring the whole of the commonwealth to bear against its diverging parts? It may be that, for the time being, youthful unconcern helped him over all the difficulties and allowed him to keep all his options open. Approximate answers to at least some of these questions can be found if one considers his subsequent career.

Caesar served his quaestorship in 69 in the province of Hispania Ulterior (southern Spain), where he was responsible for conducting court proceedings in parts of the province. Suetonius tells us that when Caesar was in Gades (Cádiz) he saw the monument to Alexander that stood near the temple of Hercules and was overcome by the thought that he had reached the age of thirty-one without achieving anything remarkable, whereas Alexander had by that age conquered the world. The story is certainly not incredible, but neither is it unique. Comparisons with Alexander were in the air in contemporary Rome.

It is also in this period that Suetonius places Caesar's dream of raping his mother; since the mother supposedly stood for the earth, interpreters of dreams saw this as presaging world-domination. Plutarch, who places the dream in the night before the crossing of the Rubicon, may well have had better sources – at least for Caesar's recounting such a dream. It may nevertheless be true that Caesar was not content just to think of Alexander, but was overcome by impatience and, fearing that he was squandering his time, returned to Rome early. There may after all be something Caesarian in such decisiveness. Yet it is equally possible that the report of his hasty departure is wrong. Accounts of Caesar's early career are not necessarily reliable. There may be a confusion here with his premature departure from his Spanish governorship in 60, which is well attested.

Caesar took the land route through Gaul, stopping for a while in the northern Italian) province of Gallia Cisalpina. The towns there were largely romanized, but did not yet enjoy Roman citizenship. This caused much discontent, which Caesar seems to have encouraged. He is unlikely to have wished to persuade them to rebel;

that would have been highly irresponsible. It is more likely that he hoped they would exert massive pressure and he would be able to support them. He wished to gain power as their champion and so win a large number of clients at a stroke. We know that the citizens of the province, where all the prominent families had been admitted to citizenship, exercised substantial electoral influence in Rome. But if any proposal was actually made in favour of the Gauls it was thwarted by the Senate.

In Rome Caesar remarried. His second wife was Pompeia, the granddaughter of the two consuls of 88, Sulla and his ally Quintus Pompeius Rufus. This curious union can scarcely be interpreted as an attempt to forge a link with the Sullans; perhaps it was a love-match. For there is no indication that Caesar changed his political stance. Or had he learnt to distinguish between Sulla and the Sullans, so that he now found himself more exposed to the cruel spell of the dictator?

In 67 Caesar supported Pompey when the question arose of his being given overall command against the pirates through a popular law. The pirates, with their large fleets, now dominated large parts of the Mediterranean. They even dared to enter the Roman port of Ostia in order to plunder it. For safety's sake the coastal towns, trading companies, shipbuilders and shipowners were forced to pay them tribute. A new order prevailed, controlled by the pirates. The greed for booty and the silent longing for anarchy – for adventure, freedom and licence – was rife, and so the pirates were joined by fresh forces.

Meanwhile Rome was increasingly afflicted by shortages, which seem to have been deliberately compounded by the withholding of existing stocks. Only when the situation became really serious did the Senate take countermeasures. But its defence posture was nervous and inappropriate. The pirates, moreover, operated with such mobility that they could hardly ever be pinned down. And above all, no one believed that they could be quickly brought to heel.

Pompey's hour had come. He was indeed, as events proved, the only man who understood war on a grand scale, on both land and sea. Aulus Gabinius, the tribune of the people, petitioned, at first without naming his master, that a large proconsular command should be created and conferred on one of the *consulares*. He should be given a large fleet and authority to levy soldiers and sailors at his discretion.

A huge amount of money was voted to him, as well as credit with all the public treasuries. He was to have command over all the land within fifty miles (75 km) of the coast – an authority equivalent to that of any local governor. He was to be able to call upon fifteen legates (senior officers of the Senate), each with the authority of a praetor, and be given two quaestors. His command was to last for two years. These were just the powers that Pompey had hoped for. He liked to fight his battles with superior forces and so be assured of victory from the beginning. He also knew how to organize and operate with such forces.

Everyone knew that Pompey was *cupientissimus legis* ('extremely desirous of the law'), as Sallust puts it, though he strenuously denied it, protesting that he wished to be left in peace and did not seek a command. From now on he always behaved in this manner, letting himself be forced, with extreme reluctance, into accepting all the commissions and powers he coveted. This is said to have been in keeping with his character; but it was also prudent. A Roman nobleman could claim the right to regular office; he might, for instance, publicly announce that he wished to be given a consulship, to which his ancestry and his ability entitled him. The same did not apply to extraordinary commands, however. By seeking them he incurred the suspicion of excessive ambition. Pompey at once met with fierce opposition. Opponents declared that one man was being given control over almost the entire world. He could raise a huge military force and dominate the whole Mediterranean. This aroused the worst apprehensions. Who could be sure that he would relinquish his command as readily as he had done in 71? The most influential circles in the Senate already found Pompey somewhat sinister. Eventually he too had been dissatisfied with his position, having been simply left aside in his somewhat inflated dignity. Was he, after commanding an army, to return to his former status? He was suspected of wanting to be king. If he wished to be another Romulus, said the consul Piso reproachfully, he should ponder Romulus' fate. (One tradition had it that Rome's first king was murdered and hacked to pieces by the senators.) This was certainly unjust. But the senators found something so strange in his nature and his ambition that they were bound to think him a highly contradictory figure: fear of Pompey was as understandable as it was groundless.

The real contradiction, of course, lay less in Pompey's nature than in the estrangement that had grown up between himself and the other senators. He actually understood the problems of empire and wished to solve them; they saw only his quest for power. True, he was out for glory and a position of privilege for himself, but only in the interests of the Senate; the senators, however, perceived this as a threat to oligarchic equality and the exercise of their central responsibility. Such a man had no place in the senatorial régime.

In this Pompey's opponents were right. Yet Pompey could not understand it; he saw things differently. He viewed the Senate and the *res publica* more from the outside, from the point of view of the practical tasks facing it, whereas they viewed it from within, from the perspective of an order under threat. Both were right. Yet the preservation of the Roman order had become incompatible with the solving of extraordinary problems, since the latter entailed too great an accrescence of power and because the Senate, in guarding its responsibility, was too anxious, and hence too narrow and too weak. This was the contradiction. Fundamentally the Senate and Pompey represented two realities, between which lay a widening gulf. But they did not know this. Moreover, Pompey wanted not only to solve the practical difficulties, but to please the Senate and society as a whole. This contradiction thus came to determine his conduct and penetrate his thinking. Sallust described Pompey as 'honest in face, shameless at heart'. With his honest face he sought to evince his loyalty to the senatorial regime, but his heart was dedicated to his ambition. He wished to solve the urgent problems of empire, even against the opposition, and so establish a special position for himself.

It was in 67 that the opposition between Pompey and the Senate began. Early that year, after the experience of recent decades, the Senate was resolved not to let any one man have supreme power. The principle underlying its resolve was that nothing new should be allowed to happen that conflicted with the example and the institutions of the fathers. They were indifferent to the hunger endured by the populace, the miserable failure of Roman rule, and the spread of anarchy and piracy. Far more important was the preservation of order and liberty.

The senators prevailed upon two tribunes of the people to enter their veto against Gabinius' bill. This caused indignation not only among the broad masses, but among the knights, who suffered particularly from the insecurity of the seas. Gabinius and his allies were determined that their good cause should triumph. There were tumultuous disputes, during which Gabinius had to flee from the Senate and the Senate from the mob; the consul was arrested and almost killed. Following the example of Tiberius Gracchus, the petitioner swept aside the tribunician veto; he threatened to remove the interceding tribune from office and would probably have done so had the tribune not yielded at the last minute. The other tribune then dared do no more than propose that Pompey be given a colleague with equal authority, but the general uproar was so great that he could not make himself heard and was reduced to making signs with his fingers. Whereupon the crowd is said to have shouted so loudly that a raven flying over the forum lost its balance and fell to earth. Cicero reports that the forum was packed with people and that all the temples with a view of the speaker's platform were occupied. The law setting up the command was then passed. A further law conferred it on Pompey; the number of legates was increased to twenty-four, and further levies approved. Pompey at once set about arming, and suddenly the price of grain fell; the market filled with goods. Confidence in the order and Roman power returned. All speculation on its weakness collapsed. All this was possible through the determination of one man.

Pompey now systematically hunted down the pirates. Within forty days they had been driven from the western Mediterranean and fled towards Cilicia (on the southeastern coast of Asia Minor). After another forty-nine days the eastern Mediterranean was cleared of the scourge. Pompey had defeated the pirates in their hiding places. However, he then acted with notable clemency; he did not execute the defeated enemies or sell them into slavery, but settled them in deserted or ruined towns. One of these he named Pompeiopolis.

Caesar is said to have been the only senator to support the bill. This might relate to a particular debate that took place in the absence of Gabinius and his allies. And it would be evidence of great courage on Caesar's part, for the atmosphere was highly charged. Later twenty-four senators, among them two *consulares*, put themselves at Pompey's disposal as legates, admittedly after the law had been passed.

Similarly, in 66, a comprehensive command was created for operations against Mithridates; this too was conferred on Pompey. He was put in charge of all forces in the east and empowered to wage war and conclude treaties as he saw fit. Again there was fierce opposition, but this time four *consulares* openly declared themselves for the bill; Cicero and Caesar spoke in favour of it, and the knights strongly supported it, especially as Lucullus, the previous commander, had gravely impaired their scope for exploitation by reducing the debts of the provincials.

As far as the pirates were concerned, Caesar may have had a personal interest; in any case it was important that he should put himself in good standing with Pompey. If he ever wished to pit himself against the Senate majority, Pompey would be a natural ally.

Pompey was at the same time the patron of the *populares* – those politicians who at any given time played the popular role – and the populace to whom they addressed themselves. He had restored the authority of the tribunes. Now they had raised him on the shield, and his great and rapid successes had fully justified the popular agitation and legislation. If it was part of their role to maintain a continuity of ideas, Pompey was bound to remain at the centre of popular propaganda. All the more so as leading senators prosecuted the tribunes who had supported him – except Gabinius, who had become his legate. Violations of the tribunician veto, on which they had once more been forced to rely since 70 in order to block proposed legislation, could not be allowed to go unpunished. The defence invoked Pompey, partly with success, especially among the judges from the equestrian order. It followed that Caesar, by supporting Pompey, gained favour with the *populares* and part of the urban populace. This too was no doubt very gratifying.

Probably a year later Caesar was elected curator of the Via Appia, the old highway linking Rome with Brindisi. The curators were responsible for checking the state of the road and seeing to repairs, and probably also for carrying out improvements. The Roman roads of the period were nothing like as impressive as those we admire today. Most of them were paved with large stones only in the Augustan period. Early road-building involved acquiring the necessary land, laying out roads leading straight across country, levelling uneven surfaces as far as possible, reinforcing them with sand and gravel, building bridges and introducing ferries where fords were impracticable. There were always many improvements to be made – the paving of certain stretches, the building of new bridges, staging posts, overnight quarters or feeder roads.

The office was popular; it gave its holder an opportunity to recommend himself to all travellers. The milestones bore inscriptions recording the services rendered by the curators. Little else made such a good impression on those who travelled to Rome for elections. This was another reason why there was no communal programme for improving the road system as a whole, for instance by widespread paving: any law to this effect would have conferred excessive power on whoever introduced it and carried it out. It was therefore left to individual curators to do what little they could or saw fit to do.

The performance of this office was largely a question of organization. The local communities could apparently be called upon to carry out repairs; some public money was probably available too. Moreover, public land near the roads had been allocated on preferential terms to certain individuals who were expected to help with their upkeep, and care had to be taken to see that they did so. Some ambitious curators were even willing to dip into their own pockets in order to finance certain works and so enchance their reputation. Caesar is said to have done so on a generous scale.

At this time he began to run up large debts. To some extent this was an inevitable part of a politician's career. Young nobles increasingly vied with one another in costly enterprises designed to draw attention to themselves and please the masses, and their income was often far from adequate to their needs. Hence they had to take out large loans, which were easy to obtain if their career prospects were good. For in time their income was bound to grow; the praetorship was followed by the first governorship, during which large sums could be saved and, above all, extorted from the provincials.

Caesar admittedly seems to have exceeded the normal limit. He was not concerned about the debit side of his budget. He refused to be troubled about such paltry matters. In this respect he became an important figure at an early stage – by the mid-sixties at the latest. His private consumption too was considerable; he was a keen collector of gems, fine vases and old pictures. Suetonius tells us that 'he is reported to have paid such immense prices for well-built slaves that he was ashamed and forbade the payments to be entered in the accounts books.' He liked to surround himself with beauty, luxury and elegance.

Suetonius records too that, having built a villa on Lake Nemi at great expense, he had it torn down as it was not entirely to his taste, 'though at the time he still lived modestly and was much in debt'. Nor can his gallant adventures have been inexpensive, especially as he liked to show himself generous to ladies – even if not all received such costly gifts as his beloved Servilia, Brutus' mother, for whom, in 59, he bought a pearl valued at one and a half million denarii. Much of his expenditure, however, went on his career.

In 66 he was elected aedile for 65. The aediles were responsible for policing and public order (unless it was severely threatened), market policy, the supervision of baths and brothels, the provision and distribution of grain, and the supply of water. For these purposes the office carried a measure of jurisdiction.

It also involved the organizing of the regular games and was therefore coveted by ambitious nobles. It enabled them to cut a dash if they were willing to incur the necessary expense, over and above what was provided from the public exchequer. Caesar was extremely liberal. He also staged great animal fights, either in collaboration with his colleague Marcus Calpurnius Bibulus or on his own account; this earned him more public gratitude than anyone else.

He organized gladiatorial games in honour of his father, who had died in 85. They admittedly came somewhat late, but had the advantage of being fresh in people's minds when Caesar sought election as praetor and later as consul. The games were exceedingly lavish. It was intended that three hundred and twenty pairs should meet in combat, but it is not known whether this actually happened. Either beforehand or just afterwards the Senate passed a resolution limiting the numbers. However, it is credibly reported that Caesar was the first to have all the contestants appear in silver armour. It was a great event that could not fail to impress: this young man was not one for half-measures, but did things in style. It has every appearance of ostentation and vainglory, and the appearance is not entirely deceptive. Here again Caesar wished to be something patently special. This may reflect the insecurity of the young outsider, and also a desire to build up his future electoral power on the grand scale, rather than by slow and laborious degrees. It was not that he shunned the labour, but he had no wish to be totally absorbed in it.

It is recorded that during his aedileship Caesar also built temporary columned halls not only on the *comitium* (the meeting place in the forum) and in the basilicas, but also on the Capitol, in order to exhibit part of his collections.

At this time another, more calculated, feature of his policy becomes increasingly clear: a desire to advance his career by the 'popular route' (*popularis via*). As the magistrate responsible for streets and squares, he arranged one night for the re-erection of the trophies that had been set up by Marius to commemorate his victories over Jugurtha, the Cimbri and the Teutones and were later removed on Sulla's orders. There was consternation in the Senate; Quintus Lutatius Catulus, by common consent its most distinguished member, is said to have accused Caesar not merely of undermining the republic, but of attacking it with battering rams.

Caesar defended himself successfully against the charge. Presumably, as on other occasions, he appeared unmoved, innocent and surprised; he probably knew full well what his opponents meant, but could not really understand it from his outsider's point of view, so that he gave an impression of superiority verging on arrogance. He had long since discovered how to put these dignified grandees out of countenance, not by attacking them, but by using superior arguments. He probably argued that it was time to bury the old enmity and once more recognize the merits of the war-hero. This was in keeping with Roman magnanimity; the Romans prided themselves on their capacity for forgiveness. It was an argument they could not easily counter; however bitterly they resented it, Caesar was right. The populace, however – or those who gathered on learning what had happened – expressed spontaneous enthusiasm.

The following year Caesar was presiding over a jury court when two men appeared before it, one accused of having committed a murder on Sulla's instructions, the other of having killed a number of citizens who had been legally proscribed, and of having collected the price on their heads from the public treasury. Caesar accepted the accusations; the court condemned the one and acquitted the other (the later conspirator Catiline). Marcus Porcius Cato, who was then quaestor and later became one of the most dedicated advocates of Senate policy, withdrew the rewards from those involved in the proscriptions.

Both Caesar and Cato took up extreme positions towards the

connivances of the ruling Sullan aristocracy – the former because he was anti-Sullan, the latter because he supported the law. Both were outsiders and therefore enjoyed special clarity of vision. They differed in that the one acted unconstrainedly – which accounted for his greatness – while the other felt himself rigidly bound by tradition. They resembled each other in their dissatisfaction with prevailing conditions, yet differed in that they drew diametrically opposed consequences. This explains why, at this stage, we find them at one with each other, whereas later, when both had become powerful, they represented opposing extremes in Rome.

Up to this point one might wonder whether Caesar's policy was primarily anti-Sullan or primarily popular, but in 63 he adopted an unequivocally popular stance on a number of occasions. This is not easy to explain. For if Caesar was intent on becoming consul, as we have to assume, he needed strong support within the centuriate comitia. For elections to the higher magistracies were held in the old army assembly.

Rome's decision-making popular assemblies all had a composite structure; it might vary in detail, but there was a general principle that the votes of the electors counted only within the single electoral subdivisions. The overall result was determined by the absolute majority of the subdivisions. In a consular election, for instance, what counted was not the total popular vote, but who came first or second in each subdivision; the results were reported to the returning officers. The centuriate comitia, however, were divided very unequally in accordance with property qualifications, so that the prosperous sections of the citizenry had a decisive say. The mass of the citizens had hardly any say at all and could at best influence the general mood in the run-up to elections. Insofar as Caesar's policy was directed to the broad mass, he stood to make no direct gain and risked losing support among the more affluent sections of the citizenry. Perhaps, having opted for the *via popularis*, he aimed at an indirect advantage? Or at a position that went beyond a consulship?

The Roman plebs had long endured great hardship and, so Sallust tells us, was inclined to rebellion. He calls its mentality alien (*mens aliena*) – implying that its mentality was estranged from itself and the community to which it belonged, from the civic body and the civic mind. Sallust speaks of a dire disease that afflicted the citizens like consumption. 'In any commonwealth the indigent look enviously upon the good (the able and prosperous) and honour the bad; they hate what is old and crave what is new; displeased with their own conditions, they wish to change everything. Untroubled, they are nourished by strife and unrest, for poverty can easily be endured without harm. Yet for many reasons the Roman populace found itself set on a particularly perilous course.' All the worthless elements collected there like the bilge in a ship's hold. For this reason the Roman nobility referred to the urban mass as the bilge of the city (*sentina urbis*). 'Moreover, the young, who had endured a life of poverty and hard work in the country, were prompted by private and public largesse to choose idleness in the city in preference to thankless toil . . . No wonder that these people could be of no more benefit to the republic than to themselves.'

So far as we can judge from other sources and recorded opinion, the picture of poverty that emerges here faithfully reflects the reality. The income of most Romans was very low and often supplied them with no more than the bare necessities. Work might become scarce; there were great fluctuations in the money supply, building activity, harvests and imports, and the earning capacity of the inhabitants fluctuated accordingly. Yet as a rule there was probably enough food to go round, thanks to the public and private distribution of corn and other supplies. Moreover, one could enjoy all the excitements of metropolitan life – the frequent games, the latest news, and other happenings. Yet housing had long posed a serious problem.

The city was grossly overpopulated. Even in the late republic it did not extend far beyond the 'Servian Wall' (built in the period after 387), but the population had multiplied as large numbers of migrants, especially slaves and impoverished farmers, moved to the city. In Caesar's day it may be reckoned at about three quarters of a million. While the well-to-do citizens needed large complexes for their luxurious houses, foyers and small gardens – Lepidus' house, the finest in Rome about 78, was reckoned to be quite modest a generation later – people in other parts of the city lived in large tightly packed tenements (*insulae*). Taller and taller buildings were erected, and new stories were often added to existing buildings. Augustus later decreed that the maximum height should be a generous sixty feet (nearly 21 metres). Party walls should not be thicker than one-and-a-half feet, and other walls must be thin in order to provide maximum space. The upper stories often projected outwards, making the streets narrow and dark – but cool in summer.

Developers kept building costs to a minimum. Anyone who wished could set up as a builder; some builders, according to the learned architect Vitruvius, were excellent, but many were ignorant 'not only of architecture, but of simple craftsmanship'. In order not to overload the foundations, they built mainly with wood, which was cheaper than stone. The occupiers lived under the constant threat that the houses might collapse or catch fire, especially in winter, when they were heated by large charcoal braziers. And the rents were not low.

The result was a vicious circle: the population was growing, and so the price of land increased and building became dearer. As the builders economized on materials, the houses might easily collapse or burn down. And fires were hard to contain. Thousands became homeless, and there was a demand for new houses. In one of his poems Catullus praises a beggar who has nothing to fear – not fire, not the collapse of his house, and naturally not theft. The rich Marcus Crassus made much of his fortune by speculating in property. He began by hiring about five hundred slaves with various building skills. If a fire broke out somewhere, his agents were promptly on the spot, seeking to buy the burning house and those adjacent to it at the lowest possible prices. In this way large parts of the city came into his possession; in a very short time, using his own labour, he built large new tenements that yielded handsome returns.

Nothing could be done about the unsoundness and squalor of the houses. The level of rents, however, led to agitation. If *tabulae novae* (remission of debts) were proposed, the poorer inhabitants took this to mean principally the remission of arrears of rent. Yet really serious demands for such remission arose only in exceptional economic circumstances – in times of war or civil strife. Otherwise little relief could be hoped for, and anyone who sought it found himself up against the powerful knights, who were not to be trifled with; he therefore lost credit and jeopardized his career.

It is obvious that politically and spiritually the urban plebs lived from hand to mouth. Their discontent might serve to foment unrest, but could not envisage a fundamental change in conditions. Everything we know argues against this. This social powder-keg was too damp to be ignited politically. Politics, as we know, are not a function of social conditions, and poverty was the normal background of much of ancient history. Hardship becomes explosive only when it is felt to be unbearable. The Roman populace seems to have regarded its hardship as normal, preferring the reliefs and palliatives afforded by daily life to the prospect of supporting or rewarding an attack on any of the central points of the Roman order. Had this not been so, it would have been impossible to rule the city with virtually no police, even in the difficult and tension-ridden period of the Catilinarian conspiracy. Sallust thus exaggerates – for reasons of his own – the propensity of the plebs to rebel.

Yet if one is actively engaged in politics and at the same time approaches them with a certain detachment, there is probably always a temptation to take an unrealistic view of what is possible. And at this period one was faced with a highly diffuse reality. No one knew quite where he stood. One might therefore hope that the possibilities were greater than they seemed, that by involving the Roman plebs one could conduct politics on a grander scale and advance one's career.

Rome lived in a state of uncertainty, as its most powerful politician was absent. There was no knowing when Pompey would return from the east. Meanwhile, whatever was done or left undone might give him grounds for leading his army against Rome. The situation bred all kinds of covert or semi-covert plans and machinations, suspicions and apprehensions.

We hear of a conspiracy in late 66. The consuls elected for 65 had been convicted of corrupt electoral practices and were deprived of office even before assuming it; new consuls had been elected in their place. The deposed consuls are said to have planned to murder their successors and several other senators and then to resume office. However they did not carry out their plans – if they ever seriously intended to. Crassus is said to have had close contacts with them. Suspicion later fell on Caesar too, probably wrongly. Clearly the most interesting aspect of the affair, which remained shrouded in official silence and unofficial rumour, is that a few prominent citizens appear to have hatched such audacious plans, that they were credited with them, and that, suspicions having been aroused, the whole thing petered out and everyone carried on as if nothing had happened.

There were alarming portents: lightning struck the Capitol, a statue of Juppiter fell from its base, and statues of ancestors were broken; law-tables were so heated that the script became illegible, and the statue of Rome's founder, suckled by the capitoline she-wolf, was damaged. The Senate summoned augurs from all parts of Etruria. They prophesied that unless the immortals were propitiated there would be murder and fire, the laws would be overturned, civil war would ensue, and the city and its empire would be destroyed. They recommended ten days of games in honour of the gods, and much else besides. A new statue of Juppiter should be made, larger than the one that had been destroyed, and placed on the summit of the Capitol, facing east – unlike its predecessor – towards the forum and the Senate House. There would then be some hope that the secret machinations that were in train against the wellbeing of the city and its dominions might be discovered by the Senate and people.

It is not at all clear whether the bulk of the senators, enlightened as they were, took these portents seriously and feared the worst. We cannot know. That the signs coincided with conspiratorial plans was certainly troubling. Reality was so diffuse as to make people impressionable. In any case, something was bound to happen, and they tried to do what was best. In future, politics would take place under the eyes of the supreme god. In a society that conducted its politics largely out of doors, in the forum, there were perhaps grounds for hope.

In 65 Crassus held the office of censor; his colleague was Catulus. He took it upon himself to try to enroll the inhabitants of Gallia Cisalpina as citizens. He also wished to make Egypt tributary to Rome. Both projects far exceeded the normal competence of the censors. Catulus blocked them, and when he and Crassus left office nothing had been achieved, not even the revision of the Senate rolls; there had been no civic census and no review of the knights.

At this time Caesar had links with Crassus. No details are known. It must therefore remain unclear whether he seriously believed, as Crassus evidently did, that ambitious goals could be attained at a stroke – as if Rome no longer had any institutions, as if any plan that occurred to one or was suggested by some wiseacre could immediately be put into effect, as if the crucial question was not whether one's own wishes could be realized, but the converse: why should they not?

Posing the converse of the obvious question may be a mark of the significant politician: he may recognize that what appears to be real is illusory and perceive the true reality that others ignore; he may discern ways and means that others fail to see and pursue aims that do not occur to them. Yet here it was not a question of a new perception of the real and the possible, of finding new and arduous ways of measuring up politically to existing reality, but merely of an attempt to fiddle one's way through to ambitious goals within this reality – a reality that was by no means better understood, but simply wrongly perceived – without any particular political intelligence and energy.

Marcus Licinius Crassus was one of the most curious politicians in Rome; he was extremely characteristic of the age – according to Burckhardt's criteria – though not energetic, not even specially ruthless, and certainly not a 'man of stature'. Yet he was so in tune with the situation of the mid-sixties that he could almost be said to have embodied them. He was the richest man in Rome except for Pompey, who later outdid him thanks to the spoils of war. In due course he became Pompey's rival; like him he was a contradictory figure, not because of his nature, but because of the interplay between it and the situation. Yet apart from their ambition they had nothing in common.

Crassus lived modestly – which was, at least ideologically, in keeping with the old Roman style. He even conformed with the almost obsolete custom of marrying his brother's widow and having his children by her. But in fact he was merely avaricious. He was also unusually active and industrious, always ready to help, but possessed of no real energy. He put himself at everyone's disposal and was even willing to defend those whose causes Cicero and Caesar no longer wished to espouse; he prepared his briefs with painstaking zeal and would put up with anything so long as it brought him connections and power. He even staked his immense fortune in politics; he did not even take interest on loans, but demanded punctual repayment.

To most debtors this was of course far worse than paying interest, for since the expenditure of the Roman nobles was high and their income irregular, Crassus increased their dependence on him. More-

over, he tried never to give offence and was utterly at home in the
prevailing climate of complaisance. He was fundamentally good-
natured, not to say well-meaning. In more restricted circumstances
he would have passed for a man of probity, and might even have been
one.

All his difficulties arose because he set his sights too high. It was
unfortunate that he was plagued with greed, but on the other hand he
was active, and the other Romans were not poor orphans. More
worrying was the fact that Crassus wanted to be the first man in
Rome, and nothing he did to further this ambition could make up for
his essential mediocrity.

Under Sulla he had enriched himself enormously. And he had
gained an insight that was important to him: for someone who
wanted to be the first man in the commonwealth, no amount of
money would suffice unless he could maintain an army on the interest
it yielded. In the period of the civil wars this was not entirely
mistaken, but the converse did not apply – that anyone who possessed so much money could, on that ground alone, become the first
man in the republic. Crassus lived under the misapprehension that by
amassing huge amounts of money and goodwill he could rise high
enough to be acknowledged as such. Power was to him a mere
accumulation of the means to power, just as wealth was an accumulation of money. Because countless people were beholden to him he
considered himself a great man. And because he could buy many
things he thought he could buy everything.

There is no evidence that Crassus had a single notable political idea,
and there is nothing to indicate that this is due to the inadequacy of
our sources. He only ever reacted, copied the actions of others, or
pursued wild plans. He remained a tactician and was never a strategist. In any crucial situation he was at a loss and resorted to half-
measures. He would probably have liked to be ruthless, if only he had
known how. And so he resigned himself to fiddling his way through.
He would probably have liked to be daring, if only daring had carried
no risk. His ambition was so far removed from the pettiness of his
nature that instead of galvanizing him into a state of high tension, it
merely whipped him into a frenzy of aimless activity.

This man came to occupy a position in which all politically dubious
elements sought his backing. He was therefore acquainted with the
shadiest dealings, contributed funds, and then regarded his influence

as decisive. He surrounded himself with a protective cordon, made up of all those who were in his debt. About the end of 63, when it was reported in the Senate that he had encouraged Catiline to march on Rome, there was a loud protest. Sallust records that at the mention of the name of Crassus, a man of high nobility, extreme wealth and extraordinary power, 'some found the matter incredible, while others, though believing it to be true, deemed it politic, at such a moment, to flatter the great power of this man rather than provoke it, but most were beholden to Crassus on account of private dealings.' It was decided to declare the allegation false. The Senate and the *principes* were too weak to show him the limits to which he could go; hence he never seems to have known them.

For this reason he was constantly devising new plans. The wildness of his plans was consonant with the indeterminacy of contemporary reality. The standards he did not possess within him could therefore not be imposed on him from without. The wisdom he imbibed from outside led him to believe that the possibilities were endless. And this was as true as it was false.

When working on his obscure plans, Crassus probably thought that various courses were open to him – either that he could take the lead in a particular cause or distinguish himself by opposing it. In the elections for 63 he supported Lucius Sergius Catilina, a somewhat deranged patrician who seems to have had a certain louche charm. Catiline had formed an electoral pact with Gaius Antonius, the corrupt old Sullan whom Caesar had once prosecuted. Crassus supported them munificently. Caesar too seems to have canvassed support for them. Such pacts met with disapproval in Rome. In this case the blatant buying of votes caused such outrage that the relevant laws were made more stringent. As many senators feared a joint consulate of these men of questionable honour, Cicero, a *homo novus*, was able attract such support that he surprisingly topped the poll and was elected by all the centuries; Antonius became his colleague; Catiline failed to be elected.

Sixty-three was to be a particularly eventful year. Various initiatives by Crassus, Pompey and others kept Rome in suspense; in the end everything was concentrated on the Catilinarian conspiracy and then on the execution of five of its leaders; meanwhile the impending return of Pompey cast long shadows. The outcome of the conflicts of 63 created a situation so pregnant with consequences for the future course of Roman politics that few years of the late republic were as decisive as this. Caesar now played a significant part and succeeded in making a breakthrough to the front rank of Roman politicians. For the first time he came into conflict with Marcus Porcius Cato, who was soon to be his most important opponent and established a leading position for himself in this year. For Cicero, now consul, it was a year of destiny.

At the beginning of the year the tribune Publius Servilius Rullus brought in a land law. This had two unusual features: first, it provided for large-scale settlements in Italy and the provinces; secondly, it proposed a ten-man commission that was to be invested with extraordinary powers, indeed military authority, for five years, in order to buy up and allocate land. Its powers exceeded those of earlier land commissions, just as Pompey's great commands exceeded those of earlier generals. The ten commissioners were to be given large staffs of officials, clerks, bookkeepers, heralds and architects, two hundred land surveyors from the equestrian order, bodyguards, and large quantities of equipment. In addition, they were to have huge funds at their disposal and determine what was private and what public property. Their powers were formulated so broadly as apparently to include even the annexation of Egypt on the basis of a dubious will made by a former king, who had been murdered in 80.

The bill included an extensive social programme: large parts of the impoverished plebs in the country towns, and probably in Rome too, were to be provided with land. This reforming impulse may have come from Caesar. The commission was also to provide for Pompey's soldiers. This would have been an extraordinary affront to the general, for he could not be a member of the commission. The whole bill was clearly designed to create a strong counterbalance in favour of Crassus, together with Caesar and eight others.

Yet although many recognized its merits, the bill had no solid support, and there was probably no determination to secure its passage. Cicero was thus able to block it, with the support of a coalition made up of advocates of Senate policy and followers of Pompey. Caesar probably began to realize how little he could achieve in collaboration with Crassus. At all events, from now on he paid more attention to Pompey and the *populares*.

A second bill was introduced in favour of the old Marians, whose political rights were to be restored. Caesar strongly supported it, but it too was blocked by Cicero, who believed that it would overthrow the whole of the civic order that Sulla had restored. A petition for remission of debts and an amnesty for convicted politicians was likewise unsuccessful.

Then, with Caesar's help, the tribune Titus Labienus pushed through a measure under which the priests were no longer to be co-opted, but elected, as they had been in the last decades before Sulla.

In addition to this – again with Caesar's support, perhaps even at his instigation – Labienus arraigned the aged senator Gaius Rabirius for the murder of Saturninus, the famous tribune of the people, in 100, thirty-seven years earlier. At that time Marius had simply obeyed a *senatus consultum ultimum*. Saturninus and his followers had surrendered to him and been held in the Senate House. A group of young men had then come with ladders, climbed on to the roof, dismantled it, and massacred the captives from above. Although this was not part of the consul's police action, it had clearly had the approval of the leading senators; at all events they must have seen to it that the killers were not pursued. Labienus' decision to take the matter up again, after a lapse of nearly forty years, may have been connected with the fact that an uncle of his had been killed at the time; it is clear, however, that the prosecution was prompted chiefly by current considerations.

For even if the murder was not covered by the letter of the *senatus consultum ultimum*, it accorded with its spirit: there was an obvious interest in eliminating legislators who had forced through important measures that the Senate opposed. Yet this presupposed that certain citizens were willing to lend themselves to such actions. Their willingness was bound to abate if a man like Rabirius was convicted, or at least exposed to the risk of conviction.

If Labienus and Caesar sought to deal the Senate such a blow and thereby blunt its ultimate weapon, they presumably acted with the imminent return of Pompey in mind. For upon his return there was likely to be a substantial programme of legislation, and this weapon might become important. Moreover, by supporting the basic liberties of the people, which were usually violated by the execution of the *senatus consultum ultimum*, they stood to win great popularity.

They opted for a particularly ancient procedure, a hearing before a two-man tribunal. The two men were chosen by lot; one of them was Caesar. He had to pronounce sentence. Rabirius was found guilty of high treason and sentenced to death by crucifixion on the Campus Martius. Cicero, however, quashed the sentence; whereupon Rabirius was arraigned before the popular court, before the centuriate assembly. They were prevented from reaching a verdict by the novel use of an ancient, but not yet obsolete custom. Since early times, when Rome had to reckon with hostile attacks from its immediate environs, the old military assembly could meet only if a guard was posted on the Janiculum. A flag was flown to indicate that there was no danger. If it was lowered, the assembly had to be adjourned. On this occasion the praetor adjourned the hearing. It is not clear whether he wished to prevent a conviction or an acquittal. However, given his political position, the general inclination of the centuriate assembly to support the Senate, and the fact that the case was never resumed, it seems likely that the assembly favoured an acquittal and that he wished to obviate the public failure of the prosecution.

The revival of this ancient form of trial and punishment – at a time when the death penalty had been all but abolished in Rome – and above all the lowering of the flag on the Janiculum may well have been products of Caesar's fancy.

Caesar supported another bill introduced by Labienus, which would have allowed Pompey to wear the magistrate's toga at theatrical performances and his triumphal insignia at circus games, in both cases with the laurel wreath. Having supported Pompey in 67 and 66, he probably made contact with Quintus Metellus Nepos, who had been sent from the east to prepare the way for Pompey's return to internal politics. Mithridates was dead; finding himself in a hopeless situation, he had had himself stabbed to death, having been systematically immunized against the normal poisons. Thus ended the war against the wily Hellenistic potentate who had repeatedly gathered new forces in order to extend his territory. The preparations for Pompey's return and the attempt to transfer as much of his power as possible into Roman internal politics coincided with the discovery of Catiline's conspiracy.

Before this, however – presumably in the middle of the year – Caesar won an extraordinary success: he was elected to the office of *pontifex maximus*, the supreme priest of Rome. Since the third century BC the election had taken place in a popular assembly of seventeen tribes. The geographical division of the votes was unequal and worked to the disadvantage of the urban plebs. Inequality of wealth played no part, except insofar as the richer citizens could more easily travel to Rome for the elections. All the priests were eligible, but as a rule it was one of the oldest and worthiest who was elected. The others were probably inclined – or persuaded – to acquiesce in this. Catulus (consul for 78) and Servilius Isauricus (consul for 79) put up against each other; the office would normally have gone to one of them. But Caesar did not see why this time things should follow the normal course.

In his favour were his reputation with the plebs and his support for the law to re-institute elections for all other priestly offices. In 102 the sponsor of the law introducing the elective principle had immediately become *pontifex*, then *pontifex maximus*, at an early age. Caesar was certainly supported by many friends and followers, but this would not have sufficed to outweigh the authority and prestige of his highly respected fellow contenders. He therefore borrowed money on an unprecedented scale and bribed the electors. Financially he was so stretched that Catulus hoped, by offering him a large sum, to persuade him to withdraw. Caesar, however, raised his stake to the limit of his credit. On leaving home on the morning of the election he told his mother that he could only return as *pontifex maximus* or not at all. He was elected with an overwhelming majority.

By normal standards his stake was excessive. He had achieved the almost inconceivable. He had staked all on one card – not only a great deal of money that was not his, but his political existence. It seems insanely daring, and one wonders whether it was not an act of desperation; at least it was an act of wilfulness.

However unusually Caesar's career had begun, he had made his way satisfactorily through the regular sequence of offices; in the middle of 63 he was elected praetor for the following year. But if he had wished to achieve something extraordinary, he had little or nothing to show. He had been able to prosecute and provoke the Sullans, but his attempts to win power or authority – however seriously or light-heartedly he had embarked on them – had all failed. In political terms the alliance with Crassus had scarcely been rewarding; Caesar does not even seem to have taken much money from him

– and it would have been unwise of him to do so, as he would have been unable to pay it back very soon. Now he had brought off his first big success.

As *pontifex maximus* he had a chance to play an important role in politics. Even more important were the prestige that he acquired and the pretensions that he could now make quite plain; for he now moved from the Subura to the house in the Via Sacra reserved for the supreme priest, where he would in future reside. At thirty-seven he was in the front rank of Rome's senatorial aristocracy.

Now, at the latest, it becomes clear how much Caesar differed from others of his generation. Among these there was undoubtedly a good deal of dissatisfaction with existing conditions, and a good deal of rebelliousness. The Senate régime was anything but convincing, with its endless deliberations and vacillations, its insistence on complaisance and consideration, its time-wasting and obsession with trifles, and above all its utter refusal to countenance anything new. The political order was full of absurdities, which made sense only because society still believed in them. Yet what was so maddening was society's increasingly rigid attachment to the past. Where governmental and administrative efficiency were concerned, the Romans may seem to us to have been relatively undemanding, but even they must have found some things hard to endure.

Marcus Cato, Caesar's later opponent, and his junior by about five years, who still most closely resembled him in intensity of temperament and strength of character, seems to have felt much the same. Yet he drew quite different conclusions: he wanted senatorial rule to be decisive, consistent and vigorous. Instead of being merely backward-looking, the senators should actually return to the ways of the fathers. But Cato was made of quite different stuff and had grown up in the bosom of the ruling oligarchy.

And if Cicero, Caesar's senior by six years, tried to give firmness and direction to senatorial policy, this too was an expression of criticism and dissatisfaction. True, Cicero was a *homo novus* and so full of admiration for the old Senate that his first concern was to do things better himself. In most of the younger nobles, however, rebelliousness and disrespect for the fathers manifested itself in a general attitude of youthful disdain, frivolity and cynicism. They had no clear notion of what they wanted to do, and there was no opposition in society to which they could ally themselves. They therefore sought an outlet for their ambitions in wild festivities and private adventures. The 'rich old men', as the sassy Marcus Caelius once called them, did not expect the young to die of respect for them; the young could be allowed their debts, their luxuries, their amours. They could even risk a little political side-stepping – so long as they gave the impression of being prepared, in the end, to don the old tightly fitting garments that became the guardians of tradition and consider them the most splendid in the world. And most of the younger nobles were prepared to do so.

According to Nietzsche, 'any living thing can become healthy, strong and fruitful only within a horizon.' The Roman horizon seemed to many to be full of holes – or far too extensive. They could do this or they could do that; they had no *point de repère*, no direction, and so in the end they conformed. It was rare for anyone to break free, like Catiline and his fellow conspirators, out of defiance and indignation, indebtedness and hopelessness. Some of the younger men found this defiance fascinating and were at first enthralled.

Lucius Sergius Catilina, so Cicero later declared, 'possessed, as you will recall, many traces . . . of excellent qualities. Never, I think, has the world seen such a strange being, such a mixture of disparate, divergent and mutually contradictory needs and passions. Who would have been more pleasing – for a while – to the most respected men? Who was more closely associated with the lowest characters? . . . Who showed himself baser in his wiles, more steadfast in his endeavours? Who was greedier in taking, more generous in giving? And what was most amazing about this man was that he was able to win many friends and bind them by his loyalty; whatever he possessed he shared with them all, and he stood by his own in all adversities, with his money, his influence, his personal commitment, and if need be with crimes and reckless stratagems; he varied his behaviour, adapting it to the circumstances, twisting and turning it to all sides; among older men he was sedate, among younger men sociable, among the unscrupulous daring, among libertines licentious.'

Like Caesar, he came of old patrician stock that had long since ceased to be important; he was highly gifted and uncommonly strong-willed. Aristocratic self-assurance combined in him with the robustness of the self-made man. He had served as one of Sulla's henchmen. He had always been ruthless and had the 'plebeian' habit, not uncommon in outsiders, of conforming with the ways of society at large, which appeared corrupt, but ultimately was not. His forcefulness and ebullience contrasted agreeably with the narrowness and rigidity, the anxiousness and caution, of his peers. Yet he was unable to turn his talents to anything positive, being by nature too much of an anarchist.

Of all who found it difficult to conform with tradition, Caesar thus stands out by virtue of the bold and uncompromising will to succeed that he evinced as a candidate for the office of *pontifex maximus*. He may have been prompted partly by desperation, but success brought him the reassurance he needed. It was perhaps only then that he reached the decision that saved him from reckless ventures like Catiline's. He discovered a path of his own, on which he could not only preserve himself from the blandishments of society, from conforming with the normal and the comfortable, but avoid losing himself in opposition, protest and desperate enterprises. He could hardly have inclined to conformity, but it is impossible to exclude the possibility that he was in danger of lapsing, in one way or another, into what Bismarck called a 'Catilinarian existence'.

Now, if not before, his will began to harden. His energy and daring, his imagination and intellect, were concentrated on the prospect of the consulship, and probably also on what he hoped to achieve during and after it – though at this stage he cannot have reckoned on obtaining a province like Gaul. From now on there is no evidence of actions or plans that could be construed as pointing to a lack of seriousness. True, he remained opposed to the leaders of the Sullan aristocracy, and a certain resentment surfaced now and then. Yet at the same time a distinct 'achievement ethic' came more and more to the fore. The youthful insouciance of earlier days was gone. He now needed something firm to hold on to, and this he increasingly found in service to the *res publica*. The better he discharged the duties that were properly incumbent upon all Roman nobles, the more he excelled – as Pompey did too, though in a different way.

What may initially have been a rather negative disposition could thus be turned to his own advantage. The change cannot be traced to any single event. The most that can have happened in 63 was a switch from quantity to quality; certain motives that were already at work were consolidated, strengthened and given a fresh momentum; and this coincided with his decision to pursue a course of his own choosing.

Caesar had immense inner resources, and the strength and skill to organize them. Hardly any Roman was so unremittingly active as Caesar, or brought such intelligence to bear on his activity. One striking detail that is always cited is his ability to dictate four important letters to his scribes simultaneously, and as many as seven unimportant ones. It is also reported that he was the first person to correspond with friends within Rome itself, as the size of the city often made it difficult to keep appointments punctually.

Although we do not know to what period these reports relate, they indicate Caesar's capacity for effective action and his ability to call everything to mind. He increasingly mobilized all his abilities, in order to exemplify the old Roman aristocratic ideal in novel fashion. In doing so he found a role for himself and created a new horizon that could be constantly enlarged and allow the full development of his potential.

At the same time his political position became clear: he saw that in future he must above all side with Pompey, though this did not mean that he wished to sever his links with Crassus which would have meant forfeiting some of his independence. Yet if Caesar was to become an important ally of Pompey, Pompey must rely on the people. Caesar could do little for Pompey in the Senate. However, he could be useful to him through his popularity with the urban masses and his good relations with those who operated the mechanism of popular politics. The important question, then, was how the various forces in Rome stood in relation to Pompey and he in relation to them..

Pompey had instructed his envoy Nepos to secure for him the right to apply for the consulship of 61. This required a dispensation from the normal rules; under these he could not have become consul until 59, ten years after his first consulship. He also sought leave to conduct his candidacy *in absentia*. As on his return from Spain, he intended to move straight from his military command to the consulship, in order to ensure acceptance of the demands arising from his campaigns – provision of land for his veterans and ratification of the settlements he had made in the east. His wishes were presumably to be submitted to the Senate in the form of a petition; if this failed, they would be put to the people. Nepos accordingly sought to become a tribune of the people for 62. Pompey must also have considered another eventuality: if there was any reason – or pretext – for armed entry into Rome, as there had been in 71, it should be acted on. He probably had to leave all the details to be decided on the spot. He was still in Asia Minor, attending to local affairs.

Nepos' success was of crucial importance for the future course of events. The Senate's readiness, or reluctance, to make concessions could determine whether Pompey felt obliged to march into Italy and whether after his return – however it happened – there would be conciliation or conflict.

The Senate's attitude, however, might be strongly influenced by Pompey's supporters in Rome. They might opt for diplomacy or impudence, staking more on accommodation or more on rejection. Even if Pompey favoured an amicable settlement, at least some of his friends probably preferred to pitch their demands high, preferring failure to compromise. In doing so they may have been mindful of what they took to be the interests of their patron and ally; above all, however, they themselves stood to gain from any conflict between Pompey and the Senate, in which he might, if necessary, rely on the support of his legions. In the course of such a conflict he would be forced to rely on them, and his friends could then bargain for concessions on their own account. This was at any rate in Caesar's interest, and we may presume that he tried to use his influence in Pompeian circles, though no details are known.

The petition relating to Pompey's candidacy was obviously not presented in 63, either because the time was not ripe or because Caesar and others delayed submitting it. From the middle of the year another possibility emerged – that Pompey and his army might be summoned to deal with an emergency. The problems surrounding his return were complicated by the Catilinarian conspiracy.

In view of the grave consequences for Rome that flowed from this complication, it might for once be justifiable to use the word 'fateful'. This of course applies only to the effects. For the element of pure coincidence should not be overlooked, especially the part played by a number of protagonists – among them Caesar himself. At this time he played an extremely significant role.

★   ★   ★

Catiline, having failed in his second bid for the consulship (for 62), now joined with others in planning a *coup d'état*. They intended to murder Cicero, seize power in the city, eliminate a number of opponents, and take over the chief magistracies. The conspirators also planned a remission of debts, which would benefit not only certain nobles who were deeply in debt and on the verge of bankruptcy, but many common folk too.

There was no extensive background to the affair. The plan of the conspiracy was not unlike that of 66, or Lepidus' rebellion of 78/77. Once again the discontented throughout Italy banded together in Etruria and elsewhere. Catiline's agents tried to form armed contingents everywhere; near Faesulae (Fiesole) over ten thousand men assembled. Stores of weapons – which ultimately proved inadequate – were laid down; fighting units were organized and drilled.

The idea was probably to march on Rome and conquer it, as Sulla had done before. How Catiline proposed to defend himself against the returning Pompey remains unclear. It would probably be unfair to consider the plans in too much detail. On the whole we are dealing once more with a product of the 'unreality' of the contemporary situation; no one saw the need for strict rules, and the conspirators were not only lured by imaginary possibilities, but driven by adversity and pressing need. Moreover, the extent of the general indebtedness was itself symptomatic of the fact that in Rome anything was deemed possible. Crassus is believed, perhaps rightly, to have been linked with the conspiracy, though when it came to the crunch he distanced himself from it.

Caesar too came under suspicion. He may indeed have encouraged the conspirators, if only to disrupt the life of the city and provide Pompey with an occasion for leading his army against it. Sallust writes: 'All who were not on the side of the Senate preferred to see the commonwealth lapse into disorder rather than lose some of their own importance.'

Unluckily for the conspirators, one of them confided in his mistress, who then passed the information on to the consul. She was a married woman of the high nobility whose passion was now less ardent than her lover's; moreover, he had such large debts that he could no longer afford to give her handsome gifts. When she showed signs of rejecting him, he hinted darkly at future riches and, on being pressed, revealed details of the conspirators' plans. 'She was unwilling,' writes

Sallust, 'to keep so great a danger to the republic secret.' She may also have made a sober assessment of the conspirators' prospects.

Cicero paid attention to what was happening and showed perhaps more than the necessary zeal. As a new man he was anxious not to neglect his duty, but at the same time, for all his nervousness and excitement, he saw a chance to achieve something great. On 21 October he persuaded the Senate to pass the *senatus consultum ultimum*, which invested him with quasi-dictatorial powers.

On 27 October the rebellion erupted in Etruria. The Senate ordered levies and dispatched troops to various parts of Italy. The conspirators decided that Catiline should join the army in Etruria while the others prepared for the rebellion in the city. It was to begin with the elimination of Cicero. But when the conspirators and their retinue turned up for his morning reception on 7 November, they found the house heavily guarded and had to leave without achieving anything. Catiline nevertheless appeared in the Senate, having re-mained in Rome to demonstrate that he had nothing to do with the rebellion in Etruria. He may also have wished to keep his options open and await developments. But the senators received him coolly; none wished to sit next to him, and Cicero attacked him so vehem-ently that he decided to leave the city. In Etruria he donned the consular insignia and took command of the rebel army, which he organized into two legions.

Even now it was difficult to prove anything against the conspir-ators in the city. In the Senate Cicero was already being mocked for his zeal in constantly referring to the imminent dangers he had heard of. Finally, letters were intercepted in which the conspirators offered concessions to a certain Gallic tribe in return for a promise of military support. They had also sent messengers to urge Catiline to call even the slaves to arms – they were probably thinking of the gladiators – and to march on Rome as soon as possible. It was planned to start fires at various points in the city and to kill many of the citizens in the ensuing confusion. Catiline was to be ready to join the conspirators in the city.

Then, on 3 December, the five leading conspirators were sum-moned to the Senate, found guilty, and detained at the houses of various senators. On 4 December the Senate continued its delibera-tions. A man now appeared who declared that Crassus had sent him to Catiline. It was probably during the same session that Catulus and another *consularis* accused Caesar of complicity. This may have been in revenge for his election as *pontifex maximus*; it may have been an attempt to compromise this troublesome colleague and get him out of the way. The leading senators were prepared to believe anything of Caesar. They found him a sinister figure, unlike all the others; he seemed to want no part in the normal, though diffuse, reality of Rome, but chose to pursue all kinds of aims, and unlike Catiline he could not be pinned down. However, Cicero and most of the senators wished to isolate the circle of the conspirators; Cicero may also have known that Caesar was not implicated.

On 5 December the Senate met to decide what to do with the five prisoners. The meeting was held in the temple of Concordia, at the foot of the Capitol. It stood – and its ruins still stand – immediately to the right of the steps leading down to the forum. The temple was easily guarded and adjacent to the Mamertinum, the prison where the captives would be executed if sentenced to death. It was felt, moreover, that the merciful power of the goddess of concord should preside over the session and spread from there to the people. Meetings of the Senate perforce took place in hallowed precincts – the Senate House, the curia or one of the temples. And the choice of venue often depended on practical or symbolic considerations.

Cicero had assembled a large body of knights to guard the forum and the Capitol. The knights had been alarmed by what they had learnt of Catiline's plans from Cicero's rich friend Atticus.

These precautions were all the more necessary as clients and former slaves of the prisoners were already trying to recruit men among the plebs to free them from captivity. The atmosphere in the city was tense. Rumour was rife; sensational reports circulated, some true, some false; messengers ran to and fro. One claimed to know this, another that; much was surmised, much alleged to have been seen or heard. Crowds of curious, interested and distrustful citizens – potential trouble-makers – congregated in the streets, probably at some distance from the well-guarded temple of Concordia. Such gatherings could easily turn into angry demonstrations.

When the senators arrived for the meeting they had to force their way through the crowd and were exposed to shouts and questions. The consul, having performed the ritual sacrifice and obtained the goodwill of the gods, opened the session from the rostrum. The senators, having risen to greet him, resumed their seats on the long benches placed in rows parallel to the side walls, with a gangway in the middle; the senators did not usually sit in a fixed order. After delivering his report, the consul asked the assembled fathers, in order of rank and seniority, what they proposed regarding the fate of the Catilinarians. The first, a consul designate, called for the extreme penalty (*ultima poena*). Everyone was bound to take this to mean the death penalty. The other consul designate and the *consulares* concurred. This was no doubt pre-arranged. The accused were to be given short shrift and exemplary punishment; this was customary at times of internal emergency and always had the desired effect. It was the best way to prevent the Catilinarians from gathering further support, to halt the spread of disorder in the city, and to forestall any attempt to free the prisoners. It also offered a prospect of quelling the rebellion before Pompey returned from the east. It all seemed cut and dried; however, the senators were swayed not only by rational arguments, but by the excitement, tension and outrage caused by the plans that had come to light. All this seems to have combined to inflame the general sentiment. The senators reassured one another of their common cause by showing their determination to attack.

The question was then put to Caesar, the praetor designate. He rose and made a lengthy speech, which is said to have been delivered with consummate skill and extreme concentration, in his high voice, with his characteristically lively gestures, and no doubt with all the elegance for which his speeches were famous. He must have radiated calm, perhaps even coolness.

In the main Caesar concurred with the earlier speakers. He roundly condemned Catiline and his followers. He spoke of the need to punish them severely; he even declared that no penalty was severe enough to match their crime. He also held that the Senate was entitled to inflict any punishment it thought fit. He did not question its claim to decide, if need be, upon anything that would save the *res publica*.

However, having so far agreed with the *consulares* – no doubt to the growing surprise of his audience – Caesar raised certain doubts. He embarked upon a philosophical discussion of the meaning of death, which the immortal gods had ordained as a natural necessity, a relief from affliction and adversity, but probably not as a punishment. Moreover, death brought a sudden end to everything. Since men no longer believed that criminals had to suffer in the underworld after they died, the end was no longer to be feared. Caesar asked the senators to consider that if there was in any case no adequate penalty, it would be better to abide by the laws, under which no one might be sentenced to death without due process of law – and which also offered him the alternative of exile. He reminded them how unpopular severe sentences were and how much the people were attached to its most important civil liberty; he also hinted at the agitation – perhaps even accusations – that the consul and the Senate would have to fear if they passed and executed this resolution. It only needed someone to come and espouse this popular cause effectively – at which many must have thought of Pompey, who was still under arms. According to Sallust, Caesar invoked the ancestors, who had abolished the death penalty. They were after all superior in ability and wisdom to anyone present, 'for from small beginnings they created so great an empire that we can sustain it only with difficulty, after it was won with such energy.' Moreover, the death penalty, though wholly justified in this case, would serve as a precedent for others who were not faced with such an emergency. He for his part proposed something much harsher than the death penalty, namely life imprisonment in various towns in Italy.

Custodial sentences were not known to Roman law; public prisons were used at most for the temporary detention of dangerous persons. What Caesar suggested amounted to preventive detention. He also proposed heavy penalties for the towns in question, should their prisons prove insecure. Cicero said later that Caesar wished to have the prisoners surrounded by terrifying guards and deprived of all hope. No plea for mitigation or pardon should be permitted. Finally, their property should be confiscated.

The speech is not recorded, but certain clues allow us to reconstruct the arguments, though not the sequence in which they were advanced. Given the excitement over the resolution to impose the death penalty, the difficulty in which Caesar found himself, and the success he achieved, it seems most likely that, having covered himself by declaring his basic agreement with previous speakers, he first launched into an attack designed to undermine their certainty, then exposed the fear that lay behind their determination, and finally stressed the severity and harshness of his own proposal, showing that it fully accorded with their interests: the death penalty was mild in comparison with a life sentence; it would expose the consul and the Senate to great danger, and it would bring no benefit.

Caesar was arguing from a difficult position, having been suspected of complicity in the plot. Others in a similar position – Crassus for one – were not even present. Now the *pontifex maximus* was distancing himself from the Catilinarians, while doing his utmost to save them. For the prohibition on a subsequent pardon counted for little. He acknowledged all the rights to which the Senate laid claim through the *senatus consultum ultimum*, yet tried to prevent its exercising them. While proposing severe penalties, he adhered strictly to the law, which did not forbid permanent imprisonment. In doing so he was comparatively mild, supporting the civil liberty of provocation; this was entirely consonant with popular agitation and his own previous policy. By frustrating the effect of exemplary justice, he was helping to prolong the conspiracy and increasing the chances of Pompey's being recalled with his army. While speaking against the consul and the Senate majority, he appeared to be defending their interests in a particularly ingenious fashion. And the more he defended them, the more he embarrassed his opponents by fostering the fear that they were needlessly endangering the consul and themselves. This fear, already present, was now compounded by their alarm at having almost made a false move.

What Caesar said carried conviction. What he proposed was seductive. And at the same time it amounted to a severe blow against a right that the Senate had hitherto claimed – the right to act with extreme severity in an emergency, if necessary in contravention of the law. It was also a blow against the Senate's attempt to demonstrate at last – and in a convincing cause – its determination and rigour.

Caesar must have enjoyed recommending to the Senate, in such statesmanly fashion, a course of action that – according to previous criteria and the conviction of its leaders – ran counter to its interests. And this in a speech that testified to a high sense of responsibility and was delivered with a demeanour of innocence, confounding the opinions that the leading senators had hitherto entertained and expressed about him! It was altogether irritating, especially as Caesar's reasoning seemed faultless and the ostensible convictions behind it unassailable. Cicero remarked that one saw here the difference between irresponsible popular speakers and a truly popular spirit, concerned for the people's wellbeing. Whether he said this at the time or wrote it later, it brings out the stimulating, surprising and provocative combination of *popularis* and statesman that accounted for Caesar's strength. For the bulk of the senators did not necessarily share the views of the *principes*. Most were probably prepared to be impressed by Caesar. And he must have delivered a performance of amazing virtuosity.

The effect was immense. Those who were questioned after Caesar voted almost without exception for his proposal, against the solid authority of the *consulares*. This had probably never happened before. They were glad not to have to reach a definitive decision or display resolution in such a difficult matter.

The consul now intervened in order to reply to Caesar's concerns and reservations – and probably also to the anxious, furtive glances of many of the senators. He told them that they should not be concerned for him, but let themselves be guided solely by the public interest. For he naturally related everything to himself and the dangerous position he occupied. Yet he too was seized with fear and uncertainty; even his brother was unsure whether he really meant what he said. There were interjections; doubts spread as to whether he would be able to implement a resolution that day, and whether it was prudent to reach any decision when everything was so uncertain. In these circumstances Cicero then began to question the senators again. The first announced that in speaking of the *ultima poena* he too had of course meant life imprisonment. And although some of the *principes* disagreed, most of them, and most of the praetorians, including Cicero's brother, voted for Caesar's proposal. Support for the death penalty proved so weak that one of the praetorians, Tiberius Claudius Nero, ventured to go a step further. He proposed that any decision should be deferred until Catiline was killed. The consul should then make a fresh report, under armed guard. This would have been to carry the impression of senatorial irresolution, feebleness and lack of leadership to the extreme and done everything to sustain and encourage the conspiracy. Nero had recently served as a legate under Pompey and obviously represented Pompey's interests. This proposal too met with approval; the consul was powerless. The advocates of the death penalty seemed to have been decisively defeated.

It was only the vote of one of the youngest senators, a backbencher, that swung opinion round. This was the thirty-two- year-old tribune designate, Marcus Cato. He too made a great speech, taking Caesar and the whole house severely to task. The fathers were mercilessly put to shame. For Cato the issue was quite clear. And he championed it accordingly: they must not tolerate such an outrageous plot; they must punish it with the utmost severity and set an example. There must be no more pussy-footing. So vehement were his remonstrances that the senators let themselves be swayed. In accordance with senatorial custom they rose, one after another, and went to stand beside Cato. Those who joined him became a majority. It was resolved to execute the five Catilinarians. One piquant little scene must be mentioned in passing. As Cato was speaking, accusing Caesar of complicity with the Catilinarians, a letter was handed to Caesar. Instantly suspicious, Cato accused him of receiving messages from the enemies of the common-wealth, even in the Senate. He demanded that the letter be read out. Caesar, with an amused smile, handed him the compromising missive – a *billet-doux* from Cato's half-sister Servilia.

Caesar seems to have become involved in a fierce altercation with Cato; excitement was so tense that some young knights of the guard intervened and nearly killed Caesar. Some sources state that this scene took place near the exit. Caesar's life is said to have been saved by a hair's breadth, when Cicero interposed himself. Some of the senior senators later held this against the consul.

This was the first time, as far as we know, that the two strongest characters of the late republic confronted each other; one of them, after this speech, became the most dedicated champion of Senate policy; the other was soon its most dangerous opponent. It was the only occasion before the end of the civil war – with one exception, in 59 – that Caesar came close to winning over the Senate, and in a matter of vital public interest. Yet we should not overlook the significance of the fact that things were on a knife-edge: there was a choice not just between Caesar and Cato, but probably between two different possibilities for senatorial policy. This becomes clear when one considers how pregnant with consequences Cato's victory was – though no more so than Caesar's intervention, which made this victory possible. The course of events, the starting positions for Pompey's return and for the future of the republic – in the decade still left to it – were far from predestined, but determined by the great performance of this extraordinary man. The outcome was very soon to become clear.

In his discussion of this occasion, Sallust compares and contrasts Caesar and Cato. In many ways they were almost equals – in lineage, age, eloquence, greatness of mind (*magnitudo animi*) and fame. 'Caesar was deemed great because of his public gifts and generosity, Cato because of his blameless life. The one was famed for his clemency and compassion, the other for his severity. Caesar won fame through giving, helping and forgiving, Cato by wasting nothing. In the one the poor found refuge, in the other the wicked found ruin. The one was praised for his lightness and flexibility, the other for his unshakable firmness.' One might also speak of Caesar's sense of the fitting and of Cato's tenacious adherence to tradition. 'Finally, Caesar had made it his principle to be active and vigilant, to be concerned for the affairs of his friends while neglecting his own, to refuse nothing that was worthy of a gift; for himself he sought a great command, an army, a fresh war in which his energy could shine. Cato's endeavours were directed to moderation, to decency, and above all to rigour; he did not vie with the rich for riches or with the intriguer for influence, but with the brave for valour, with the moderate for discipline, with the blameless for unselfishness; he chose to be good rather than appear good. Thus the less he sought fame, the more it pursued him.'

It is both Roman and Sallustian that morality should be in the forefront of this evaluation, especially as regards Cato; but it is not wholly inappropriate. Cato made morality into a programme. Since everyone saw the decline of the age as the result of moral failure, he drew the conclusion that one must behave in the old Roman manner and make others do the same. His nature was tough and unshakeable, rooted in the Stoic philosophy, which he practised with almost ludicrous consistency, with a rock-like conviction of the rightness of his policy. He was ingenious and quite unconventional in his tactics and choice of means, but no strategist; he was inflexible in his aims, which he never doubted and none disputed: there must be no innovations; every inch of the old order must be maintained and defended.

Caesar and Cato were thus two fundamentally different representatives of the Roman aristocracy. Each exaggerated particular traits, to the exclusion of others that properly belonged with them; and in the long run they carried their one-sidedness to perverse lengths. For us Cato's position is linked with that of the senators, if only because he became the champion of the Senate and clung tenaciously to the old. Yet Caesar in no way inclined to the new: he simply did not close his mind to it. If Cato was for the republic, Caesar was not against it; he simply acted more freely, and this ultimately had consequences for its survival. Cato's thinking proceeded from within, Caesar's more from without. Cato wanted to preserve the republic, Caesar to realize his potential within it. For Caesar it was something given, whereas for Cato it was something that must be defended and restored.

It may be presumed that Cato would in any case have played an important role in the late republic. But it is questionable whether he would have begun to do so as early as this had it not been for Caesar. For it was Caesar who, by intervening so successfully and overturning the relations of authority in the Senate, gave Cato the chance to score a victory that instantly brought him such great authority. This was followed by a second big success, in which Caesar also had a hand and which also had far-reaching consequences.

Had the Senate adopted the political course advocated by Caesar on 5 December 63, it would probably have been less resolute in opposing Pompey's demands. But then the chance for Caesar's ascendancy would probably not have arisen. And in any case the weakness of the Senate was the best guarantee for the continuance of its régime. In objective terms, then, Caesar acted much more in its interests than Cato. Of course neither knew what he was doing. It was generally held that the Senate must be strong if it was to be able to rule. In the light of subsequent events, there was a profound irony in what happened. Caesar's brilliant intervention, prompted no doubt by immediate considerations, by the requirements of popular politics and Pompey's interests, not only failed, but made Cato powerful; it thus laid the ground for the conflict that made Caesar's rise possible. Since one can hardly assume that he planned so far ahead, it follows that what both Caesar and Cato ultimately achieved was precisely what neither had wanted.

As soon as the Senate had reached its decision, the prisoners were brought from the houses where they had been held. The five groups made their way through the silent crowd in the forum to the prison. Here Cicero had the prisoners strangled. Outside the prison he announced to the people: 'They have lived.' As the hero of the hour, he was solemnly escorted home; lamps and torches were set up everywhere; the tension was released. Any sympathy there had been for Catiline and his plans had evaporated when it became clear that the conspirators had intended to set fire to the city. The awesome demonstration of senatorial power had a liberating effect. It was now possible to take in what had happened.

The executions had a devastating effect on the rebels in Etruria. Many drifted away. Only a hard core attempted to break out to the north. In January they were defeated and wiped out. Catiline fell after a brave fight.

None of this could be foreseen on 10 December, when the new tribunes of the people assumed office, especially as the most important army was commanded by the consul Antonius, Catiline's old ally. However, it can hardly have seemed likely that Catiline would be able to hold out until Pompey's return.

Nepos, however, immediately petitioned that the general and his army should be recalled to Italy to restore order. In preparation for Pompey's return, he vehemently attacked Cicero for violating the laws of provocation. He apparently announced that Pompey would see to it that the guilty were punished. He probably thought that this would shake the newly established power of the Senate. But Cicero arranged for a military detachment to be stationed outside Rome, and the Senate resolved that anyone who accused those involved in the execution of the Catilinarians should be considered an enemy. The knights supported this policy. Cato persuaded the Senate to agree to a substantial increase in the provision of corn for the plebs. Nepos therefore resorted to force in order to put through his petition. His supporters occupied the forum and drove out his opponents. The voting was to take place in the temple of Castor, where he presided over the assembly, supported by Caesar; the steps of the temple were occupied by gladiators. Cato had difficulty in forcing his way through. When he interceded to prevent the herald's reading out the petition, Nepos read it out himself. Cato tore the document from his hand, and when Nepos continued from memory, another tribune clapped a hand over his mouth. Armed men then began to storm the podium. But Cato remained steadfast. His supporters were at first put to flight, but then returned, encouraged by his example. Nepos finally gave up after further altercations and after the Senate passed the *senatus consultum ultimum*. In a threatening speech he inveighed against the tyranny in Rome, then fled to Pompey. We do not know whether all this was staged in order to provide Pompey with a pretext for making himself the champion of the tribunes of the people and marching on Rome.

The mission of Nepos had failed, and Cato had won his second big victory, which virtually guaranteed that in future there would be opposition between Pompey and the Senate. Perhaps the Senate would in any case have refused to accommodate Pompey after he had won his two great commands in defiance of it. However, weakness and fear of Pompey might have inclined it to agree to a measure of conciliation. However, after being challenged so strongly by Nepos and gaining such power and confidence through the authority of the young Cato, it was no longer willing to compromise. It was above all for this reason that Catiline's conspiracy and Caesar's various powerful interventions affected the politics of the following months and years.

In the spring Pompey let it be known that he wished to return in peace – without his legions. He wanted to allay any suspicions that had been put abroad and to dissociate himself from the threats issued by Nepos. When he asked the Senate to delay the elections until his return – ostensibly so that he could support the candidacy of a friend, though it was his own candidacy that he had in mind – the majority was prepared to agree. Cato, however, prevented it. They went some way to meet Pompey – by postponing the elections until his friend arrived.

At the end of December 62 Pompey landed at Brundisium (Brindisi) and dismissed his troops, promising that he would later lead them to Rome in triumph. Mommsen's generation did not understand why Pompey should have dismissed his troops, since they took it for granted that he must have sought the crown. However, we now know that at that time a monarchy could have been set up only with brute force and would not have lasted long; the idea could probably not have been entertained at all. Marching into Rome would have made sense only as a way of exerting pressure in support of particular demands. Pompey now had to ensure the provision of land for his veterans and the ratification of his new order in the east. He does not seem to have thought it possible that his demands would be opposed; or, to be more precise, he probably considered it so unlikely that he was unwilling to risk the criticism, disapproval and hostility that he would have met with in Italy had he marched on Rome. Wherever he went he was honoured and received with generosity. But then he found himself back in day-to-day politics; he could hardly find his way around, behaved maladroitly and suffered repeated rebuffs.

All the more splendid was his triumph in late September 61. It was his third triumph, in celebration of his victories over the pirates and the kingdoms of the east. After Africa and Spain, he had now conquered Asia. There was so much to show that the procession was spread over two days. Yet even so it was not possible to display all the spoils. Large tablets caried lists of the lands and peoples he had conquered: Pontos, Armenia, Paphlagonia, Cappadocia, Media, Colchis, the Iberians, the Albanians, Syria, Cilicia, Mesopotamia, Phoenicia and Palestine, Judea, Arabia, as well as the pirates he had defeated on sea and land. Other tablets stated that through his conquests the public revenues had risen from fifty to eighty-five million denarii a year, and that Pompey would deliver twenty talents in gold and silver to the treasury, not including the soldiers' shares in the booty. Huge pictures illustrated his victories, and also the death of Mithridates. Several conquered kings and members of their families processed in front of Pompey's triumphal chariot. He himself wore a cloak that had belonged to Alexander the Great and been part of the treasure of Mithridates. This documented the fact that world-wide dominion had passed from the Macedonians to the Romans. Next came the senior senators who had served as legates under Pompey.

He had vowed to build a temple of Minerva from the proceeds of the spoils. There he later declared proudly in an inscription: 'Cn. Pompeius Imperator, having ended thirty years of war, defeated, killed or subjected 12,183,000 men, sunk or captured 846 ships, brought under Roman protection 1,538 towns and fortified settlements and subjected the lands from the Sea of Asov to the Red Sea, fulfilled his vow to Minerva in accordance with his merit.' As early as 63 Cicero had declared that the limits of Roman rule could no longer be found on earth, but were determined by heaven.

★   ★   ★

Pompey's achievement seemed magnificent, and so did his rank. This was the reality in which he had served Rome, the reality he inhabited. Accordingly he was allowed to wear the triumphal robe at public games and thereby repeatedly present himself as the great victor. We may certainly assume that the imperial pride that was manifested in all this was shared by the Roman populace, however terrible the times, however miserable its daily life.

Yet Roman reality was determined chiefly by the aristocracy, which was not much impressed by Pompey's achievements. When he first wore his triumphal robe at the games it aroused so much criticism that he refrained from doing so again; he did not even summon his soldiers to take part in his triumph. He probably feared their resentment, as nothing had yet been done to provide for them.

Some members of the aristocracy were highly distinguished and influential, others less so. But it was deemed improper for any to occupy, let alone flaunt, a position of privilege that went beyond that of *primus inter pares*. Moreover, the Senate refused to give blanket approval to Pompey's settlements in the east and insisted on examining them in detail. Pompey was reluctant to expose himself to this, as he foresaw an endless war of attrition with his opponents. Lucullus was eager to settle old scores, and Crassus kept up a fierce feud.

Opposition to the bill for the settlement of the veterans was no less fierce. It would have been possible to break it and Pompey was on the point of doing so. But he drew back; even now he dared not summon his veterans. He did not wish to be seen as a law-breaker. He had thus failed in his two pressing demands when Caesar returned from Spain in mid-60 to seek the consulship.

In 62 Caesar had served his term as praetor in Rome. As early as 1 January he had once again caused a stir, indeed great indignation. He proposed that Catulus, the senior senator, should report on the rebuilding of the Capitoline temple and that someone else should now be charged with the task; he clearly had Pompey in mind. He refused to allow Catulus to deliver his report from the speaker's rostrum. It was a gross humiliation for the dignified old gentleman to have to speak from the floor, below Caesar's feet. This was clearly Caesar's revenge for the suspicions that Catulus had voiced against him in connection with the Catilinarian conspiracy. The *pontifex maximus* and praetor was not prepared to put up with such treatment and reacted quite pettily. Abandoning the quietly superior tone of his speech of 5 December 63, he reverted to behaving like an agitator. Meeting with strong opposition from the senators, he desisted from his attack, but it had served its purpose. Catulus had done himself no good with his aspersions.

After the armed altercations attendant upon Nepos' petition, Caesar, like Nepos himself, was suspended from office. This may have been part of Cato's determined policy, but it may also have been prompted by the senators' annoyance at Caesar's increasing intransigence. He regarded it as illegal and continued to hold court. However, hearing that the Senate meant to assert its will by force, he doffed his *toga praetexta* and retired to his house. Two days later a crowd of citizens gathered outside it 'spontaneously' to protest against his suspension. Caesar pacified the crowd; this gave the Senate a welcome opportunity to thank him and restore him to office. When he was again denounced for involvement in the Catilinarian conspiracy he acted decisively by throwing one of the denouncers into prison and preventing the other from receiving the promised reward. Nothing else is known about his term of office except that he defended an African so vehemently against the claims of his king that the king's son, who represented his father's cause, pulled Caesar's beard. The court found Caesar's client liable to tribute, but when his opponent tried to lay hands on him, Caesar hid him and had him taken to Spain in his own litter. Such stories, which were often told, gave the impression that Caesar was very loyal to his friends. And the impression was not false: he was indeed a very reliable friend.

At the end of 62 he was involved in a domestic scandal. During the Bona Dea festival, to which only women were admitted, Publius Clodius visited Caesar's house in disguise, supposedly to be alone with Caesar's wife. Attempts to conceal or play down the incident foundered because Cicero and others insisted on prosecuting the perpetrator of such heinous sacrilege. Controversy arose over the constitution of the court before which Clodius should be arraigned. However, his friends first blocked a law that would have ensured a fair trial and then bribed the jury court that was subsequently set up; accordingly Clodius was acquitted. The money was supplied by Crassus, who thought to find a valuable ally in such a daring and ruthless man. Caesar declared himself ignorant of the whole affair, but divorced his wife all the same. Asked why, he replied that members of his family must also be above suspicion.

In 61 he took over the governorship of Hispania Ulterior, which had its centre in southern Spain. He wished to leave in haste, even before the relevant financial arrangements had been made. But his departure was delayed by creditors. He had no recourse but to seek help from Crassus. Crassus stood surety for 830 talents; this was an immense sum, amounting to about an eighth of Crassus' fortune – and he was the richest man in Rome.

While crossing the Alps, Caesar and his companions found themselves in a wretched little village and fell to discussing whether there too men contended for office. He is said to have declared in all seriousness: 'I for my part would rather be the first man here than the second man in Rome.'

Soon after arriving in the province he moved against a number of tribes in what is now Portugal and northwest Spain, partly with newly levied troops. Acting with great energy and skill, he took much booty and made many conquests. After one battle the soldiers, in accordance with established custom, acclaimed him as 'imperator'. This was the title of every Roman general, but to acclaim him as such was to honour him as victor in a particular battle. On receiving Caesar's report, the Senate decided to accord him a triumph.

While he was attending to provincial administration and jurisdiction, Caesar found the solution to an old conflict between creditors and debtors: the debtors should pay yearly no more – and no less – than two thirds of their income until the debt was discharged. By cancelling a special tax that the Senate had imposed on the province in the seventies, Caesar became its patron. During his governorship he must have accumulated a great deal of money – by begging, according to one source. He is also said to have plundered a number of settlements. At all events his finances were in a much better state on his return than they had been before.

In the spring of 60, he departed for Rome in great haste, without waiting for his successor to arrive. Having heard that Pompey had been unable to put through the two demands that were so important to him, he saw the chance of taking them over in exchange for a high reward. For he could now seek the consulship for 59. His opponents, Cato and his allies, viewed the prospect with apprehension.

Having been granted a triumph, Caesar found himself in a quandary.

If he wished to celebrate it – and he began preparing for it as soon as he returned in early summer – he would have to remain at first outside the sacred bounds of the city, which he was not allowed to cross until the day of the triumph. Yet he had to declare his candidacy for the consulship in the city at the beginning of July, and his triumph could not take place so soon. Should he forgo his triumph? To make the consulship less attractive, the Senate had decided that the provinces for the following year's consuls should be the 'woods and pastures of Italy' – two sinecures that could yield little profit. We know of no Senate decision remotely resembling this. On one occasion the Senate had sent a consul to a province in order to prevent his putting through a bill to which it objected. But that was an honourable commission, and the Senate was merely reacting. In the present case it was taking a preventive or deterrent measure – imposing a punishment, as it were, on anyone seeking a consulship.

Caesar was not to be prevented or deterred. He sought the Senate's leave to stand for office *in absentia*. Most of the senators seem to have been ready to accommodate him, but Cato embarked on a filibuster that did not end before sundown, so that no decision could be reached. The following morning Caesar waived his triumph and announced his candidacy to the election officer. It is curious that the Senate majority was prepared to resolve upon the 'woods and pastures', but not to reject Caesar's request to be dispensed from the usual procedure. Had the situation changed in the meantime? Could they now see more clearly what prospects Caesar's candidacy held out? Were they reluctant to be on bad terms with him? Or had Caesar's friends become stronger in his support? Pompey probably had nothing against the decision regarding the provinces, as it was likely to make Caesar dependent on him, but he had a great deal against delaying his candidacy.

It is in any case difficult to see how anyone could imagine that this kind of petty chicanery would deter Caesar from pursuing such promising prospects. What beckoned was the chance of a lifetime. Pompey needed him and could surely be induced, after so many defeats, to mobilize all the support he commanded in order to win success in the popular assembly with Caesar's help. And he had to offer something in return. It became clear, however, that Caesar wanted a province from which he could make great conquests. There was no other way to gain great power, fame, riches and a large following of veterans, no other way to prove his worth, far from all the irksome restrictions of the oligarchy, all the pettifogging of city politics. If Pompey's demands were pushed through, it must also be possible to push through an appropriate provincial law.

Had Cato and his allies no inkling of Caesar's plans? It would admittedly have been something quite novel to set up an important special command when there was no war. Hitherto special commands had always been preceded by wars. Yet it surely cannot have been beyond the wit of Caesar's opponents to conceive such a possibility.

Yet whatever thoughts Cato may have had about Caesar's intentions, he relied on the various means of obstruction and the superiority of the Senate, which in the past had nearly always proved effective in the end. True to his programme, he was determined to defend every inch of the old order. And he did not want to be overtaken by events, but to forestall them. He could expect nothing good of Caesar. He therefore opposed him systematically. For his own part, and on behalf of the Senate, he forced him to remain an outsider. Rome's aristocrats had always tried to outdo one another. But as a rule they had kept within certain bounds. There was competition for every position. But in principle everyone had an equal chance of attaining it. It was not the business of the aristocracy to devalue the positions themselves for the sake of an individual. Yet now, in defence of the Senate régime, one of the prerequisites of this basic class solidarity had been violated. Perhaps it was the attempt to defend the régime that provoked the attack on it.

Reality hardened and the quarrels took on a new seriousness. Cato at least made it clear that the fate of the *res publica* was at stake, and that the game was being played against both Caesar and Pompey.

As Cicero wrote at the time, Caesar had a good following wind. He made an electoral pact with Pompey's wealthy friend Lucceius, who promised to distribute money to the centuries on behalf of them both. Crassus and Pompey came out in support of Caesar. It was not uncommon for men who were on bad terms, or even at enmity, with one another to support the same candidates at elections. The number of candidates was consequently limited, more or less in accordance with their year of birth. Only someone who commanded substantial electoral influence in his own right had any real prospects. Support from such rivals was then determined by their obligations; moreover, they might compete with one another for the candidate's favour. Crassus probably mobilized large numbers of knights for Caesar. Caesar, moreover, could rely not only on his clients and friends, but on his reputation. At least he was well known, had given splendid games – with gladiators in silver armour – and used his good offices on behalf of many people. He was the *pontifex maximus* and had been granted a triumph. To the bulk of good Roman society he seems to have been an extravagant, restless aristocrat, impressive and intelligent, but in no way dangerous. An *enfant terrible*. If he went his own way now and then, leaving the beaten track and exasperating his elders and betters, this simply made him interesting. No doubt his aloofness from the others was noted. He was not easy to fathom and did not fit into any obvious mould. But whatever objections this might have occasioned were obviously mitigated by his social bravura and versatility. He was no simple *popularis*, but played the 'popular' role in a superior fashion – persuasively, with brilliant arguments and therefore, it must have seemed, responsibly. He was presumably all the more friendly to the younger generation, the less he saw eye to eye with their elders. Because he offended the most powerful citizens, he may have felt it prudent to treat the others with consideration. And by doing so he probably succeeded in making them feel honoured. Most were bound to be impressed and captivated by his superb skill as a speaker, his insouciance and charm, his aristocratic bearing and assurance, and perhaps by even his arrogance. And the masses were already on his side – however little that signified.

His opponents strongly supported the third candidate, Marcus Calpurnius Bibulus, Cato's son-in-law. He was a serious, stern and somewhat limited man who could steer a straight course. Ever since they were aediles he had competed with Caesar. The leading senators behaved with a cohesion that was rare at elections. As a rule each acted in accordance with his personal obligations, but this time it was a question of high politics. They joined forces and persuaded Bibulus to promise the electors the same sums as Lucceius and Caesar had offered. Even Cato, for all his scruples, agreed to this, as the cause of the republic was at stake.

★   ★   ★

It was probably in early July that the election was held on the Campus Martius, roughly where the Pantheon now stands. Rome was crowded. Many citizens, accompanied by servants, had repaired to the city and were staying with their Roman hosts. At dawn the assembly of the centuries was announced by the heralds with tuckets of trumpets. The consul responsible for conducting the election took the auspices in a tabernacle near the Campus Martius. It was important to obtain the goodwill of the gods, so that the successful candidates might be worthy successors of the long line of past magistrates. The formalities were probably still strictly observed. The consul, as returning officer, then went to the place where the voting was to take place. The citizens streamed towards it in groups of varying size, escorting the candidates to whom they were specially beholden from their houses or joining their growing retinue. The crowd probably ran into several thousands.

On the open field where games were usually played or practised, a platform had been set up. From here the returning officer opened the proceedings with a *contio*, a kind of unordered popular assembly. He first spoke a solemn prayer beseeching the gods 'that this matter may end well and happily for me, my conscientious endeavour, my office and the people and plebs of Rome'. He then announced the names of the candidates. These, together with other magistrates, had taken their places beside him on the platform, wearing their whited togas. Their names may have been posted somewhere in large letters. There were no election speeches – at most a general exhortation by the returning officer. He then sent the citizens away to vote.

To do so they had to enter the enclosures (*saepta*) or pens (*ovilia*), made of wooden scaffolding and ropes, which were set up at election times; these were long, narrow structures lying side by side, each large enough to accommodate the members of one voting division, and numerous enough to accommodate all the divisions taking part in one ballot. At the entrance the elector's membership of the division was checked; voting tablets may have been handed out here (for it was a written election). At the exit everyone passed separately over a narrow bridge, so that he could hand in his tablet without being influenced. The handing in of the tablets was supervised by officials appointed by the returning officer and the candidates. At this end there must have been another platform from which the returning officer could observe the proceedings. The candidates stood here too. On leaving, the voters had to pass in front of them.

Before the voting began, one century from the first class was chosen by lot to be the *centuria praerogativa*; it voted before the others, and the result was publicly announced by the herald. It was meant to act as a suggestion to the assembly. According to Cicero, the candidate whom the *praerogativa* placed first was always successful. He therefore picked up many of the uncommitted second votes (each elector had as many votes as there were offices to fill). The practice of calling on one century to open the voting tended to ensure uniformity. After this the remaining sixty-nine centuries of the first class, together with twelve equestrian centuries and a few special divisions, were called to vote. When their ballots had been counted, the result was read out. There then followed a separate ballot for the six senior equestrian centuries, in which senators' sons voted. It had obviously been instituted so that these centuries, having heard the results of the first class, could vote solidly for the most successful candidates; this would again give later voters the impression of a uniform vote carrying special authority. It was a striking manifestation of senatorial solidarity. We do not know whether the institution still functioned in Caesar's day as it had in the past. At all events, the intention was that the various forces should show their relative strength in the first ballot and then demonstrate their solidarity in succeeding ballots. It was now the turn of the other classes, from the second to the fifth. As soon as the required number of candidates had absolute majorities, the voting was terminated. As there were a hundred and ninety-three centuries, this stage could be reached in the second class. It was considered important that the result should not be decided by the lowest classes and that they should not be able to give any one candidate a specially large majority. A candidate could distinguish himself only by winning the support of all or most of the centuries by the time he had gained a majority. The first candidate to be declared elected had certain privileges over the other.

The assembly lasted many hours, and by the time the second class came to vote it must have been very hot. We do not know whether the enclosures had awnings to protect them against the sun. One can imagine that candidates felt tempted to provide them. Anyone who had cast his vote could canvass among the others or repair to the shade of one of the buildings nearby.

Finally the successful candidates were 'proclaimed' by the returning officer in a *contio*. Only then were they duly elected: the retiring magistrate formally 'created' his successor. In 60 Caesar was the first to attain an absolute majority; according to one source he carried all the centuries. Bibulus came second. Probably many citizens voted for both Caesar and Bibulus, because they had obligations to both and because they were not bound by the political oppositions, being unable to perceive or feel them as keenly as Cato and his followers. They probably tended to view the phenomenon of the outsider in social terms.

In the next few weeks Caesar had long negotiations with Pompey and Crassus and forged an alliance with them. The aim of this 'triumvirate' was to realize jointly certain objectives that they had failed to achieve alone. Pompey was to have his two demands met, Crassus to obtain a long-sought remission for the tax farmers of Asia, and Caesar to be given a province. As for the longer term, there was only the vague formula that nothing should be done in the commonwealth that was displeasing to any one of them. As each wanted to be the first man in Rome, it was hardly possible for them to agree on anything beyond the present moment.

The alliance had important consequences. A generation later, Asinus Pollio began his history of the overthrow of the republic with the year 60 BC. Cato later declared that it was the unity of the three, not their disunity, that destroyed the republic. Yet we must ask whether it was really the unity of the three men that was decisive, rather than the way in which Caesar carried out their policy.

At this time the country was visited by a great storm. Many trees were uprooted and many houses destroyed; ships sank in the Tiber and its estuary; the wooden bridge over the river was destroyed, and a wooden theatre built for a celebration collapsed. Many lives were lost. Roman historiography tells us that great events were widely believed to be reflected in nature. Here was confirmation of this belief.

For a time Cicero had deluded himself into thinking that he could draw Pompey on to the side of the Senate. In 60 he wrote to his sceptical friend Atticus that he was thinking of making Caesar see reason (literally 'making him better'). He spoke also of a 'medicine that heals the sick parts of the commonwealth instead of excising them'. As always when viewing great decisions with hindsight, one wonders whether things could have been ordered differently; one considers the possibilities and probabilities immanent in the situation. This is merely another way of asking how it was that things happened as they did. 'Happy is the man who can say "when", "before" and "after"! He may have suffered misfortune, he may have writhed in pain, but as soon as he is able to recount events in the order in which they occurred he feels as well as if the sun were shining on his stomach.' This, says Musil, is 'the law or narrative order', the 'proven way of shortening the perspective of understanding'.

We must resist any such shortening of our perspective if we wish to appreciate the full complexity of the events narrated here. It is easy to see how the Senate came to oppose Pompey, but was it not exceedingly short-sighted? For he wanted to be on good terms with the Senate and in fact had great respect for it. In view of its weakness, did it not need all the help it could get? If the *res publica* was in danger, was not an alliance with Pompey an obvious precaution? Was it not very imprudent to oppose him? And did it really make sense to pursue such a petty policy towards Caesar?

There are many other questions. How was it possible, for instance, that the once so powerful Pompey, having had such great successes, had become so weak? How was it possible for Cato, at the age of only thirty-five, to prevent the assembled fathers, partly against their inclination, from reaching an accommodation with Pompey, and above all with Caesar? For the resulting conflicts were clearly no longer binding on the bulk of affluent Roman society. Cato, incidentally, filibustered for weeks in order to frustrate the remission of debt for the tax-farmers.

What was it that made for power or impotence? What was decisive in creating oppositions? Since the oppositions governed the scope of possible positions and possible points of contact, this also raises the question of the opportunities open to Caesar in the Roman society of his day.

9

Crisis and Tensions: Cato's Authority, Pompey's Difficulty, Caesar's Problem

THE DISCREPANCY BETWEEN EVERYDAY
POLITICS AND CONSTITUTIONAL
POLICY · CATO AS CHAMPION OF THE
SENATE · WHY POMPEY WAS OPPOSED ·
THE RESPONSIBILITY OF THE
LEADING SENATORS · DISINTEGRATION,
NOT A CRISIS OF LEGITIMACY · NO
PLACE FOR OUTSIDERS

THERE WAS AN ELEMENT of schizophrenia in the politics of
these years. On the one hand, relations based on mutual
obligations still obtained, as did other traditional motives of
all kinds; splendid games, for instance, could predispose the electors
in favour of a particular candidate, and politicians continued to
respect one another's claims. This was the stuff of everyday politics,
with their constantly changing groupings. Pompey and Caesar were
naturally locked into the various mechanisms of social relations.

On the other hand there was the sphere of constitutional policy, in which the Senate ultimately acted with relative cohesion, however this might be achieved at any given time. This too had a long tradition. As the Roman order was not strictly defined, but rested largely on example and precedent, and as the power relationships within it were generally viewed as part of the 'constitution', it was obvious that the ruling aristocratic body must pay careful attention to the way in which the order was affected by new political facts. When in doubt, the senators preferred to make practical concessions rather than submit to 'new unheard-of precedents'. They also did everything they could to prevent any individual from becoming powerful enough to be able, if necessary, to override the Senate. In this regard the senatorial oligarchy had always achieved unity in the end. However flexible it might be in other respects, in this it had always remained firm. This solidarity had guaranteed its leadership.

Everyday politics and constitutional policy had always complemented each other, but the latter only rarely found its way on to the agenda. Much the same was true of the Senate's handling of the great questions of war and foreign affairs. What was new in the late sixties was the weakness it showed in its conduct of constitutional policy. It is the pettiness and rigidity of this policy, and its incongruity with everyday politics, that gives the impression of schizophrenia.

From early times, constitutional policy had been primarily in the hands of the *principes*. It was their duty to safeguard the inherited order and ensure that any differences were resolved within its framework. This was an established part of their role and consonant with their interests. For their authority rested largely on the fact that they represented the cause of the whole house. Old experiences, deeply etched into Roman ways of thought and feeling, had taught them that the pretentious Roman aristocracy, with its world-wide empire, needed strict control and discipline. No magistrate could provide these. The *principes*, however, had been able to do so on various occasions, and practice had made them remarkably adept. Rome was thus able to pull off the trick, exceedingly rare in world history, of finding, within a class that was full of powerful tensions and particularist interests, a group in which the overriding interest of the whole was institutionalized in a non-partisan way – if the outcome of a long series of skilful attempts can properly be called a 'trick'. Of course, the individual *principes* might often be highly partisan. It was only as a group, when performing their common role – and maintaining a fine balance between the freedom to differ and the limitation of differences – that they exercised this function, in which the power of their patronage worked to their advantage.

This practice continued into the late republic – though the tensions had by now become so much fiercer that it could no longer operate as impartially, convincingly and definitively as before.

It had of course been clear for some time that there was always one highly authoritative personality who espoused the cause of the Senate. To do so required perpetual diligence and painstaking effort, the expenditure of much time, energy and responsibility, as well as confidence and endurance. Since the seventies Catulus had displayed these qualities and for this reason been universally respected as the first of the senators. On his death in 61, Crassus is said to have tried to take over his role. None of the *principes* would have offered any competition. But young Cato did. Cato then became the most committed, resolute and tireless advocate of senatorial policy.

His only problem was that he did not receive enough support from the *principes* or that they were too weak to dragoon the bulk of the senators into consistently pursuing a common line. Rather, there arose within the ruling class a movement of mutual repulsion and rigidification, between the indolence and resignation of the old and the orthodox, defiant rigour of the young. Various elements may have come together: relative impotence led to resignation, and resignation to relative impotence. There was a tendency to withdraw into private life. Cicero characterized the grandees of the time as 'fish-ponders': nothing was dearer to their hearts than their fish, nothing gave them more pleasure than the red mullet that lived in their ponds and ate out of their hands.

The Senate was hard to govern: its members, of whom there had been twice as many since Sulla's reforms, were now confronted for the first time with the problem of supporting a strong, resolute policy in the interests of the house. At the same time a mood of complaisance had taken root. After all, Pompey's demand that land should be provided for his veterans was not altogether unjustified. And why should Caesar's triumph founder on the prohibition against conducting his candidacy *in absentia*? There were many impediments to a decisive senatorial policy. But a new realization that they must unite with Pompey rather than oppose him was not one of them.

Cato was all the stronger, his authority all the greater. For he was able to advocate forcefully and with utter conviction what the *principes* and the Senate represented only feebly and half-heartedly: first, that the Senate was responsible for the commonwealth, and secondly that everyone must conform with class discipline and no one become too powerful. Cato's cause corresponded to the old tradition of constitutional policy, to which the senators generally subscribed. So what objections could there be to him?

One might think of two: Pompey's goodwill and the present situation of the republic. Yet however well-disposed Pompey might be to the traditional order and however genuine his desire to serve the Senate régime, he was an outsider and could, if necessary, bring great power to bear; if his demands were met he would become significantly stronger, and this would inevitably result in repeated conflicts between him and the Senate majority. Even though Pompey had represented the interests of countless clients, his reputation as a patron was bound to suffer if he could not assert himself. Whatever his plans, he could not evade the demands that went with a position of privilege.

Moreover, he was unreliable. It was to be expected that he would seize the first opportunity of getting his way with the support of the popular assembly. Cicero wrote at the time that there was about him 'nothing great, nothing outstanding, nothing that is not low and popular' – and Cicero was still better disposed to him than anyone. What this really meant was that Pompey was everybody's man, and above all reluctant to spoil his relations with the unruly *populares*, who might cause the Senate much trouble. What else could he do in his position, one might ask, given his ambitions, which were based on achievement? Yet it was this that so exercised Cato and others. In the end it was immaterial whether he was opposed because of his particular interests or because of the political methods he used in pursuing them. Yet in view of the strong moral component in Roman political judgments the reproach of unreliability probably weighed most heavily. Politicians tended to be categorized as good or bad. The most important criterion was whether their actions were oriented more towards the Senate or more towards the populace. Cicero later expressed this view when he said, 'The good is what pleases the good.' For the Senate majority, then, morality and politics were coterminous.

It could be said that the more powerful Pompey became, the more necessary it was to oppose him. And the more indolent the Senate majority showed itself to be, the less it could be relied upon in an emergency and the more necessary it was to fight for every inch of the order. This was probably why it seemed desirable to take preventive action against Caesar, which would incidentally have been justified only by his plans for a great command that would not result from a war, but precede it. These plans had to do with power relationships in which a man like Caesar might take the view that he was not dependent on the Senate. And they were the same power relationships in which the Senate majority was unreliable and which determined Cato's policy. Cato's authority was thus unusually great, but by no means unquestioned. The harder he had to work, the greater it became. Had he not worked hard he would not have been able to assert himself. The pettiness and rigidity of his policy is therefore symptomatic not only of the power that this solitary figure wielded as the champion of the Senate, but also of his weakness, which was essentially that of the Senate.

Yet just because the Senate was weak, we should not overlook the strength of its cause. This strength went into the role that Cato was able to play so incomparably well, thanks to his philosophical convictions and his tenacity. It derived from the continued attachment of Roman society in general to the inherited order. The Senate was at the heart of this order; it was the centre of power, the place where all the forces were held in balance. This disposes of the second objection that modern commentators repeatedly raise against Cato's policy: that he failed to recognize that Rome's aristocratic régime was superannuated, that the republic was in a profound crisis, that Pompey's power resulted simply from the Senate's inability to deal properly with the practical problems of the commonwealth and from the inadequacy of Rome's old institutions when faced with the new realities of empire.

It is probably out of the question to demand that the ruling class in a community should simply abdicate its power. To do so would be irresponsible, inasmuch as power is associated with a consciousness of responsibility. Its members do not stand outside their world, but within it; their perspective is therefore restricted. If they are not only to have occasional doubts about themselves, but become totally disoriented, they must be forced into surrender by opposing forces. It is probably as well not to overrate the powerful to the point of crediting them with the absence or the weaknesses of an opposition. The power of the powerful should not immediately serve as an alibi for the impotence of the weak. They are only parts of society, which is responsible as a whole for the state of the community – through power and weakness, action and inertia.

Quite apart from this general consideration, it would have been especially difficult in Rome at this time, given the make-up and preoccupations of Roman society, to conceive of the voluntary abdication or capitulation of the nobility. On the one hand, it was not prey to modern uncertainties – to the consciousness that history is a great process of change in which everything is relativized and the old becomes 'superannuated'. Ancient thinking about social structures was static. It would have been inconceivable that present conditions might be out of date, that the simultaneous might be non-simultaneous. Nor was it possible to imagine that everything could be totally different, that present conditions were arbitrary and fortuitous.

The nobility, after ruling Rome for centuries, could not assume the innocence of mere functionaries. Rather, it had existential ties with the republic. And it could not evolve a sense of state weakness, because there was as yet no divorce between state and society. The commonwealth was constituted by its citizens, though there were admittedly substantial gradations among them.

Finally no one, even outside the ruling class, seems to have been in any doubt that this class had responsibility for the commonwealth. It had no competitor. Nor did anyone envisage an alternative order.

This sounds odd when one considers all the ills that afflicted the republic. Yet the oddity was hardly noticed, or at least hardly pondered or emphasized. Nowadays we are inclined to postulate a crisis of legitimacy, which we deduce from the misery and discontent of the broad masses not only in Rome, but in other towns and in the country. It was here that Rome's soldiers were recruited. Attention is drawn to their willingness, after Sulla's march on Rome, to follow their generals even against the Senate.

Yet for the Roman republic of about 60 BC it is possible to postulate a crisis of legitimacy only if one unthinkingly – and erroneously – credits the Roman republic with certain features of the modern state.

The Roman commonwealth had not evolved an autonomous state apparatus. Innumerable functions that the modern state has arrogated to itself or evolved over time – and that are no longer feasible without it – were performed by the citizens acting among themselves: they thus had no need of a bureaucracy, a public prosecutor, a police force, a public education system or a postal service. Even the need for public order was normally met by individuals, assisted by neighbours, clients or slaves. The Romans relied infinitely less on public services than we do. They therefore expected much less of the community. Conversely, they paid no taxes. Only until 167 BC had there been occasional levies based on wealth, but only when the need arose – usually in times of war – and these were repaid whenever possible.

There was therefore none of the pressure for legitimation that goes with the development of the state. In modern times the individual citizens have given up basic rights, including the right to self-help, paid much in taxes and become subservient to the state, in exchange for the promise – indeed the guarantee – of protection, care, welfare, and ultimately total provision for existence. And they needed these things. Max Weber speaks of the 'growing need for order and protection ("police") felt by a society accustomed to absolute security' in all areas of life. To this were added all the organizational requirements of an increasingly specialized economic society.

When the state detaches itself from the whole and becomes so powerful, it may appear, from the perspective of the church and later of society, as an opposing force. This creates quite new perspectives, attitudes and demands. And where society can be shaped and changed by the state, it begins to seem as though the order one belongs to is largely at the disposition of others. Add to this the modern belief in progress, and it becomes clear that today, given the huge expectations we have developed, the legitimacy of political systems can easily become a problem. This is unique in world history. There seem to have been certain remote parallels in the monarchic high cultures of non-European history.

In classical antiquity, however, we find free communities rather than states and monarchic régimes. Their members were citizens to a far greater extent than elsewhere. To put it in extreme terms: they did not *have* constitutions – they were constitutions. Evidence for this is afforded by the Greek word *politeia* ('polity') which means both 'constitution' and 'citizen body'. Or to put it more precisely: the citizen formed part of the order in which he lived. Whereas in Greece democracy and oligarchy could still alternate, there was something inescapable about the Roman order: it had to retain its inherited form or cease to be an order altogether. The opposite of *res publica* was not monarchy, but *nulla res publica* ('no republic') or *res publica amissa* ('lost republic'), that is to say, disorder. An order made up of the citizens, however, represented part of their identity. It hardly needed any justification and could hardly be viewed from outside.

Within such a community one might of course feel the ruling class to be an opposing force; one might think that it ruled badly or selfishly, that the Senate and the magistrates were inadequate and no longer bore comparison with their predeccessors, that the republic was no longer what it had been. Yet these were political and – above all – moral criticisms; they did not call the system into question. And Cato at least would have agreed with them.

In such a commonwealth there might be unrest, rebellion and civil war, which undoubtedly arose from social hardship and discontent. It could be seen that the Senate no longer had full control of the commonwealth, that the republic might in certain circumstances be exposed to serious disturbances, that its existence might even be threatened from within. All this was a sign of disintegration, but it did not amount to a crisis of legitimacy.

To distinguish between disintegration and a crisis of legitimacy is not just to quibble. The distinction is in fact extremely important. The manifest discontent was not generalized into criticism of the system. At no point were the opinions and interests of those suffering hardship objectivized as a cause; at no point did their objections and new ideas interact with current needs in such a way as to generate a political alternative. The lower classes were probably largely indifferent to the existing order. But they did not question it, even though they might periodically revolt.

This was crucial for the existing power relationships, and for politics as a whole. Social discontent did not produce a separate force that could press politicians into its service. The outcome was that from time to time, when some politician found it advantageous or feasible, the urban populace or the soldiers and veterans would be mobilized politically and at best granted a share in any gain that accrued to the republic; all were concerned simply to improve their lot within the existing system.

Hence political conflicts were largely confined within the aristocracy. Extraneous causes for such conflicts usually arose from problems connected with the empire, initially within Italy itself. After the period of the Gracchi, most of the social questions that came on to the agenda resulted from wars or civil wars and concerned the settlement of the veterans. Other controversial matters were taken up only if individual nobles found it opportune or if powerful groups such as the knights pressed for it. Even civil wars resulted from decisions made by individual nobles. The troops followed, but they never insisted on following – or at least not until the forties.

Cato and the Senate majority could thus see the crisis as essentially confined within the nobility. If Cicero arrived at a different judgement, this was because he was rather more sensitive and had personal reasons for feeling that the republic was in constant danger. As a *homo novus* he had a livelier conscience as regards the tasks facing the republic than other members of his class, who merely wished to represent it. There was a more important reason, however: as the one who had had the Catilinarians executed, he was in danger of being attacked by the *populares*. When the *senatus consultum ultimum*, and hence the republic, were attacked or defended, he was in the crossfire. He therefore wanted everyone, in his daily actions, to be as concerned for the republic as he was for himself. He felt that the whole of the order was permanently on the agenda; accordingly, all who cared for the Senate régime and the republic should form a united front. From his point of view Pompey belonged on the side of the Senate majority. Cicero thus drew the dividing lines differently; but he gave no more thought than anyone else to the social ills and the problems of empire, and it certainly never occurred to him that social conditions had made the Senate régime superannuated. As a good intellectual he merely related things primarily to himself and judged Pompey on the basis of his opinions, not of his interests.

If there was no alternative to the old *res publica*, an attempt must be made to restore its previous efficiency. And to this end the Senate must first of all regain the leadership of the commonwealth. Hence Cato was doing what was necessary. He was more keenly aware than most of the Senate's failings.

Cato differed from the others only in the intensity of his personal commitment, in the consistency and seriousness of his concern for the whole. According to Mommsen, this made him one of the conservatives 'who basically conserved the republic to death'. Objectively this is correct – except that where there are no 'progressives' the term 'conservative' can have only limited validity. The rapid collapse of the republic was not due to Pompey, but to the struggle against him. We know this with hindsight, but at the time the crisis was inexplicable. It was possible to yield, but impossible to argue that this would have been anything more than mere resignation.

Nor could the outsiders, Pompey and Caesar, change Cato's mind. Their difficulty, their great weakness, lay in the fact that while they might change their conduct, they could not redesign the commonwealth.

Whenever Pompey was needed, he had many allies, such as the knights who supported him in 67 and 66. On the other hand it was not in their interest that he should become too powerful. They too were attached to the Senate régime, unless they happened to be in dispute with it over particular issues. True, they did not think it desirable that the Senate should as a rule be strong. Indeed, it was its weakness that endeared it to them. Yet in extreme cases they always sided with the Senate – especially as it never failed to accommodate them. And there was no other cause to which Pompey could have attached himself.

He could therefore never free himself from his inner ties to the old republic. He wanted to serve the Senate régime, but if he wished to assert himself he was bound to oppose it. This contradiction was too much for him. What he was – and could not help being – was unacceptable.

The opposition between Cato and Pompey was thus asymmetrical. Cato could defend the republic, but Pompey could not attack it. The Senate was itself a party, but at the same time it stood above the parties. Pompey was only a party. The Senate had a cause, but Pompey could only point to his personal ability and personal achievements. There were possibilities for him to rise, but no place that he could occupy when he had risen.

In 61 he made a characteristic attempt to overcome this unsatisfactory state of affairs by proposing to Cato that he and his son should marry two of Cato's nieces, daughters of Caesar's friend Servilia. Mother and daughters were thrilled, but the girls' uncle rejected the proposal out of hand.

The opposition between the two men was doubtless inevitable. And it was equally inevitable that both sides should make such a disagreeable impression: Cato petty, intense and dogmatic, Pompey half-hearted, anxious and hypocritical. By setting his face against anything new, Cato elevated the old sceptical attitude to innovation to a principle – as always happens in critical situations when one takes one's measure from the past or an imagined future. He put on the armour of ideology, as it were, to protect his thinking. When Pompey honestly declared that he disapproved of the attacks that his helpers mounted against the Senate, he became a traitor to his friends. And his 'honest face' was still not wholly convincing. Yet by conducting such a vacillating, shady, unreliable and cowardly policy he could at least retain a minimal degree of trust, which might later provide a point of contact. He did not destroy his bridges. He remained reliant on the approval of the 'good'. This was the best way to retain the prospect of one day establishing a power like his in Rome: when the city needed him, when the situation had become sufficiently difficult – with a little help from him, though this should if possible not become obvious.

Only if the problems of empire, which Pompey understood so well, and the crisis itself had come on to the political agenda – and not just indirectly, through their various side-effects – and if a permanent grouping had taken shape, might things have been different; only then might Pompey have gained real strength and direction.

However, the crisis could not be compassed politically. The parties avoided tackling the major problems of the republic head-on. The contradiction between the two disparate realities of Rome did not become an opposition – especially as the victims of hardship had no voice. This again strikes the modern observer as astonishing. When faced with a crisis, we always think first of solving it through political action, either by the government alone or by the main parties arguing it out. It may take time, but in the end something is bound to be done. We have the same expectations of our rulers as we have of the state. And we have grown accustomed to the efficacy of the modern party system. Yet parties of this kind are a product of modern times, fathered by the established state, which mediatizes its citizens. Only the state makes it possible to view the parties as something positive. Only within the state can there be party programmes, involving the idea of representation and social change initiated by the state. Yet even today it is not clear to what extent the modern capacity for opposition can still cope with the problems of our world – to what extent we are capable of solving the crises of our own making.

In any case the Roman republic had reached a point where it could have coped with its problems only at the cost of a substantial loss of freedom and scope for personal development. For things were ultimately moving towards monarchy, though no one could have known it. Yet when in fact does a society regard such a political price as unavoidable – at least if it is politically and existentially as attached to its order and as unconcerned with efficiency as Roman society was? This was possible only after it had been worn down by the protracted civil wars of the forties and thirties.

Yet if no force arose in opposition to the existing order and if outsiders had no points of purchase from which they could objectify their claims within a wider framework, this meant that Caesar's opportunities and the limits within which he could act were strictly determined.

He was in the same situation as Pompey, but in a different way. He too could justify his extraordinary claims only on the basis of his ability. And among the old ideals of the Roman aristocracy he was intent on exemplifying that of service to the commonwealth. Yet he was obliged – and able – to do this more forcefully and single-mindedly than Pompey. Pompey was moved by a conscious desire to serve the Senate régime; he would hardly have had the strength to strive for a position of privilege had he not been convinced that it was necessary and meaningful according to conventional criteria. Caesar seems to have needed no such *rationale*. He did not necessarily think differently, but he had less respect for the republic and was therefore less reliant on the approval of the 'good'.

Both had been strongly influenced by Sulla. Yet while Pompey had been accepted by him, Caesar had felt repelled by him. Pompey wished to emulate him by performing extraordinary services for the Senate and the republic. Caesar was fascinated mainly by Sulla's extraordinary personality, his audacity, and his determination to take matters into his own hands when they were set on the wrong course. And apart from the fact that he hardly had an opportunity to commit himself to the Senate, personal fascination was to him of overriding importance – because he did not want to be like the others, and because he had such high standards.

If Pompey was vain, Caesar was proud. From his youth onwards. This accounted for his early opposition to the ruling circles. And for the spirit of contradiction within him. And because he had demonstrated this spirit so clearly at an early age, the merits he brought to his 'popular' career were enough to dispense him from the need to woo the urban populace. The trouble he had with the leading senators was thus more valuable than any advantages he could have gained by compliance.

It meant a great deal to Pompey that he had won fame in his early years. Caesar, however, still had to make good his claim. He felt certain of accomplishing great things in the future, and this made him to some extent independent of his surroundings. Yet he thereby incurred an obligation to himself. And in time this may have made his need for ultimate recognition – to which he had at first been indifferent – all the stronger.

Much in his early career seems to indicate that Caesar felt sure of rising step by step. But he probably took little thought for the future. He may often have been in danger of coming to grief like Catiline. Yet he knew how to calculate; above all he could count on his amazing superiority and his genius for captivating others – and not just women. Even if the leading senators sized him up fairly accurately and opposed him, even if they recognized that he was really far more dangerous than Catiline, he was able to present himself in all innocence as a victim of gross injustice. He could provoke his opponents to the utmost and simultaneously put them in the wrong. This no doubt predisposed most people in his favour. Justice required it of them, and in any case they were inclined to complaisance. Caesar's generosity, courage and enterprise contrasted pleasingly with Cato's rigidity, reservations and inflexibility. Cato was enervating, Caesar refreshing – if at times excessively so.

He had so far won every round in his great game and been elected consul. The question was now what would happen if he espoused Pompey's demands. Would he succeed in persuading the majority of the senators to act against the Senate's cause? The case was far from open and shut.

For the present, however, the only certainty was that Caesar had found his own path, the path he had always sought. Since the contemporary oppositions afforded the outsider nothing that would have induced him to take up a firm, objective position, he had to find his point of reference, his criteria, within himself. With no cause to take up, he had to develop his personality freely and without ties. He wanted to demonstrate his *virtus* – the manliness so admired by the Romans – through deeds. His pride, his awareness of his own superiority and the self-confidence that grew with every new success made him certain of achieving the goals he had set himself.

Having distanced himself from everyone else and deliberately set himself apart from his peers, he could begin to show his true greatness, knowing that he was free to realize all his rich talents without too much consideration for others, and convinced that he possessed incomparably more strength, skill and insight than all of them. This may be aesthetically pleasing. Yet it is far from clear that it was to be a blessing for Roman society. And it was a problem for Caesar too, in that a great gulf might open between himself and society. Was it perhaps because he found himself in this position that he came to attach such overriding value to his achievement ethic?

10

The Consulship (59 BC)

ASSUMPTION OF OFFICE · THE SENATE
DEBATES THE LAND LAW: CATO IS
ARRESTED · THE LEADING SENATORS
ADOPT NEW TACTICS · LATE APRIL
AND THE SECOND PACT WITH POMPEY ·
OPPOSITION · CAESAR IS OFFERED A
COMPROMISE: HIS RELATIONS WITH
THE POLITICAL ORDER · A YEAR OF
GREAT CHANGE

T HE CONSULS BEGAN their year in office by taking the auspices. Just before dawn each took up a post from which to observe the sky. If any light that could be interpreted as lightning appeared on the left hand in the morning twilight, it presaged good fortune for them and the city during their term of office. It even sufficed if one of the competent officials, the *pullarius* (the guardian of the sacred chickens), reported to them that he had seen lightning. If what he said was untrue – as it usually was – the responsibility was his. The magistrates might believe it with a good conscience. And no ill would befall the commonwealth.

Good auspices were necessary in Rome, for no important actions could be taken without them. Yet for practical reasons important actions often could not be deferred. It had therefore become customary to help the auspices along. Yet they still had to be taken. At least there must be evidence of lightning. It may seem to us a mere formality, but it was part of tradition. By observing the formalities one was acting as men had always acted since the founding of the city. This was the proper way of doing things, and if propriety was abandoned all would be lost. Special care was called for in the religious sphere. No one had ever wished to break with religious observances. Nor did Caesar. Perhaps the *pontifex maximus* was particularly conscious of putting on an act, with the sense of parody that one would like to ascribe to him. How else could he steel himself to the many impositions he had to endure? Moreover, Caesar was determined to show no chink in his armour. His opponents were on the lookout, and they should have no occasion to find fault.

The fateful year of the consuls Gaius Julius Caesar and Marcus Calpurnius Bibulus thus began, as far as the latter was concerned, like any other – and probably more correctly than any other. But the auspices were deceptive.

Three weeks earlier, Caesar's helper Publius Vatinius, a tribune of the people, had announced that he would have no truck with the reports of the augurs and all the pretensions of their college. In other words, he rejected the whole business of augury. This was the other side of the coin – the open threat that the auspices posed to any proposed legislation. For if a magistrate discerned a flash of lightning in the morning sky – or had one reported to him – no popular assembly could pass any resolution that day. Lightning was regarded as propitious for any undertaking except in the *comitia*. But whatever Vatinius said, he was not Caesar, not the *pontifex maximus*, not the consul, who was responsible for the conduct of Roman politics during the year just beginning.

Having taken the auspices, Caesar donned the purple-edged *toga praetexta* in the house on the Via Sacra where he resided as *pontifex maximus*. The toga was probably newly made, shining and chic. The lictors arrived, as well as countless friends and acquaintances. In festive mood, and acccompanied by a large retinue, Caesar set out to walk the short distance across the forum to the Capitol. On a dais stood his official chair, the *sella curulis*. Here he took his place, presumably beside his colleague. Thanks were rendered to Juppiter for protecting the commonwealth during the past year; white cattle were sacrificed to him, as the previous consuls had promised a year earlier; and the same sacrifice was promised for the new year.

The senators now proceeded to the temple of Juppiter Capitolinus, the Best and Greatest (*optimus maximus*). The new consuls held their first session. Caesar led the proceedings because he had been elected first. He had to report on the state of the commonwealth (*de re publica*), dealing first with religious, then with secular matters; he probably mentioned the principles he would observe and the objectives he hoped to achieve during his term of office. He may have outlined the programme agreed by the triumvirate – but cautiously, for he wanted to bring everything up for debate in the Senate and attend to any objections and proposals. He also appealed to his colleague Bibulus to join him in conducting their consulate in a spirit of harmony, to the benefit of the commonwealth and the general good. Again he acted with an air of superiority. If any conflict erupted, the blame was not to be laid at his door. And since his opponents took a very basic view of everything – precise and petty, divisive rather than conciliatory – he could play the magnanimous and responsible role that he so much enjoyed.

Then the questioning began. It was eagerly awaited. For the consuls customarily questioned a few *consulares* before the rest, paying no heed to seniority. They usually adhered to this order throughout the year. In this way they indicated their wish to collaborate especially closely with the senators in question, who were chosen on grounds of reputation and personal friendship. Caesar called first upon Crassus, not Pompey. He had after all long-standing contacts with Crassus and may have been moved by the consideration that Crassus came off worst in his intended programme of legislation.

On the same day, or soon afterwards, the consul usually made his first speech to the people, in which he thanked them for his election and spoke of himself, his achievements, and his ancestors, so that it should become clear who he was. Caesar probably made a programmatic statement about the way in which he intended to conduct affairs and about his plans – undoubtedly with a very statesmanlike air, in keeping with convention, but in such a way as to give expression, here as elsewhere, to his personal style and distinction.

He can hardly have denied the fact that he was a true *popularis*, concerned with the welfare of the whole commonwealth.

He at once introduced an innovation: daily reports were to be compiled and published on proceedings in the Senate and the popular assembly. This was a contribution to political objectivity, a practical measure in line with an old tendency of popular politics; though not necessarily to the senators' liking, it could be justified as a means of protecting Senate debates from misrepresentation. Presumably it also reflected, like Caesar's later campaign reports, his zeal for documentation. He always had an eye to posterity, the more distant horizon within which his achievement ethic operated.

It was probably in the early days of the year that Caesar introduced his land bill. It had long been known that he was planning it. It provided for two forms of land acquisition. On the one hand, all lands still in communal ownership, except those in the Campania, were to be divided up; on the other hand, part of the proceeds of Pompey's booty was to be used to purchase land. The sale of land was to be voluntary, at a price equivalent to the value entered in the census assessment lists. In settling these lands, priority was to be given to Pompey's veterans. Any residue would go to others. Caesar seems to have pointed out that there were far too many people living in the city who would benefit from acquiring land of their own. This would strike at the root of much unrest. The land commission was to be fairly large, with a membership of twenty. Caesar himself declared that he did not wish to be favoured in any way by his own law; one provision precluded the sponsor from membership of the commission.

Caesar submitted the text to the Senate and declared himself ready to consider any proposed changes. On a number of occasions soldiers had been given land after serving in campaigns. Most of them were impoverished country-dwellers. Sulla himself could serve as a model for the proposed policy. Yet how much milder Caesar's law was: it robbed no one! There was only one fundamental objection, and this related not to individual provisions of the expertly drafted bill, but to its effect: it would substantially increase the power of Pompey. Hardly anyone liked to say this openly. There was certainly strong opposition, not least from Caesar's colleague Bibulus. On the whole, the fathers seem to have made heavy weather of the debate; most were probably opposed to the law, but unable to raise any substantive objections. Some spoke in its favour; the rest refrained from negative comments. They tried to prolong the debate. The outcome could not be foreseen. There may have been a real prospect that Caesar's eloquence and cogent arguments – in the absence of compelling counter-arguments – would win the majority over. Then once again, as in the Catilinarian debate, Cato spoke out in opposition. Without beating about the bush or going into detail, he baldly insisted on the principle that in all things one must adhere to convention and not transgress it in any way. He launched into a long speech on the subject. The day was short – the session had to be adjourned before sundown – and there was no end in sight. Cato clearly intended to monopolize all the remaining time. For a while Caesar seems to have listened patiently. Then his patience snapped. He ordered an official to arrest Cato and take him to the prison. Did he hope for a favourable decision in Cato's absence? Or had he suddenly realized that he would get nowhere with his policy of superiority and consideration? Or had he merely been looking for a pretext to put an end to what had meanwhile become a charade? Was he dropping his mask?

Viewing these events with hindsight, we should not be tempted to conclude that matters were more clear-cut than they probably were. There was much to be said for Caesar's doing his utmost to win Senate approval for the land law. Cato's filibuster seems to suggest that he expected the majority to support it. Some may have welcomed it, and many, impressed by Caesar's determination and the threats of his helpers, probably feared a defeat that they wished to avoid. They were bound to be struck by the fact that Crassus, Pompey's rival, declared himself in favour of the bill; this is admittedly not attested, but may be presumed. And whatever the Senate finally decided, it was to Caesar's advantage that he had demonstrated as clearly as possible the practical justification of the bill, the superiority of his own arguments, and his wish to collaborate with the Senate. Pompey too should see – and perhaps wanted to see – to what lengths Caesar would go to reach an amicable solution. At least he had tried everything. At least the rigidity and intransigence of his opponents had become evident.

Whether Caesar really hoped that the house would agree is not clear. If so, he would presumably have had to be satisfied with a more modest province and a more or less normal continuation of his career.

For only if it became necessary to assert himself against the Senate by violent and illegal means would he have any prospect of winning the extraordinary command for which he saw himself destined. There is no way of knowing which of the two possibilities he preferred. We do not even know whether he saw the question in this light; he may have decided to see how things went. Reason may have counselled him to try for the conciliatory but more modest solution. If so, he was not wearing a mask, but would admittedly have had to be prepared to expend much effort in long and fairly fruitless discussions. For quick, resolute action was much more to his taste, and his whole consulship was directed towards action. Thus, whatever his preference, he cannot have been unhappy when Cato frustrated his diplomacy. His determination and anger erupted, and he gave full vent to them.

Plutarch reports that Cicero mistrusted Caesar's friendliness as he mistrusted a calm sea. He felt it deceptive and feared 'the monstrosity of Caesar's nature concealed in his gay and friendly manner'. Plutarch uses the Greek word *deinótes*, which designates anything monstrous, awesome and violent. The underlying adjective (*deinon*) is used by Sophocles in the famous chorus in the *Antigone* to describe the whole range of man's potential, his huge capacity for good and evil. A fearful will, immensely compelling in its controlled strength, must have been discernible within – not behind – Caesar's arrogantly superior gaiety. It was not simply masked: here was a man who had trained himself to project an outward gaiety that derived from his aloof and disdainful inner self and to conceal the awful depths of his soul. This required concentration, discipline, and enormous effort – the style of the great statesman, the brilliance of the superior personality. At this crucial moment, however, when he had to lay the ground for his extraordinary future career and could get no farther by adhering to his accustomed style, his awesome will broke through the surface and manifested itself publicly for the first time.

Many senators rose and left the hall with Cato. When the consul pointed out to one of them that the session was not over, he replied that he would rather be in the prison with Cato than in the Senate with Caesar. Whereupon Caesar had Cato released, probably by asking a tribune to object to his arrest. As he adjourned the session he remarked that if they did not wish to consider the bill with him he would put it to the vote as it stood. He was no longer disposed to sue for a majority.

An alternative scenario had long been planned and was now quickly staged. Pompey had summoned his veterans to Rome for the vote on the land law. They and others were organized into teams that were able to terrorize the public life of the city. Their leader was Vatinius. They soon controlled the streets.

At one meeting Caesar insistently questioned Bibulus. Had he any objections to the individual provisions of the bill? When Bibulus could only reply that there were to be no innovations during his year in office, Caesar began to plead with him and appealed to the crowd to join in. They would get their land law if only Bibulus would consent. Bibulus replied, 'You will not get the law this year, even if all of you want it.'

In another altercation that cannot be dated precisely, Vatinius had the consul arrested. Previous tribunes of the people had arrogated such rights to themselves. They were part and parcel of the repeated trials of strength between the Senate and the tribunes. Vatinius had prepared everything in advance. His helpers had placed rows of wooden benches in front of the court tribunals, so as to form a long, narrow gangway from the speaker's platform to the prison. Along this Bibulus had to run the gauntlet to the accompaniment of jeers from the crowd. At last other tribunes of the people intervened and rescued him. Caesar was naturally in no way involved in the doings of his most important helper. He was after all a mild and generous man.

Caesar then called a popular assembly in the forum, at which Pompey and Crassus were invited to speak. In measured tones Pompey explained all the points in favour of the land law and the well-deserved provision for his soldiers. He went through it clause by clause. Finally Caesar asked him whether he would support him if those who opposed the law resorted to violence. As he did so he surveyed the crowd and assured himself of their support. Pompey replied, 'If anyone comes with the sword I shall bring both my sword and my shield.' It was like a flash of lightning from a clear sky. The cautious, reserved old gentleman, so anxious to distance himself from his own helpers and so jealous of his prestige, was ready to identify himself with open violence from the start! Caesar had probably worked hard on him. But he was no insignificant tribune whose bill might succeed or fail. If he was willing to push through their common interests he could not fail, and then he need show no consideration; but it had to be clear that he had Pompey's backing.

In the night before the voting Caesar's supporters occupied the forum. When his opponents, Bibulus and three tribunes, arrived with their retinue, they had difficulty fighting their way through. The sources are not altogether clear. However, it seems that they reached the temple of Castor, where Caesar was to conduct the voting from the platform. But they were unable to present their intercession, as they were thrown down the steps. The lictors' fasces, symbols of consular authority, were broken. A basketful of dung had been kept in readiness for Bibulus and was tipped over his head. He was to be made a laughing stock. Two tribunes of the people and several others were wounded. During the scuffles the group was forced to the east along the Via Sacra. Bibulus is said to have bared his neck. He wanted to die and call down the curse of murder on Caesar. But his friends dragged him to safety in the shrine of Juppiter Stator (near where the Titus Arch now stands).

Caesar did not let himself be disturbed by all this. He was apparently in the middle of a speech when his fellow consul was forced from the steps to the temple. He may have paused briefly or raised his voice. In any case the session took its expected course; everything was perfectly organized, and the law was ratified.

Next day Bibulus called the Senate together. He complained about the disgraceful violence. But when he questioned the senators, none had any proposal to make. What were they to decide? A *senatus consultum ultimum* was unthinkable. Only if there was a prospect of superiority could the citizens be mobilized for a police action. The senators were outraged, but they were also paralysed by fear. And there was no legal basis for repealing the law. The fact that Caesar had foiled the intercession might be a ground for a prosecution. Yet because no intercession had been possible the law was valid. This meeting of the Senate clearly served only to write the illegal defeat of the *res publica* on the wall and initiate a new phase in the opposition to Caesar.

An entirely new tactic had been devised – a boycott on politics.

Bibulus retired to his house and refused to leave it. Vatinius and his gangs besieged him several times; on one occasion he wanted to have him dragged out by an official. But Bibulus was well guarded, and Vatinius did not wish – or was forbidden – to go to extremes. Three other tribunes of the people stopped discharging their official duties. Probably the majority of the senators stayed away from the Senate, even the dutiful Cato.

Their aim was to demonstrate that law no longer prevailed in Rome, that no one could feel safe, that there was no longer any freedom. In view of the terror spread by Caesar this was not wholly untrue. They believed that they could turn their defeat to advantage and gather support. Earlier, when they were considering whether Bibulus should intercede against the land law, he had been advised to rely on the reputation of having been defeated rather than on that of inattention and dereliction of duty. Now they were dramatizing their inability to act, counting on the citizens' deep-rooted attachment to the old order, on its strength and evident rightness. They calculated that this was the surest way to fuel universal indignation. After earlier defeats – in the days of the Gracchi, for instance – they had at first simply bided their time; now they hoped that time could be used in a quiet but effective bid for support.

Another new tactic was tried. Whenever a vote was due to be taken on a bill, Bibulus observed the sky and sent word to the official conducting the ballot that he had seen lightning. Unlike intercession, religious objections of this sort remained effective, even if they were ignored. They involved dealings with the gods. There were precedents in which the Senate had repealed laws because they had been passed in disregard of the auspices. The aim was thus to have all the legislation annulled. In earlier times there had been perhaps one intercession against major legislative programmes, but if it was unsuccessful no further action was taken, lest the instrument of the veto should be blunted. The opposite procedure was now adopted, obviously under Cato's influence. An attempt at obstruction was made, even though they knew that Caesar and Vatinius would ignore it. Being so sure of their cause, they deliberately incurred a high risk. In earlier times they had normally used the procedure of intercession, not 'obnuntiation' (which relied on signs observed in the sky). Intercession required one to appear in person, but one could 'obnuntiate' through officials. This was another advantage, but there was a concomitant disadvantage: the exclusive use of obnuntiation as a means to block legislation meant that no open confrontation occurred between the legislator and the sky-gazer – which made the whole business somewhat ludicrous.

In addition, Bibulus was constantly issuing proclamations, highly polemical commentaries on the doings of his fellow consul. Otherwise there was hardly any opposition. It was too dangerous, and this was precisely what had to be demonstrated.

For a time Caesar was unmoved by all this. He would have preferred to get his way by legal means, but this was not possible, and he probably did not grieve too much over the hallowed institutions. If his opponents wished the constitution to be violated, then they should have their wish. In future he would do everything alone. As jokers observed, they were living under the consuls Julius and Caesar.

In his land law Caesar, following precedent, inserted a clause requiring all senators, on pain of exile, to swear an oath to honour and respect it. Cato and others hesitated for a long time, but finally yielded to bitter necessity. The members of the land commission were appointed; foremost among them were Pompey and Crassus.

By the beginning of April the agreed legislative programme had been put through. Pompey's arrangements in the east had been ratified. At this point the aged Lucullus once more ventured to raise an objection. But Caesar threatened him with prosecution and railed at him so vehemently that he threw himself at Caesar's feet and begged for mercy. The tax farmers of Asia received a one-third remission, but the consul warned them not to overdo things in future. At Pompey's request Ptolemy XIII was recognized as king of Egypt, in return for large payments. Many other individual concessions were made. Finally, Vatinius brought in a law granting Caesar the province of Gallia Cisalpina, together with Illyria (the Dalmatian coastal strip), for five years. This command had now become more necessary than it had perhaps at first appeared. Having so often infringed the law, he had to be offered security from prosecution in the immediate future, as well as an opportunity to make conquests, so that he could eventually return to Rome in safety. There might have been rich pickings in Illyria. But as the region was at peace, Caesar would have had to start a war.

In March, Cicero made a speech in court in which he openly complained about current political conditions. Three hours later Caesar had Cicero's enemy Clodius translated from the patriciate to the plebs. Members of the patriciate were debarred from becoming tribunes of the people (*tribuni plebis*), but this is what Clodius had sought ever since the Bona Dea scandal of 62/61. He had various motives, one of which was a wish to get his own back on Cicero, who had questioned his alibi at the time and treated him with scorn. Although Cicero had not succeeded, Clodius still bore him a grudge.

Publius Clodius Pulcher had admittedly scant respect for the ideals of the republic and the proud but often ineffectual old gentlemen who represented it. He was no more averse to scandal than his egregiously generous and much-loved sisters. Yet at the same time he was extremely touchy. He might do whatever he liked to others, but he was put out if he met with unfriendliness.

He was certainly highly talented, open-minded, energetic and ambitious; but he lacked stability and any readiness to exert himself consistently, any real will or ethical principles. His anarchic temperament expressed itself without constraint, as it met with little resistance and aroused much admiration; the youth of Rome found him fascinating, especially as he dared to do what most of them only dreamt of doing. Yet in the end this temperament became increasingly self-consuming. Ambition apart, his most important motive seems to have been hatred, or rather boundless aggressiveness. He had an eye for anything that would cause a stir and give offence. He somehow needed to lash out. His aims were secondary and might change. They did not matter. In this regard he was free. But activity and aggression were vital to him. To this extent he was not free. He lived on negation. And he had a flair for potential sources of protest. In 68 he had served as an officer in the army of his brother-in-law Lucullus; feeling that he had been slighted, he had made inflammatory speeches that led to a mutiny. In 58 he was to discover quite new and unsuspected opportunities for popular politics; he discerned strong and largely untapped resentments among the Roman populace and made them his own. He was a *popularis*, but far from well-intentioned. An enigmatic and dangerous character.

Cicero, writing in June 59, remarked appositely: 'He rushes around, behaving like a madman; he still does not know what he wants; he threatens many and appears to want to strike if chance will give him an opportunity. Seeing the bitterness caused by present conditions, he pretends to want to attack those who have caused them. Then, considering their strength, their financial and military power, he turns against the good.'

Translation to the plebs required a special act of adoption that took the form of a law. The application had to be scrutinized by the college of the *pontifices* and then posted publicly. Three weeks later an archaic and by now purely formal meeting of the people could be held at which the applicant submitted himself to a plebeian as his son; the assembly then gave its approval. In Clodius' case the *pontifices* were probably not consulted, and there was certainly no three-week delay. The *pontifex maximus*, acting on his own authority, staged the whole proceedings as a demonstration of his contempt for tradition. The only 'father' that could be found for this son of well over thirty, a scion of the highest patrician nobility, was a nondescript man of twenty; he of course at once emancipated his 'son'. In this farce Pompey had to play the role of augur.

The haste with which the proceedings were conducted suggests that the adoption was not originally intended. Perhaps all three rulers – or at least Pompey and Caesar – had been against it. Or perhaps the resistance came only from Pompey, who, in his worthy way, did not like this unpredictable, ill-mannered young man and also did not want him to act against Cicero. If so, the event is easily explained. Caesar exploited Cicero's complaints as a means to take Pompey off guard. It was after all quite clear that they needed a strong man in the tribunate who would defend Caesar's laws when his consulship came to an end. Otherwise one would have to assume that Cicero's open criticism was in itself too much for Caesar. In this case he was seized by a fierce resolve not only to put down his opponents, but to keep them down, displaying the inordinate fury of someone who was determined to use every means to achieve the impossible. It may be that Caesar's action was partly prompted by both these motives, especially as he may have been infuriated by Pompey's continued caution. Moreover, Pompey extracted a promise from Clodius that he would do nothing against Cicero.

At the beginning of April the power relationships seem to have been reversed. Cicero wrote to Atticus: 'Believe me, the wheel in the commonwealth has turned gently and with less noise than I expected, in any case more quickly than it should.'

Then came the recess. Between 4 and 24 April no laws could be passed. Most of Pompey's veterans seem to have left the city. One could breathe freely again and reflect on recent events. There was widespread displeasure over what had happened. Even Pompey seems to have remarked how much his reputation had suffered. He probably began to rue his recent actions. Clodius, having at first toyed with the idea of making a lucrative ambassadorial journey, now thought of putting up immediately for the tribuneship, with the intention of opposing Caesar's laws. Politically, this seemed the more profitable course.

At the end of April, however, everyone was suddenly startled by the news that a fresh land law was planned. It was even intended to split up the Campanian lands – the apple of Rome's eye, as it were. Citizens with three or more children were to be settled there. Immediately after this it became known that Pompey wished to marry Caesar's daughter Julia, who was about thirty years his junior.

The two had obviously made new agreements. Caesar seems to have convinced Pompey that there was no turning back. At all events, Pompey executed a sharp turn. Everything he had done in the first few months had been dictated by necessity: he had to get his laws through, and Caesar had him in the palm of his hand. All the same, he had maintained a certain distance between himself and his unruly ally. And this aloofness seems to have worked to his advantage. Now he acted on his own initiative and set out much more openly to augment his power. The chief beneficiaries of the land distribution were his veterans. He himself took a leading role, but above all he was now wholly on Caesar's side.

Caesar must have pulled off a diplomatic master-stroke. Perhaps he incited Clodius to oppose his laws in order to arouse Pompey's apprehensions about their future; in any case he made it clear to him that an accommodation with the Senate would no longer be easy to achieve. Caesar's charming daughter may have had a hand in all this.

Under the new agreements Caesar managed to obtain the province of Gallia Transalpina (corresponding geographically to Provence and Languedoc). It presumably afforded substantially better opportunities for conquest than Illyria. There was unrest and certain movements of population in neighbouring regions; a *casus belli* would not be far to seek. Numerous minor laws were passed, under which the allies conferred gifts on various people, the aim being to benefit friends, to win new friends, and to boost their own power. Caesar's overriding concern was to prevent an understanding between Pompey and the Senate, which even now could not be ruled out.

There is no evidence that Crassus had any part in the new arrangements. Nor can one see what he might have gained from them. In May Caesar broke with senatorial custom by changing the order in which the senators were called upon to speak: from now on he called Pompey first. In July we find Crassus among Pompey's opponents. He had probably not followed the change of course that took place in late April.

Caesar had his new province conferred on him by the Senate – or the rump that still met. The senators believed that at least the power of decision lay with them and not again with the people. Pompey himself was the proposer. He was supported by Gaius Calpurnius Piso, whose daughter Caesar had recently married. Cato inveighed against the horse-trading in daughters and provinces and warned that the Senate was itself installing the tyrant in the castle. A few days later Caesar boasted that he had achieved what he desired, against the will of his opponents and to the accompaniment of their groans; now he could dance on their heads. At this point one of the senators, in a veiled allusion to Caesar's alleged relations with Nicomedes, is said to have interjected that this was not easy for a woman. Caesar riposted that Semiramis had once ruled Syria and the Amazons had possessed a large part of Asia. This was in June or thereabouts.

At about this time there were more and more demonstrations against the new rulers, whose laws benefited few of the citizens. The withdrawal and silent opposition of many leading senators created an increasingly powerful impression. People thronged round Bibulus' public notices. The terror, intimidation and arbitrary action employed by the rulers had induced a change of mood, even among the masses. 'Nothing is so popular as hatred of *populares*,' wrote Cicero. Indeed, the ability to take pleasure in the ascendancy or even the tyranny of an oligarchy in the name of the people had scarcely evolved in ancient times. Not only lies, but ideologies too, were quickly seen through. In the day-to-day life of Rome it soon became evident who wielded the power and used it to do whatever he pleased. And there was nothing to be hoped for from impalpable advantages. Caesar had done nothing to benefit the urban populace; there was no new corn law, no games. Only talk – and the pressure exerted by the veterans, the gangs organized by Vatinius.

Cicero goes on to say, 'No one approves of what has taken place; everyone complains and is indignant; people all agree, speak their mind openly and do not hesitate to inveigh; but no one knows what to do. If we resist there will probably be bloodshed; but everyone sees that constant acquiescence can only lead to ruin.' Whistles greeted the rulers wherever they appeared in public. In July many citizens came to Rome for the games. When Caesar entered the theatre, everyone remained silent. But the entrance of young Curio, his one passionate and open opponent, met with the tumultuous applause that Pompey had enjoyed in his better days. Caesar felt grossly insulted; he complained and threatened the knights who had risen from their special seats to join in the clapping; in his impotent rage he even threatened to abolish the people's corn rations. In spite of all this, an actor won such applause with a verse directed at Pompey the Great – 'Through our misery you are great' – that he repeated it several times. When Pompey wished to reply openly to the reproaches of Bibulus he was so helpless that people almost felt sorry for him. Slowly the wheel seemed to be turning again. Yet power still lay entirely with Pompey and Caesar.

On the one hand, their opponents hoped that the next elections would bring in suitable men who could energetically support the Senate's cause in the following year. To gain time, Bibulus postponed the elections from July to October. On the other hand, they seem to have realized that it was no longer possible to cancel all the results of Caesar's consulate – on the basis, for instance, of a *senatus consultum ultimum*. They therefore hit upon the idea of putting a compromise solution to him: he should reintroduce all his laws, but this time in the proper manner, paying due respect to the auspices. They were obviously ready to refrain from obstruction. They would accept all Caesar's laws if only he would reintroduce them and thereby admit that he had acted illegally, that laws introduced in this manner were invalid, and that the traditional institutions must be unconditionally respected. In return for repairing the breaches of the law they offered to guarantee Caesar's legislation and naturally to grant him indemnity.

★　　　★　　　★

The leading senators were now showing remarkable flexibility, in contrast with their former rigidity, and were prepared to jump over their own shadows. They realized that they had carried their opposition too far. They were probably helped by the reflection that their offer amounted to a continuation of earlier senatorial policy under new circumstances: they were willing to make practical concessions in return for the maintenance, restoration, and reinforcement of established rules and precedents. Their proposal must have been enticing to all the beneficiaries of Caesar's legislation, especially Pompey, who was growing more and more uneasy. For Caesar too there were great advantages: he would gain everything he wanted, and all his illegalities would be written off.

True, he would have had to eat humble pie and acknowledge that he had acted unlawfully. He would also have had to concede that Bibulus' ludicrous sky-gazing had been justified. Not that anyone in Rome is likely to have believed the daily reports of lightnings. The efficacy of this means of obstruction depended not on religious conviction, but on the senatorial sanction behind it, the existence of a power that could punish any infraction. Caesar was invited to recognize and strengthen this power, to help ensure that in future the means to obstruct legislation would function more effectively. From now on it was not to be so easy to override the Senate.

These two reasons alone doubtless sufficed to make Caesar reject the offer, which was put to him several times. He probably did not trust his opponents. What would happen if he was prepared to reintroduce all his legislation and they failed to keep their side of the bargain? It would be interesting to know what weighed more heavily with Caesar – his distrust, his pride, or his aversion to any strengthening of the Senate's authority. On this occasion in particular one wonders about his attitude to the Roman order. First as a matter of principle: did he believe that his scorn and disregard for hallowed institutions had opened a serious breach in their walls? Then in pragmatic terms: how did he assess it in relation to his own future? How did he envisage his future within the Roman republic?

As far as we can judge, Caesar was not in principle opposed to the Roman order. If he acted against it he was simply putting his own interest above respect for its rules. He may have been motivated also by anger at the restrictions they imposed. The fact that the hallowed rights of intercession and obnuntiation could be used so arbitrarily –

as it was bound to strike him – and that they could be simultaneously valid and invalid – as he now found – must have diminished his respect for them. He had little appreciation for institutions altogether. This must have been due in part to deep-seated impressions gained in youth, which made it impossible for him to distinguish between the institutions and their representatives. Assertion of his own claims, friendship, opposition, and practical achievement – these were the categories in which he thought. What united the citizens, the legal order to which they felt bound, the vital element in the republic – its institutions in other words – failed to impress him as having an independent reality. His sharp gaze passed straight through them. Hence Caesar did not understand the Senate's constitutional policy, especially as he was one of those against whom it was directed (he probably did not understand this either). The means of obstruction and the policy of Cato and his allies were to him all one. Cato's strength was his weakness, Cato's weakness his strength. For Caesar this was the nub of the matter. If he wanted to assert himself over any matter in future, his opponents would only impede him again.

In short, he probably had nothing against Rome's traditional order. He may have been an outsider, but not in any fundamental sense. And there was after all no cause to which he could have attached himself in striving for a new order. He simply did not want his opponents to have the power to frustrate him whenever they chose, let alone over such practical and reasonable demands and in such a ludicrous fashion. To this extent he must have been pleased with the effects of his victory on the Roman order.

To this extent, any going back must have been far from his mind. Having trounced his opponents, was he now to make it all up to them? Probably others would have done so. Pompey would have been happy to have such golden bridges built for him; he may have urged Caesar to cross them. One should of course beware of inter-preting this period of his career by extrapolation from its later phases; this was the period when the points were being set. However, the way in which he chose his course – unwaveringly, without yielding to temptation, purposefully and with the utmost commitment, wil-fully embracing the role that had been forced on him – strongly suggests that he was determined to go his own way, at first alone, in his province; then he would have to take stock. He was ready to render great services to Rome, and he must have hoped to benefit from them. Just as he had little understanding for institutions, so he had little for the power that his opponents could still summon up within the framework of the republic, in defence of its laws. After all, he had at first soundly beaten them.

In 59 Caesar demonstrated his ability to serve the republic by introducing, in August or thereabouts, his *lex repetundarum pecuniarum*, a law that regulated the whole area of provincial administration, not in a fundamentally new way, but more rigorously than hitherto. This brilliant piece of Roman legislation remained in force throughout the imperial period.

Soon after the middle of the year Caesar found himself politically in such straits that he resorted to highly dubious means to extricate himself. He commissioned Lucius Vettius, a notorious informer, to spin an intrigue. Vettius wormed his way into the confidence of Curio, Caesar's fervent young opponent, and revealed to him one day that he and his slave were going to make an attempt on Pompey's life. It was clearly his intention to have himself seized in the attempt. Curio, however, reported the matter to his father, who warned Pompey. The matter came before the Senate. Vettius was summoned to appear and accused many young nobles of having instigated a plot. Among these was Servilia's son Brutus. He said that Bibulus had sent his private secretary to him with a dagger. It was all highly improbable, especially as Bibulus himself had warned Pompey of a possible attack. Some statements were patently false and contradictory. The Senate therefore had Vettius arrested. The next day Caesar brought him before the people so that he could make a full statement. Some of the names he now gave were different. He did not mention Brutus, whom he had at first strongly implicated. Cicero comments: 'Only one night had intervened, during which someone had used his good offices on his behalf.' He feared a whole series of political trials. However, the case was so threadbare that Caesar dropped it and had Vettius killed in prison; naturally it was suicide.

In October Caesar somehow managed to get two of his friends elected consuls – Lucius Calpurnius Piso, his father-in-law, and Aulus Gabinius, Pompey's old supporter. This virtually ruled out the possibility of a *senatus consultum ultimum* against him when his own consulship ended. Who would have executed it? Clodius was already a tribune of the people. In the praetorian elections, however, two ardent opponents of Caesar were successful, Lucius Domitius Ahenobarbus and Gaius Memmius.

It may be presumed that Caesar had again pulled all the levers of political pressure. This is suggested by the reluctance of the praetors to charge Gabinius with electoral corruption. When the accuser in a popular assembly called Pompey an 'unappointed dictator' he was almost lynched.

During his consulate, Plutarch tells us, Caesar behaved like the most insolent tribune of the people. Perhaps things were even worse than in the days of the Gracchi and Saturninus. In any case Caesar, unlike any tribune of the people, was assured of a provincial command. And he could be almost certain of obtaining it with impunity.

Caesar had managed something quite unprecedented: he had put through a large legislative programme – ignoring all attempts at obstruction – and escaped senatorial punishment. He had dealt the Senate a heavier and more lasting blow than anyone else. No longer was it the supreme authority, which ultimately, in case of doubt, could ensure order in Rome. This was an enormous change.

Caesar had been able to bring about this change because he proceeded with unexpected ruthlessness and, unlike all the others, had no particular respect for the Roman order. Another precondition for his effectiveness, however, was the fact that the leading circles in the Senate did their utmost to oppose him and were not prepared 'to wait for the storm to pass', as Cicero formulated the old precept.

The situation was new, and so were Caesar's methods. Not that he was an innovator in any modern sense. Yet everything he did – and above all the manner in which he did it – ran counter to previous custom. So did the actions of his opponents. The crisis of the republic had produced the possibilities that Caesar now exploited. The fact that Pompey was needed to attend to foreign problems, that he was powerful – and that the more powerful he became, the more he was opposed – justified Caesar's actions and gave meaning to his victory. The way in which he had emerged from the crisis determined the way in which he now went to work. The crisis thus perpetuated itself and became ever more intense.

Fifty-nine BC was thus an important watershed in Roman history and in Caesar's biography. He now had every chance of winning exceptional power. On the other hand he had ruined his relations with Roman society. He now had the Senate majority and large parts of the equestrian order against him. Cato had been proved right: everyone saw how Caesar had until now deceived so many and who he really was. This is how it must have seemed, even if it was only in 59 that he really became the man he now was.

Yet what society sees and what it does are two different things, especially when it has suffered a severe blow. How Caesar would in the long run be perceived by the Romans, how he would present himself to them, how he would find his way back among them, may still have been a matter of future politics.

At all events, from now on he had to work towards being once more accepted in Rome; to do so he had to strain every nerve, but not necessarily to make concessions that would have cost his pride too dear. He can hardly have thought it hopeless, and perhaps it was not impossible. It was not yet time for the question 'Caesar or the republic?' – though his opponents already knew that the one precluded the other. Yet Rome did not consist solely of Caesar and his adversaries.

We do not know how anxious Caesar was about his future. But he must have been aware of the problems he faced, even though a certain nonchalance and his contempt for his opponents armed him against an immediate perception of their full gravity.

11

Achievement in Gaul

CAESAR'S GOVERNORSHIP was not to be just a means to attain internal political objectives. On the contrary: nowhere could he demonstrate more clearly his passion for achievement. And nowhere, it seems, was he as happy as he was with his soldiers, during his campaigns. This increasingly became his true element.

Yet Caesar could not let internal politics out of his sight for a single moment. His future depended on helping to shape events in Rome, securing his rear, doing everything to restore his political credit, making sure that his opponents did not recoup their strength and above all that they did not reach an accommodation with Pompey.

Caesar was thus simultaneously involved in warfare and politics. Many of his strategic decisions had a political aspect. And what he achieved in Rome was often a function of his military position. Above all, though physically absent from the city, he was in a way constantly present.

Indeed, Caesar's political strength derived in no small measure from his ability to be present in many places at once and so to multiply his influence. He maintained a dense and well-organized courier service that rapidly informed him of the latest turn of events in Roman politics and conveyed his directives to Rome. He wrote innumerable letters, either during marches, as he was carried on a litter, or in camp. He was thus constantly in touch with many men – and women – in Rome; he knew of their needs, offered them help and generous loans, sent them exotic gifts, and established or strengthened many vital connections. For confidential communications he used a cipher in which the letters were rearranged in a manner agreed with his secretaries Oppius and Balbus.

Caesar saw to it that he had friends among the magistrates in office at any given time, and he was prepared to pay any price for such contacts. At the same time his interests were represented by an admirably organized office in the city.

Caesar was ever present in Rome through his ideas, his orders and advice, his gifts and his requests. There was no comparison between him and other governors, who also were keenly interested in what went on in Rome, but for the most part remained spectators. He wished to have a hand in affairs – and indeed he had to. It was vital to him that he should be consulted on all important matters and be able to make known his opinions and wishes; his allies were not to take any action without first clearing it with him. They were to have no advantage for which he was not compensated. Accordingly they often sent agents to treat with him, and he is unlikely to have left them in any doubt as to his precise wishes and demands. Given his ability to calculate so many things in advance and take up a stance on them, as well as his rapid access to information about current events, he could demand to be heard. No one could pretend that he had been unable to obtain Caesar's opinion. He had the advantage of being able to spend much of his time in Gallia Cisalpina, relatively close to Rome. Nor was he at first in a great hurry to set out for his province.

The Period before his Departure

CONFLICT OVER CAESAR'S LAWS · CICERO

AND CATO REMOVED FROM ROME ·

CICERO · CLODIUS' POPULAR POLITICS ·

ATTACKS ON CAESAR'S LAWS

SOON AFTER the beginning of 58, the two anti-Caesarian praetors reported on his laws. They considered that an investigation was necessary to determine whether the laws had been introduced in the proper manner; if not, they were to be declared null and void. Caesar took part in the debate and declared that the Senate should decide. But the fathers lost themselves in endless deliberations. After they had talked for three days the proconsul left the city; that is to say, he took the auspices for his departure and, together with his lictors, donned his war-gear. The trumpets sounded, and Caesar, escorted by friends and curious onlookers, crossed the hallowed bounds of the city and formally entered upon his proconsulate. But he did nothing further. He remained outside Rome, or more precisely in Rome, but outside the city limits.

It is hardly likely that Caesar feared the Senate debates; he merely found them tiresome. Above all, he felt it beneath his dignity to devote himself patiently to dealing with all the futile reservations, objections and misgivings. After three days of debate it must have been clear to him that he could not convince the senators – or only at a price that he was too proud to pay. In the end he would only have railed at them. To the speeches of the praetors he replied in writing, and with extreme asperity.

When one of the tribunes summoned Caesar to appear before the popular court, he appealed to the college of tribunes and obtained a ruling that while he was absent in the public interest he could not be prosecuted. One would like to assume that such impotent attacks were made for the sake of principle. But the situation may have been more complicated than it at first appears. Caesar at least did not think that matters could be left to take their course. Otherwise he would not have remained for months outside the city.

As early as 3 January Clodius had pushed through a number of laws. At first he secured a decision that the distribution of corn to the populace should in future be free of charge. He was also very generous in determining who might benefit from it. The *plebs urbana* thereupon increased considerably, partly through immigration and partly through the emancipation of slaves; emancipated slaves of course at once acquired civil rights. In 56 a fifth of the public revenues is said to have been spent on the distribution of grain. Another law once more permitted the formation of urban associations. As these had often been instruments of unrest and corruption, the Senate had suspended them in 64. The new associations organized by Clodius were widely employed as armed gangs that repeatedly terrorized the city. In a third law he restricted the scope for religious obstruction based on observation of the sky. He presumably ruled that from now on the signs were to be reported not by an official, but by the magistrate himself, who, like any intercessor, would have to expose himself to direct pressure from the popular assembly. This made obstruction more difficult, at least in the case of bills that had overwhelming support. It is interesting that even Clodius did not tamper with obnuntiation as such. A fourth law made it more difficult for the censors to remove senators from the Senate roll; this benefited a few of the more controversial figures in the Senate.

The laws went through unopposed, though one tribune of the people had offered to oppose them. Clearly no one wished to provoke Clodius. Cicero, for instance, convinced himself that it was to his advantage that the law on the associations should go through. Was it hoped that Clodius would be against Caesar's laws rather than for them? It is by no means inconceivable. But there may also have been a fear that the means of obstruction would once again be blunted.

These four laws look like the beginning of a big legislative programme, as though the weapons were being forged for passing more important measures. There had never been any other reason for passing such a generous corn law.

Also in January, probably towards the end of the month, a bill was published under which anyone who killed a Roman citizen without due process of law or had done so in the past was to be outlawed. The principle of *nulla poena sine lege* is not Roman. There had in fact been a law to this effect since the time of Gaius Gracchus. Clodius may have done no more than reformulate it. His aim may have been to assure himself of a compliant court. On the same day another bill was posted, under which the two consuls were to be provided with particularly desirable provinces. They were also given moneys that had been earmarked for land settlement under Caesar's laws; at least one of them is reported not to have taken it with him, but to have invested it at a favourable rate of interest. Clodius was thus hoping to forestall consular opposition. For on 1 January the consul Piso had honoured Cicero by calling on him to speak third, after Pompey and Crassus, and Gabinius was a supporter of Pompey.

Gabinius now came forward and declared that the time had come to punish those responsible for the execution of the Catilinarians, adding that there was no reason to believe that the Senate still had anything to say. Piso confined himself to saying that he was not a brave man. Cicero should withdraw; he would soon be allowed back.

The law in itself did not pose a new threat to Cicero. Only a court could condemn him. Yet he at once put on mourning and persuaded senators and knights to support his cause. Clodius' gangs pursued him, shouting vulgar abuse and pelting him with dung and stones, and the consul Gabinius threatened the knights who spoke up for him. Clodius invoked the understanding with Caesar, Pompey and Crassus.

Pompey had withdrawn to his country house. He told a deputation that as a private person he could do nothing against an armed tribune. If there were a *senatus consultum ultimum* he would take up arms. This was an obvious hint that the Senate could reach an agreement with him, but it did not cost him much, as there was little chance of the Senate's passing such a resolution. Pompey admitted to Cicero that he was powerless to act against Caesar's will. His promises of protection were forgotten. He had to forget them; otherwise he would have been bound to seem either a coward or a traitor. Crassus privately promised help, but sided publicly with Clodius.

Caesar, however, gave his opinion at a popular assembly held by the tribune on the Campus Martius, in the Circus Flaminius. He pointed out that from the beginning he had declared himself against the execution of the Catilinarians, but added that he was not in favour of retrospective legislation that imposed penalties for past deeds. As nearly always, he appeared statesmanlike, noble, magnanimous, and full of innocence, thus belying the picture his opponents painted of him. Once again he probably irritated a number of people.

At the beginning of March, Cicero voluntarily left the city and went into exile. Shortly afterwards Caesar set off in haste for his province. He had received reports that the Helvetians were planning to march through it. This Celtic tribe wanted to leave its settlements in what is now Switzerland and seek a new home in western Gaul. Caesar's presence was urgently required. He ordered the legion stationed in Gallia Transalpina to march to Lake Geneva and hurriedly levied confederate troops. At first, however, he made no use of his main force of three legions stationed in Gallia Cisalpina, near Aquileia. They were to remain in Italy, *ante portas* as it were, in order to exert pressure on Rome. Only at the end of April were they given their marching orders, so that they could quickly be thrown against the Helvetians.

Probably by now another dangerous man had been removed from Rome – Marcus Cato. The occasion for this was a matter of foreign policy. Clodius had decided to annex the kingdom of Cyprus. The island actually belonged to Egypt and was ruled – on the basis of a disputed will – by a member of the Ptolemaic dynasty. The tribune bore him an old grudge, because when he had fallen into the hands of pirates the king had been miserly with the ransom, sending a large, but not sufficiently large sum. The pirates, feeling they were being mocked, rejected the money and generously freed Clodius. He, however, could not get over the challenge that he thought Ptolemy had issued. Moreover, the republic's coffers needed replenishing after Clodius had made over so much money to the consuls. He had a law passed by which Cyprus was to be summarily annexed and another entrusting Cato with its execution. Cato, being a law-abiding man, obeyed. Clodius informed the assembly that Caesar had sent a letter congratulating him on ridding his tribunate of Cato and at the same time gagging him, for now that Cato had accepted an extraordinary commission he was in no position to polemicize against extraordinary commands.

It is doubtful whether the removal of Cato and Cicero was really necessary in order to safeguard Caesar's laws of 59. It was of course asserted that the three rulers could not afford to fall out with such a popular tribune when their own interests were so much at risk. Yet what would their opponents have tried to do? In fact it looks very much as though Caesar wanted above all to deal the Senate further heavy blows and simultaneously weaken Pompey. He was planning farther ahead: his opponents must not recoup their strength, and they must not ally themselves with Pompey. Any stabilization of the situation in Rome must be prevented.

Marcus Tullius Cicero posed some danger to Caesar. Though not really powerful, he was an important, rousing and persuasive speaker. True, he was rather on the periphery of the leading senatorial circles, but highly esteemed by the majority; and his services against Catiline were not forgotten. He was not a brave man but in the heat of battle he could be impelled by a sense of duty or indignation into making passionate protests. He was unpredictable. And he was an exceedingly loyal supporter of the Senate. Being a *homo novus*, a countryman of knightly extraction, he had a rather old-fashioned attachment to republican ideals; he was also extremely diligent, efficient and skilful. He had been fortunate in managing to rise to the rank of consul, to top the poll and be elected by all the centuries – what is more, at the earliest possible moment. The consulate had been exactly the right place for him. Seldom had there been such a perfect match between what a man was and what was required of him. A supporter of the Senate who was also bound – with due moderation – to the urban plebs, the son of a knight, and a consul: this was the very combination that was called for at the time. Whether in opposing popular legislation or in the struggle against Catiline, he was able to commit himself to broad coalitions. He regarded this as important, for he was averse to acting in a partisan fashion, except against men whom he regarded as enemies of the republic. He was all for harmony, between the classes – the Senate, the knights and all the 'good' – and between the Senate and Pompey. This attitude resulted from his situation and from political theory. As a *parvenu* he thought it essential to be able to rely on other forces against the Roman grandees, who were unwilling to treat him as an equal. Being an outsider, he felt a special affinity to Pompey. And since he had not only studied philosophy, and rhetoric, but needed them to legitimate and balance his identity, he preferred to adhere to the teachings of political theorists. While the other *principes* represented the republic, he was chiefly concerned to find solutions to its problems. Whereas they perceived an internal threat to the traditional exercise of power, he saw the whole of the republic threatened more from without. All the more necessary, then, was a philosophy that embraced the whole of the commonwealth. And this attitude became stronger, the more he identified himself with the republic after the execution of the Catilinarians.

He was borne up by his vanity – or what normally passes for vanity. This quality is after all not something absolute, but derives from a sense of indigence. Cicero's vanity developed to such a marked extent because in him a high degree of personal sensitivity coincided with real indigence, that is to say with the lack of a secure and respected position in the political establishment. His need for praise was linked with his isolation, which arose from the fact that he often found himself supporting the cause of the republic single-handed; this produced a syndrome that we perceive as vanity, and it became all the more marked as it found repeated political confirmation.

This placed Cicero in a quite abstract political position, in which he was at once secure and insecure. He made the wrong distinctions. A popular horror like Catiline was for him quite beyond the social pale, whereas an aristocratic horror like Caesar could perhaps be 'improved'. Yet as soon as Caesar set about putting the triumvirate's plans into effect, there was nothing left to Cicero but his honour: he had to remain true to himself. His victory over Catiline must not appear to have been a one-off affair. Hence, he had to oppose Caesar – though only as far as seemed reasonable. Yet because he was sensitive, nervous and vacillating, because he thought he represented the republic better than all the others, and because he tended to let them see this, Caesar had to be prepared to see Cicero one day in the front rank of his opponents.

Moreover, Cicero's ties with Pompey had never been wholly severed. He could encourage him, when the opportunity presented itself, to resume negotiations with the Senate. Moreover, the fact that Cicero was being sent into exile caused Pompey considerable embarrassment. His unreliability and weakness were becoming all too apparent.

Finally, and most importantly, it was now clear that the Senate

231

could not protect the man who had executed the *senatus consultum ultimum* in accordance with its wishes. Another grave defeat was added to those it had already suffered in 59.

Publicly Caesar held back. He did not use his influence, but he treated Cicero kindly and offered him a post as a legate in his province; and he pretended to be in despair over Clodius, who admittedly could not be restrained. It is curious that Clodius did not bring in the bill until the end of January. Perhaps he needed another occasion to show Pompey how vital it was to remove Cicero and Cato. Perhaps he himself was uncertain whom to attack. Then Caesar, whom his opponents saw as his 'aider and abetter', had to force him to act. In any case, Caesar no doubt represented himself to Pompey as the clever one who would have so much liked to spare Cicero. What he engaged in during these weeks was a shameless piece of politicking.

He could now safely leave Rome to its own devices. Clodius and his gangs controlled the streets. He was the hero of the mob, perhaps even of the citizenry at large. They enjoyed their free rations of corn, as well as the way in which Clodius took it out on the wielders of power. He developed novel ways of articulating the 'popular will'. Violence had previously been used as a means to get laws through the popular assembly, but now it served a kind of 'popular justice'. First against Cicero, in the name of civil liberties, and then against others. Repeated assaults and demonstrations were staged; the forum was frequently occupied and the actions of the magistrates frustrated. Popular 'liberty' took the form of violence. And the ways in which this happened suggest that it was not due entirely to Clodius' zeal for tough action, but partly to the people's desire to satisfy certain needs. Whereas Clodius needed scandal, the populace wished to see its dissatisfaction with its living conditions translated into deeds – not into resolutions.

When engaged in agitation, the tribunes of the people had always invoked the power of the Roman people, which supposedly ruled the whole of the Mediterranean world. And parts of the urban populace had always been involved in political demonstrations: politics took place before their very eyes, they were interested in them and took part in them. Yet at the same time it was clear that many lived in poverty and that the real rulers of the Roman world were the Senate and the magistrates. The citizens could accept this, having long been inured to it; it was after all traditional and therefore natural; only occasionally did they rebel. Yet this pattern of rule and exception presupposed that the Senate and the magistrates could regularly make their authority and superiority felt.

The new element in the present situation was the fact that protest and 'popular anger', superbly stage-managed by Clodius, could be indulged in with impunity. The forces that might have opposed them were too weak or held one another in check. In many ways the effects of Caesar's consulship were still felt. The Senate was beaten, the magistrates' hands were tied, and there was no police. There were admittedly some forces of order, but they were employed merely to keep the streets safe from thieves and vagabonds. Anything else was a matter for the magistrates, who sought help from friends and clients, and in extreme cases formed a kind of special constabulary. This system, however, presupposed a degree of energy and security that could nip disorder in the bud.

As such energy and security no longer existed, Clodius had a free hand. He could indulge his passion to the full. And the urban populace enjoyed the power it seemed to possess, some by exercising it, others by identifying themselves with it. The disintegration of Roman society was proceeding apace. For the time being the gangs could be opposed only by setting up other gangs; this happened in 57. And until 52, when Clodius was murdered, his chief adversary banished, and the Senate's authority restored with Pompey's help, street battles flared up in Rome again and again.

Not long after Caesar's departure, Clodius and Pompey came up against each other. When Clodius was setting up an assassination attempt or something that looked like one, Pompey withdrew from public life, as Bibulus had done a year earlier. On 1 June the Senate, no doubt with Pompey's approval, accepted a proposal to recall Cicero from exile. A tribune of the people interceded. A little later Clodius took the bold step of attacking Caesar's laws. He led Bibulus before the people and had him testify that he had in every case observed the sky. Hortensius and other *principes* made common cause with him in the hope of achieving by popular means what the Senate was unwilling to risk. It was a curious reversal of the fronts, a strange alliance that suddenly brought together the leading senators and the popular terrorist. But two important tasks confronted them – recalling Cicero and reversing the defeat inflicted by Caesar. The former had to be shelved for the time being because Clodius was opposed to it, but the latter could be carried out with his help. It was presumably intended that Clodius should first annul Caesar's laws and then have their content legally ratified. This would have greatly increased his power. He even declared that if the Senate repealed Caesar's laws he would carry Cicero back to Rome on his shoulders. This would have been a fresh scandal, but this time to the Senate's liking. But the Senate majority was not prepared to countenance it. To conduct constitutional policy with a gangster against Caesar's breaches of the law was probably too bold and its consequences too unpredictable to tempt the senators out of their reserve. Perhaps they were also too strongly opposed to Clodius because of the exile of Cicero.

Pompey then turned the knife. Speaking in the Senate on 1 January 57, he openly supported the recall of Cicero. Scarcely anyone but Clodius could speak against him. For the first time he had the majority of the Senate on his side and could perhaps hope to win it over permanently. This would have brought about a fundamental change in the internal politics of Rome. After some initial reluctance Caesar agreed to Cicero's recall – though on condition that Cicero promised to behave loyally towards him.

The First Gallic Campaigns (58/57 BC)

CAESAR'S COMMENTARIES, written in 51, begin with the sentence: *Gallia est omnis divisa in partes tres* . . . ('Gaul, in the comprehensive sense, falls into three parts; one is inhabited by the Belgae, the second by the Aquitani, the third by peoples who in their own language are called Celts and in ours Gauls.' This describes the whole country whose subjection is described in Caesar's commentaries. And its subjection was intended by Caesar from the start. It is significant that he begins with space and not with time, with which modern historical works tend to start. But he is describing 'contemporary history'. And his task is dominated by geography.

He had no instruction to make conquests, no authority to do so. For there were laws – including his own *lex repetundarum* – that forbade a governor to make war on his own initiative. Caesar's instructions may have permitted armed operations beyond the borders of his province if Rome's interests required them. Roman governors probably had to be given such latitude, but it can have related only to intervention in isolated trouble spots, not to the conquest of whole countries, let alone a territory as large as Gaul.

How then did Caesar come to do so? How was it possible for him to embark upon a conquest far greater than had been undertaken by any Roman general before him, without instruction or permission, and indeed quite needlessly, in contravention of the principles of Roman foreign policy and in spite of the fact that Rome had long suffered from the size of her empire? The conquest of Gaul greatly extended the borders of the empire, pushing them as far as the Atlantic, the North Sea and the Rhine, bringing Rome into conflict with many brave tribes, making her a neighbour of the Germani, and opening up to ancient culture a gigantic land mass that no longer faced the Mediterranean, much as Alexander's conquests had opened up parts of Asia to Greek culture. How did a Roman proconsul come to engage four legions, then six, then eight and finally ten, along with auxiliary troops, in a war undertaken on his own initiative? Four of these legions had been provided by the commonwealth, but the rest were high-handedly levied by Caesar himself.

Statements on the subject should be sought first in Caesar's own work *De Bello Gallico*, for the other sources, while containing judgments on the war, say nothing about its origins. Cicero praised Caesar in a speech he made in 56. Having just changed sides politically, admittedly with a bad conscience, he was resolved to see some good in Caesar's enterprise. He declared to the Senate: 'The Gallic war, assembled fathers, is now for the first time being waged under the command of Gaius Caesar; formerly it had been merely contained. Our generals saw it as their task to confine the local peoples permanently within their borders, but not to challenge them . . . Gaius Caesar has, I observe, been guided by quite different principles. He believed it his duty not only to fight those whom he found already armed against the Roman people, but to bring the whole of Gaul under our authority. He has thus most happily conquered the various tribes of the Germans and the Helvetians in great battles, and the rest he has intimidated, repulsed and subjected, teaching them to endure the dominion of the Roman people. Regions and tribes of whom we previously knew nothing through writings, oral accounts or hearsay have been traversed in all directions by our commander, our forces, and the arms of the Roman people. Until now, assembled fathers, we possessed only the rim of Gaul; the other regions were in the hands of tribes that were hostile to our rule, unreliable or unknown, or at all events terrible, warlike barbarians. It never occurred to anyone to conquer and subject these peoples. For as long as our empire has existed, everyone who reflected wisely on our commonwealth has believed that no country posed such a danger to our rule as Gaul. Yet because it was the home of so many powerful tribes we have never before waged war against them all; we have offered resistance only when attacked. Now at last we can say that our rule extends over all these territories.'

Cicero made no mention of Caesar's reason for going to war. Yet two years later, in his *De re publica*, he stated Rome's principles in the matter: 'Unjust wars are those that are undertaken for no reason. For a war can be deemed just only when it is a question of taking vengeance on enemies and repulsing them, not otherwise.' Here, on the other hand, he presupposes a long-lasting 'Gallic War' that did not exist and pretends that it was simply a question of how to continue a war against hostile tribes.

Cato, on the other hand, criticized Caesar openly. Traces of his criticism are to be found in Suetonius, who states that Caesar neglected no opportunity to wage a war, however unjust and danger-ous, but took it upon himself to challenge both allied and hostile tribes. Unlike Cicero, Caesar's opponents were unable to see a single long-standing Gallic war, but only the reopening of a series of campaigns. And at least some had no doubt that they were unjust by conventional standards and undertaken in defiance of the rules of international law.

Caesar's own version starts with single events, by which he was drawn step by step into a great war, in the conscientious performance of his duties as a Roman governor. He gives a very detailed account of how the hostilities first arose. He derives their external causation and legitimation from certain shifts of power and movements of population that had alarmed Rome in 61 and 60 BC. Admittedly everything had meanwhile settled down.

Since the end of the second century Rome had had an 'official district' to the west of the Alps, known as the province of 'Gallia Transalpina' or 'Narbonensis'. It was ruled directly by Rome and used as a base from which to observe the immediate and more distant approaches. The Gauls were organized in tribes that seem to have had a loose sense of affinity based on a common language and a shared religion. A close connection obviously existed among the priests, the druids, who met once a year at a sacred place in the middle of Gaul. Here numerous disputes were settled.

The political and social order was aristocratic, the power structure apparently unstable. Wars between tribes, and alliances between nobles from different tribes, often caused unrest, but most of these seem to have had only local significance. The system was occasionally disturbed from the outside as a result of tribal movements beyond the Rhine. Pressure from the north and east, for instance, caused the Helvetians to move from southwestern Germany to what is now Switzerland. Larger or smaller groups of Germans often entered the country, invited or uninvited, to make conquests or merely to take booty. Yet such incursions seldom affected large areas of Gaul. Hence, what went on in this part of the world need not as a rule concern the Romans, even if the Gauls occasionally tried to involve them in their affairs.

True, Rome had friendly relations with many Gallic tribes and prominent nobles within them. A specially close friendship – what the Gauls called blood-brotherhood – bound Rome to the Haedui, who enjoyed a certain hegemony among the tribes who inhabited the regions bordering the Roman province and beyond; this hegemony was disputed by their neighbours the Sequani.

The Sequani, wishing to displace the Haedui, had sought the help of a German prince from the tribe of the Suebi; he had assembled a large following and come to their aid, to be rewarded with land in Alsace. This prince was called Ariovistus. He had defeated the Haedui; and in 61 the Romans considered whether they should intervene in favour of their friends. The Senate could not make up its mind and decided on a delaying measure. It instructed the governor of Transalpina to 'protect the Haedui and the other friends of the Roman people as far as is possible without disadvantage to the republic'. This meant that he could do anything or nothing. Moreover, during Caesar's consulate, the victorious Ariovistus had been acknowledged as 'king and friend'. If, as we must assume, Caesar had a decisive say in this, he may have wished to prepare the way for a conflict between the Haedui and Ariovistus that could later be exploited. In any case such acknowledgments were regularly linked with gifts.

★　　　★　　　★

At first, however, another reason for intervention arose. At the end of March the Helvetii wanted to march through the Roman province, in order to conquer new territory in western Gaul (on the Bay of Biscay). They had spent two years preparing for this diplomatically and militarily. They now broke up their settlements and made ready to leave. The most convenient route lay through the Roman province, which they intended to enter near Geneva.

When Caesar arrived there after long forced marches, the Helvetii asked his permission to pass through Roman territory. He asked for time to consider the matter and told them to return in mid-April.

He had already destroyed the bridge over the Rhône. Now the soldiers he had summoned in haste had to build a wall between the lake and the mountains, some thirty kilometres in length and nearly five metres in height, with trenches and watchtowers. When the envoys returned, Caesar declared that according to the custom and precedent of the Roman people he could not allow anyone passage through the province. Any attempt would be frustrated by force.

The Helvetii then sought to reach their destination through the territories of the Sequani and the Haedui. Caesar concluded that this posed an extraordinary danger to Rome and hurriedly augmented his forces. He sent for the three legions stationed at Aquileia. He also raised two more in Gallia Cisalpina; these presumably did not consist solely of Roman citizens, as they should have done. However, as early as 65 he had sought Roman citizenship for the inhabitants of this province. It is true that the raising of legions really required a Senate resolution, but things were now pressing. The Helvetii had already laid waste the land of the Haedui. Caesar felt obliged to come to the aid of Rome's friends. He was determined to act on the Senate's resolution requiring the governor of Transalpina to do so.

In a forced march he led his army across the Alps and surprised the Helvetii as they were crossing the Saône. He wiped out the last quarter of the tribe, which was still on the left bank. He then had a bridge built. In one day his army crossed the river. The Helvetii, having spent twenty days in the same endeavour, were greatly alarmed and sought an accommodation with him. They sent an envoy to say that if Rome would make peace with them they would settle wherever Caesar told them. If not, they would defend themselves. The envoy proudly reminded him that they had beaten the Romans before: Caesar's surprise victory by the river should not mislead him into underrating them, lest the name of the place where they now stood should come to be known as the site of a fresh Roman defeat.

Caesar, for his part, enumerated the offences committed by the Helvetii, in the past and recently. He found it insolent that they should boast of their previous victory and feel secure. 'For the immortal gods,' he said, 'are accustomed at times to grant favourable circumstances and long impunity to men whom they wish to punish for their crime, so that they may smart the more severely from a change of fortune.' Despite this he was ready to make peace, but not on equal terms; they must give hostages and make reparation for the harm they had done.

This was good Roman argumentation: the opponent was always in the wrong, having at least done something to the detriment of Rome's friends. For Rome had friends everywhere, far beyond her borders. As Cicero wrote, 'Our people has already gained control of all lands by defending its allies.' Here, as elsewhere in his commentary, Caesar felt it important to present himself as Rome's proconsul, attending to her interests in the traditional manner, in accordance with senatorial policy. At the same time he ignored the fact that the war he intended to wage did not accord with the will of the Senate majority.

The Helvetian envoy replied that his people had a tradition of taking, but not giving, hostages. The Roman people could bear witness to this. He then departed, and on the next day the tribe began to move westwards. For about two weeks Caesar and his soldiers followed close behind. He was not prepared to do battle until he had a good chance of defeating the Helvetii. One well-devised battle plan failed as a result of false intelligence, but battle was finally joined near the town of Bibracte. 'Caesar first had his own horse and then all the others led out of sight, so that the danger would be equal for all and no one could think of fleeing. After encouraging his soldiers he joined battle.' The Helvetii put up stiff resistance. When they withdrew on to an eminence and the Romans followed, other parts of the tribe intervened and tried to encircle the Romans. Caesar states that not one Helvetian fled during the battle, which raged for half a day; even the camp-followers defended themselves courageously. When the battle was over the Helvetii moved off. For three days and nights, without pausing, they marched northwards to the territory of the

Lingones. The Romans could not follow because they needed just as much time to tend the wounded and bury the dead. When Caesar threatened to attack the Lingones if they provided the Helvetii with corn or anything else, the latter finally capitulated. They were obliged to hand over the hostages they had taken from other Gallic tribes, their weapons and any deserters; they were ordered to return to their territory and rebuild their settlements. Caesar instructed the neighbouring tribe, the Allobroges, to supply them with corn. For they had destroyed all their stocks. He added that he did not want the Germans to be attracted by the fertile areas that the Helvetii had vacated and so become neighbours of the Roman province. The Helvetii were thus to remain outside the area of direct Roman rule.

In their camp precise lists are said to have been found, showing that in all 368,000 people had taken part in the migration. According to another count ordered by Caesar, 110,000 returned to their old territory. We may presume that these figures were grossly exaggerated, as was usual in reports by Roman military commanders. Yet it is quite possible that the Roman army, consisting of six legions and auxiliaries – 35,000 men at most – was inferior in numbers. Caesar gives no details of his own losses, but makes it clear enough that they were substantial.

At the same time he gives the reader to understand that this victory cancelled out the defeat that the Helvetii had inflicted on the Romans in 107. The background to this is that Rome always saw the Celts and the Germans as particularly dangerous opponents. Only the Gauls, in 387, had succeeded in conquering Rome and burning the city. The Romans still recalled the *vae victis* with which the Gallic commanders had rejected the Romans' laments. Then, in the last decade of the second century, the Cimbri and Teutones had inflicted heavy defeats on them and struck terror into the city, until it was finally saved by Marius. Now Caesar had defeated one of the bravest peoples, parts of which had been involved in the Cimbrian campaign, for all its superiority in numbers. This spoke in favour of his campaign. The question of the *casus belli* could take second place.

The defeat of the Helvetii forced the rest of the Gauls to reach an accommodation with the proconsul. The leading men of most of the tribes came to congratulate him. They explained that they knew his chief purpose had been to punish the Helvetians for their earlier transgressions. But they considered him to have acted in their interest too. For the Helvetii had intended to bring the whole of Gaul under their dominion. They then asked him to convene a meeting of representatives from the whole of Gaul. By paying Caesar a kind of homage they probably sought to convince him that he could leave the Gauls to their own devices. The meeting took place, but Caesar tells us nothing about it. It may be presumed that it did not produce the desired result. He writes simply that afterwards the same men met him secretly in order to ask for his protection against Ariovistus.

For Ariovistus was bringing more and more Germans across the Rhine and demanding more and more land. There was a danger of a large invasion. Moreover, the king was an unpredictable barbarian, quick to anger, whose rule was intolerable. He had also taken many Gallic hostages, especially from among the Haedui. If Caesar did not help them they would have no recourse but to emigrate, like the Helvetii, and find themselves a new home far away from the Germans. Caesar promised to take up their cause, declaring himself confident that Ariovistus would be discouraged from further incursions. While the Gauls feared the worst, he trusted that the king would be reasonable, that he would respect his authority and the friendship that Rome had shown him.

Caesar then reports what had prompted him to warn Ariovistus to keep within bounds. Because of the long-standing friendship between Rome and the Haedui it seemed to him intolerable that the latter had come under the control of the Germans. In view of the size of the Roman empire, it was a disgrace both to himself and the republic. This reflection implies a criticism of the Senate, which, having so often set its seal on this friendship, had now let matters come to the present pass. Moreover, it seemed as though the stream of Germans crossing the Rhine would never end. Were these savage barbarians ever to occupy Gaul, they would invade his province and enter Italy, as the Cimbri and Teutones had done. It was therefore vital to oppose them as quickly as possible. Finally, Ariovistus had shown himself insufferably arrogant.

This local affair, involving a few Gallic tribes and a German prince thus assumed the proportions of a major threat to Rome. It is not clear whether Caesar really believed this. And it is quite unclear how he knew about Ariovistus' arrogance. For his only contact with him had been in the preceding year, when he had had him recognized as a friend and a king.

1. Caesar. Portrait from Tusculum in Turin, the only one likely to have been made during his lifetime.

2. Caesar. Colossal portrait head from the 2nd century AD. Archaeological Museum, Naples.

3. Caesar. Portrait head in green slate. Egypt, early 1st century AD. Pergamonmuseum, Berlin.

4. Caesar. Portrait in Pisa. The head corresponds to a type dating from the Augustan period and preserved in several copies; it differs clearly from the earlier portrait (ill. 1) in that Caesar's baldness is concealed by a fuller growth of hair and the high, furrowed brow and the resolute jaw are more strongly emphasized.

5. The secret ballot was one of the Romans' most
important liberties. The descendant of one of the legislators who introduced it proudly cites this deed on a
coin. The obverse is Rome, and behind her the voting
urn; on the reverse Libertas is drawn by a team of four
galloping horses; the personification of liberty is recog-
nizable by the *pileus* (*cf. below ill. 26.*) that she wields
with her outstretched right hand. Coin of Gaius
Cassius Longinus, *c.* 126.

6. Coin of Lucius Cassius Longinus (63 BC). The abbreviation III V[ir] identifies Longinus as one of the three annually appointed mintmasters (officially called *tres viri aere argento auro flando feriundo*). A citizen is seen casting his ballot into the urn. On the ballot is the letter U, short for *uti rogas*, a conventional formula indicating assent to a motion. The picture alludes to the law, requested by an ancestor of the mintmaster, which introduced the secret ballot in most proceedings of the popular court.

7. Portrait of Sulla: 'he possessed immense power of soul, was avid for pleasure, but still more avid for fame; he led a luxurious life of leisure; yet no pleasure ever detained him from his duties' (Sallust). Coin of Quintus Pompeius Rufus, *c.* 54 BC.

8. 'Honest in face, shameless at heart': this is how Sallust described Pompey. With his honest face he hoped to evince his loyalty to the senatorial regime, but his heart was dedicated to his ambition. Portrait head of Pompey. Augustan copy after a public portrait from the 50s of the 1st century BC.

9. Supposed portrait of Marcus Licinius Crassus: fundamentally good-natured, not to say well-meaning, in more modest circumstances he would have passed for a man of probity and might have been one. Torlonia Collection, Rome. Later copy after a contemporary portrait.

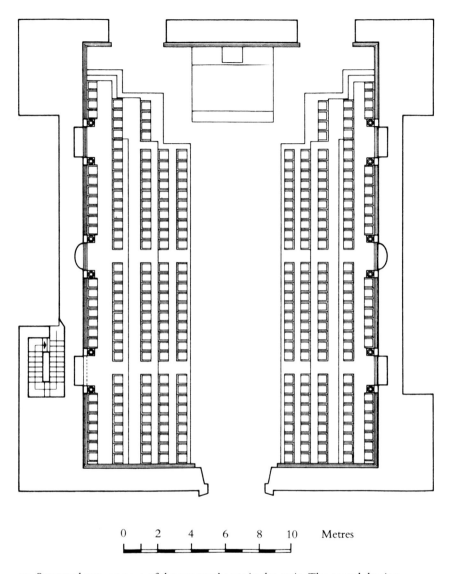

| 0 | 2 | 4 | 6 | 8 | 10 | Metres |

10. Supposed arrangement of the senators' seats in the curia. The consul, having offered a sacrifice and secured the agreement of the gods, would open the session from his podium. The senators, having risen to greet him, would resume their places on the long benches parallel to the side walls.

11. Marcus Porcius Cato: ingenious and
quite unconventional in his tactics and his
choice of means, but no strategist; inflexible
and sure of his aims, which he himself
never doubted and none disputed.
Inscribed portrait bust of the 1st century
AD, modelled on a contemporary portrait.
Archaeological Museum, Rabat.

12. Voting scene. The voter on the left receives his voting tablet from an election officer. Horizontal lines in the background indicate the barrier separating every voting division from the others. Both voters go across narrow raised walks (*pontes*); this is intended to ensure that the voter is seen to cast his vote without influence. Coin of Publius Nerva, *c.* 112.

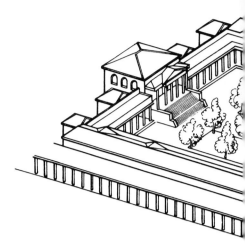

13. Marcus Tullius Cicero: he was not brave, but in the heat of the fight he could be carried away by his sense of duty or his anger to make passionate protests. Capitoline Museum, Rome. Copy from the imperial period from a portrait of the mid-first century BC.

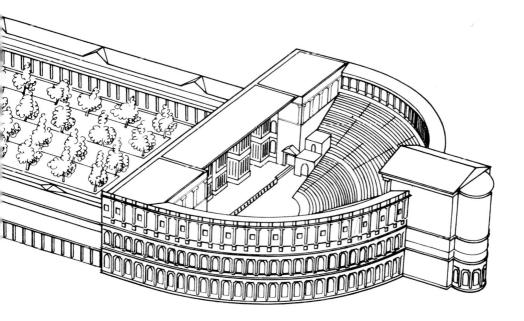

14. Pompey's theatre and colonnade: a bird's-eye-view from the northwest
(reconstruction in the Museo della Civiltà Romana, Rome, with modifications.
This complex, dedicated in 55 BC, was Rome's first theatre built of stone. The
auditorium served also as a stepped approach to a temple of Venus Victrix,
Pompey's tutelary goddess. Behind the stage-front was a public garden surrounded
by arcades; on the narrow western side, opposite the temple, was a raised hall,
particularly grand, that was used from 52 BC onwards for meetings of the Senate. It
was here, six years later, that Caesar was assassinated.

15. Coin of Lucius Hostilius Saserna. The images relate directly to Caesar's military successes. On the obverse is the head of a Gaul with the characteristic necklace (*torques*) and behind it a Gallic shield; on the reverse is a battle chariot with a barbarian fighting in the rear.

16. Pompey: he linked commanding and ruling on the grand scale and splendidly represented Rome's claim to world-wide dominion. Coin struck in the Pompeian army in Spain (46-45 BC) commanded by Pompey's son Gnaeus: Cn[aeus] Magnus Imp[erator] F[ilius].

17. Portrait of Cleopatra VII: she is said to have radiated an irresistible charm that had everyone in thrall. In her hair she wears the royal diadem, its ends falling down on her shoulders. Coin struck in Alexandria *c.* 40 BC.

18. Mark Antony: he was fundamentally an outstanding 'second man', though he clearly considered himself a 'first man'. . . Perhaps he owed his charm and his weakness to the fact that he had never quite grown up. Portrait bust in green slate. Egyptian work from the thirties of the 1st century BC. Kingston Lacey (Dorset).

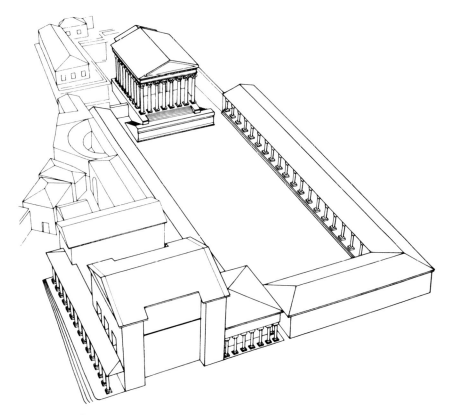

19. Forum Julium: a bird's-eye-view from the south (based on a reconstruction in the Museo della Civiltà Romana, Rome).

20. Coin of Marcus Mettius, struck in 44 BC. Caesar Dict[ator] quart[o] appears wearing a wreath; behind him the curved staff (*lituus*) of the augur, indicating his religious office. It was not uncommon in Rome for portraits of famous personalities to appear on coins, but before Caesar no one had been so honoured during his lifetime.

21. Mark Antony: aristocrat though he was, Antony thought nothing of drinking away the night with the soldiers, boasting and bragging. Coin issued in the Antonian army *c.* 33 BC: Anton[ius] Aug[ur] Imp[erator] III Co[n]s[ul] Des[ignatus] III III V[ir] R[ei] P[ublicae] C[onstituendae]. (For the third time acclaimed victor in a battle, for the third time designated consul, triumvir for the reconstitution of the commonwealth).

22. The curia still stands on the forum. Though rebuilt at the beginning of the fourth century, it presumably corresponds to the original design, except that it was originally surrounded by a colonnade. Representation of the Curia Julia on a coin struck *c.* 30 BC. The building of the new Senate House was inaugurated by Caesar and completed by Augustus. On the roof-ridge is the winged Victoria; under the gable, in large letters, is the inscription 'Imp[erator] Caesar'.

23. Coin of Gaius Cossutius Maridianus, struck in 44 BC, showing Caesar as *parens patriae*. The toga drawn up over his head, indicating that he is performing a sacrifice, relates to his religious office as *pontifex maximus*. There are symbols of other priestly offices: on the left the cap of the *flamen*, on the right the curved staff of the augur.

24. Coin of Lucius Aemilius Buca from the year 44 BC showing Caesar as Dic[tator] perpetuo.

25. Marcus Junius Brutus: 'it depends very much on what he wants; but what he wants, he wants utterly' (Caesar). Coin struck by Lucius Plaetorius Cestius from the army of Caesar's assassins, commanded by Brutus: Brut[us] Imp[erator]. (43-42 BC: *for the reverse of this coin see ill. 26.*)

26. To Caesar's assassins the Ides of March meant the day on which Rome was freed from the tyrant. On a coin struck in 43 or 42 in Brutus' army, the date (Eid[us] Mart[iae]) acquires the character of a symbolic watchword: in the middle of the image is the *pileus*, symbolizing liberty regained, and to left and right are the daggers used in the murder.

Caesar now sent envoys to request a meeting with him. Ariovistus was to choose the place – somewhere between the two of them. Ariovistus replied that if he wanted anything from Caesar he would visit him, and Caesar should do the same; he wondered, moreover, what business Caesar, or indeed the Roman people, had in 'his Gaul', which he had conquered by military force. He thus confronted the proconsul proudly, on terms of equality. Herein lay his monstrous arrogance. Caesar at once sent further envoys to say that if Ariovistus was so ungrateful for the title of friend and king as to reject his invitation to a meeting, he was obliged to convey certain demands: he should bring no more Germans into the country, hand over the hostages he was holding, and not challenge the Haedui and their allies again. Invoking the Senate's decision to protect the Haedui, he concluded that it was for Ariovistus to decide whether he wished to be a friend or an enemy of Rome.

With regard to the Haedui, the king invoked the right of the victor, adding menacingly that if Caesar attacked him he would discover 'what the undefeated, battle-tested Germans, who had for fourteen years had no roof over their heads, could achieve through their valour'.

A little later it was reported to Caesar that fresh hordes of Germans were preparing to cross the Rhine. He therefore struck camp and moved against Ariovistus. After three days he received intelligence that Ariovistus had begun to march against him. He forestalled him by occupying Vesontio (Besançon).

Here a panic arose in the Roman army. Caesar writes that it started among the junior officers. Having heard from travellers about the huge stature, valour and training of the Germans – it was said that no one could endure their looks or the piercing gaze of their eyes – they were seized with abject fear. Some had deserted, and the rest despaired. Throughout the camp wills were being written. The young officers ran to tell Caesar that the extensive woods were unsafe, that it was difficult to obtain corn supplies, and that a mutiny was to be feared. And indeed the confusion seems to have communicated itself to the men.

Caesar at once called a council of war, to which all the officers were summoned, including the centurions, the long-serving non-commissioned officers from the ranks. He upbraided them for imagining that it was for them to consider where they should be led and what strategy should be adopted. There were no grounds for fear. Even the Helvetii had beaten the Germans. However, anyone who hid his fear under the pretext of being concerned about the supply of corn or poor road conditions was acting insolently, either distrusting his general's ability to perform his duties or presuming to give him instructions. Caesar was unmoved by suggestions of a threatened mutiny. Such things happened only to commanders who lacked fortune or were motivated by greed. In his case, his whole life testified to his selflessness, and his victory over the Helvetii testified to his fortune. Contrary to his original intention he would begin the march that very night in order to see whether they were moved more by duty or more by fear. If necessary he would set out with only the tenth legion, which should be his personal bodyguard. For he could rely on it. After this address his officers were as if transformed: the whole army was at once imbued with eagerness for battle.

Caesar's descriptions of his dealings with the soldiers are sometimes striking. He often praises their bravery, experience and steadfastness. Yet again and again they display fear and regain their courage only through his intervention. He mentions various meetings of the army in which he restored their fighting spirit. After being addressed by him they showed remarkable martial zeal. He treated them in a way that makes them seem a little like overgrown children. Such accounts serve to demonstrate his own controlled detachment and superiority. Yet at the same time they correctly reflect the much greater openness with which the soldiers – and the Romans generally – expressed their emotions.

As we have known since Elias, the matter has a historical dimension. The particular emotional control that we have inherited from early modern times, which sets up official channels, as it were, between emotions – good and bad – and their expression, arose only with the modern state and the particular civilization it engendered. Fear then came to be regarded as a base emotion. In the Roman soldiers, however, courage and fear probably manifested themselves more directly; they did not need to conceal them and therefore reacted more naturally. It was thus possible to address them more openly. This probably has to do with the fact that ancient communities lived in the present, a fact that is also obvious in the sphere of public order and the performance of countless public functions. After all, a large number of soldiers who had to fight man against man in a confined space felt differently from a modern army.

However, Caesar's unflattering descriptions of the young noblemen of the officer corps is connected with the fact that they had obviously expressed political doubts regarding his war aims. According to other sources they objected to his waging war solely out of personal ambition.

Knowing that the enemy was approaching, Ariovistus sent further envoys. He was at last willing to talk to Caesar; the time and place were agreed. The armies were in northern Alsace. Ariovistus insisted that both leaders should be accompanied only by horsemen. Caesar emphasizes that he did not want the meeting to founder on this condition, yet on the other hand he did not want to entrust his safety to the Gallic auxiliaries. He therefore ordered the tenth legion to mount their horses.

The Roman proconsul and the German king met. The mounted troops took up position two hundred paces behind each. Regarding the reproaches that Caesar made to the German it is interesting that he again spoke of the generous honours that had been conferred on him. According to Roman thinking the fact that he had sought and been granted the title of king and friend made him dependent. Ariovistus failed to understand this. He thought his friendship with Rome must bring him honour and protection; if it worked to his disadvantage he would rather forgo it. Moreover, the Gauls had called him to their country and he needed German reinforcements in order to defend it. He had arrived before Caesar: hence, this part of Gaul was his province, as Gallia Transalpina was the Romans'. After all, in the recent wars the Haedui had had no help from Rome; nor had they, for their part, helped the city. Hence Caesar was using Rome's friendship with them only as a pretext for destroying him, Ariovistus. The king concluded by saying that if Caesar did not quit his lands he would regard him as an enemy. He added that he knew from leading Romans that it would suit them very well if he were to eliminate him. However, if Caesar left Gaul to him, he would find him to be a loyal confederate.

Caesar replied that Rome had previously refrained from the subjection of Gaul and thus had older claims on the country. It was the will of the Senate that the Gauls should be free. At this point Ariovistus' horsemen began to discharge stones and darts at Caesar's retinue, and the parley was broken off.

Shortly afterwards the military operations began. At first there were cavalry engagements. Ariovistus tried to cut Caesar off from his supplies. Caesar secured himself by setting up a second, smaller camp to the rear, which Ariovistus tried unsuccessfully to storm. As Ariovistus refused to do battle, Caesar led his troops right up to the enemy camp. At last the Germans took to the field; their king surrounded the order of battle with carts and waggons, on which the womenfolk stood weeping, their arms outstretched, imploring the fighting men not to let them be led into slavery.

The two armies clashed so fiercely and rapidly that they could not cast their javelins. Hand-to-hand fighting started at once. Under Caesar's leadership the left flank of the Germans was beaten, while their right flank pressed the Romans. The engagement of the Roman reserves proved decisive. The Germans fled and did not halt until they reached the Rhine, about seven-and-a-half kilometres from the site of the battle. Few escaped across the river, but the king was among those who did. 'Our cavalry,' Caesar reports laconically, 'caught up with the rest and killed them.'

In the late summer, 'having victoriously concluded two such great wars', Caesar sent the legions to the winter quarters he had allocated to them in the territory of the Sequani. He himself went to Gallia Cisalpina to hold court – and take up closer contacts with internal politics. He received numerous visitors from Rome and gave or promised all of them whatever they demanded. The proceeds of the booty already enabled him to show some generosity.

Caesar's description of the fight against Ariovistus is not quite as brief as the summary given here, but there is a striking disproportion between it and the account of the negotiations, which is about twice as long.

What is curious about Caesar's account is the way in which Ariovistus successfully counters his arguments. The much-vaunted friendship with the Haedui had indeed played no significant role up to this point, and in respecting only this friendship and not the one with Ariovistus, Caesar was acting quite arbitrarily. His account leaves out only the strongest argument that Ariovistus could have used (and must have used), namely that when he made war on the Haedui Rome did nothing for them, and that not long after he had defeated them Rome honoured him with the title of friend, thus acknowledging his victory.

It is probable that Caesar deliberately exaggerated the threat posed by the Germans. The manner in which he weaves reports about it into his account is evidence of extremely skilful composition.

Especially puzzling at first sight is the fact that on the one hand Caesar rehearses all the arguments over the legitimacy of the demands he made of Ariovistus and the war waged against him, whereas according to his chosen criteria it seems clear that he is in the right. However, the manner in which he propounds his view of the Gallic war and presents himself to the reader deserves to be treated as a whole.

At the beginning of 57 Caesar received word of an alliance among all the Belgic tribes. 'First, they feared that after the pacification of the whole of Gaul our army would be led against them; secondly, they were stirred up by certain Gauls.' According to Caesar, these people were indignant that the Romans were wintering in the country and planning to stay. This, Caesar adds, would deprive them of the opportunity to seize power in their communities. Caesar the *popularis* had serious objections to such ambitions, being generally opposed to the policy of nobles who relied on popular support to gain their ends.

This is the first mention of the pacification (which implies also the subjection) of the whole of Gaul (*Gallia omnis pacata*). True, Caesar is not reporting his own opinion, but that of others, and it is not clear whether the pacification in question had already been achieved or was still to come.

With war threatening, he raised two more legions, doubling the number provided by Senate and people. In Gaul he then ordered the tribes occupying the territories next to the Belgae to spy out what was afoot. Having made sure that adequate rations were available, he set off with his army against the Belgae. The first tribe he encountered, the Remi, submitted without a fight. As the Belgic tribes were very strong and courageous – and also reinforced by Germans – he persuaded the Haedui to mount a relief attack and so divert part of the huge enemy forces. Soon after this he seems to have come into contact with the enemy. At first he appears to have avoided a full-scale battle, preferring to test the bravery of the enemy and the courage of his own soldiers in minor skirmishes. Then he ordered his army to move against the enemy. He secured his flanks with large catapult machines – weapons whose power relied on the elasticity of twisted animal tendons. But no battle ensued. The Romans retired to camp, and after the Belgae had tried unsuccessfully to cut off their supplies, their army broke up, being short of grain. Each of the tribes retired to its own territory to obtain fresh supplies, having agreed to join forces again if any of them was attacked by Caesar.

When he moved into Belgic territory, several tribes surrendered to him. The Nervii, however, were determined to defy him. They were a fierce and audacious tribe. They lived in isolation and allowed no merchants to enter their territory; they disapproved of wine and luxury goods because they feared these would make them soft.

The Nervian levy, with reinforcements from other tribes, was encamped in vast woods beyond the Sabis (Sambre). They had heard from the Belgae among Caesar's followers that the Roman legions marched separately, each followed by a long baggage train with supplies and equipment (amounting perhaps to a thousand pack animals and some waggons). The Nervii intended to fall upon the first legion and plunder the baggage train. They hoped that the rest would then offer no further resistance.

On approaching an enemy, however, the Roman army always observed a different order of march. Caesar ordered six legions, fully prepared for battle, to march forward behind the cavalry; they were followed by the whole baggage train, which was covered by the two legions he had recently levied. The cavalry crossed the Sambre and engaged the enemy cavalry, but pursued it only as far as the edge of the woods. Meanwhile, one legion after another crossed the river and began to set up camp on a hill already chosen for the purpose. The land having been surveyed, the soldiers fanned out to collect material for the fortifications. When the baggage train approached – this was the agreed moment – the Belgae suddenly emerged from the forest on a wide front, where their army had been drawn up in battle order. They are said to have come running at incredible speed. In no time they stormed up the hill where the Romans were still working at their fortifications.

Caesar states that at this moment he had to do everything at once – raise the flag, sound the trumpet, marshall the army in order of battle, encourage the soldiers, and give the signal to attack. But for most of this there was no time. He was obliged – and able – to rely on his experienced soldiers to take the necessary action themselves. He nevertheless took care to forbid the legion commanders to leave the camp before the fortification works were completed.

Even in Caesar's later account it is clear what confusion must have been caused by the unexpected onslaught. The soldiers did not even have time to put on their helmets or remove the leather covers from their shields, let alone to attach plumes to their helmets. Caesar was able to give only the most urgent instructions before rushing to the front to encourage his men, to halt those who were giving ground, and to bring some order into their ranks.

The terrain was broken and the legions widely separated. The fortunes of battle varied. A few units had managed to push the enemy troops down to the river and pursue some of them across it; others were hard put to hold their positions. The camp was unprotected on two sides, and it was here that the Nervii now concentrated their attack. Some pressed straight towards the camp; others tried to circumvent it from the flank. Panic broke out among the Roman cavalry and waggoners.

The soldiers of the twelfth legion had crowded so closely round their standards as to hamper one another in the fighting. Many officers had fallen or been wounded. Resistance began to crumble as the Nervii relentlessly stormed forwards, threatening the soldiers from two sides. Caesar saw that the situation there was extremely dangerous, yet he had no reserves to send to their aid. He then took a shield from a soldier in the rear ranks – he himself had come without one – and forced his way through to the front line, calling to the centurions by name, encouraging the rest of the soldiers and ordering them to go on to the attack and draw the units apart. The soldiers then found fresh courage; the enemy attack lost some of its impetus.

Caesar was able to rush over to the seventh legion, which was fighting beside the twelfth and likewise being attacked from both sides. He gradually brought the two legions together and then ordered them to wheel round in order to fight back to back. When the two legions that had been covering the baggage train finally came into view and one of the successful legions brought help to the points that were threatened, the course of the battle at last changed. Many who had given up resumed the fight. The Nervii had no way of escape. Caesar praises them for having, even when there was hardly any hope of rescue, 'shown such bravery that, when the first line had fallen, the second took up position on the bodies of the fallen and went on fighting. When these soldiers fell and the bodies piled up, the survivors cast their darts at our soldiers as from a burial mound and hurled back our pikes.'

In this battle, says Caesar, almost the whole tribe of the Nervii was wiped out; only the old survived, and they surrendered. 'In order to make it clear that he had mercy on the hapless and the humble' Caesar not only gave orders that they should be spared, but instructed the neighbouring tribes to respect their territory. A few years later, however, we find the Nervii once more mounting a rebellion against him. Their losses cannot therefore have been anything like as great as Caesar claimed.

He clearly could not refrain from the old Roman custom of grossly exaggerating enemy losses. It was a strange ambition, but probably in tune with the age, in which martial fame counted for so much and human life for comparatively little.

After this Caesar vanquished the Atuatuci, the remnants of the Cimbri and Teutones, who had made preparations to come to the aid of the Nervii. They lived by the Eifel. Here there was no open battle. The tribe had taken shelter in a well fortified town. Having surrounded it, Caesar built corridors protected from above (so-called mantlets) and had a ramp thrown up against the wall. He also had a siege tower built, a tall wooden structure several storeys high, that could be moved on rollers. Below was a heavy battering ram, and at the top artillery pieces were set up; bridges were probably prepared for surmounting the wall. The besieged enemy followed these preparations with undisguised scorn, unable to imagine how the Romans, inferior to them in stature, would be able to bring such a huge construction up to their wall.

When it was rolled forward relatively quickly on the ramp, the ingenuity of the contraption seemed to convince them that the Romans were in league with the gods. Consequently they wished to surrender with all their belongings. Caesar was prepared to spare them, although they had not deserved it. But he rejected their request to be allowed to retain their weapons because of their hostile neighbours. He would, however, 'do what he had done in the case of the Nervii and order their neighbours to do no wrong to people who had submitted to Rome'. Thereupon they threw down masses of weapons from the wall into the trench. During the night, however, they attempted to break out. The Romans forced them back into the town and next morning broke open the gates. Caesar left the town to his soldiers and sold the whole of the booty. 'The buyers gave him a return of 53,000 people.' Clearly the trickery of the Atuatuci suited Caesar very well.

About the same time he received news from Publius Crassus, the son of his friend, that all the tribes of Brittany and northern Normandy had submitted to the Romans.

The whole of Gaul was pacified, Caesar declares. The formula sounds strange. There were still vast tracts of the country in which no Roman soldier had set foot. Nor do we hear that they had not been at peace, and even if they were not, we still have to ask what business it was of Caesar's. Only those tribes that had opposed Caesar were 'pacified'. In his formula they appear as disturbers of a universal peace. And Caesar appears as the man to whom the peace of the whole of Gaul was entrusted. This was a monstrous pretension, a claim to the whole, in relation to which any stirring of traditional independence was bound to be seen as rebellion, as a breach of the peace. The conqueror subsequently presents himself as having waged war merely for the sake of peace! He anticipates in his claim something that did not exist: a *pax Romana* – or *pax Caesariana* – in the whole of Gaul.

For Caesar's campaigns of 58 and 57 had not led to the subjection of Gaul. They were essentially a series of surprise strikes. Caesar suddenly penetrated many hundreds of miles into Gaulish territory, and moreover sent a sub-commander as well. He demonstrated the power of Rome, the bravery and efficiency of her army. Many tribes surrendered to him; the few who defended themselves were beaten. He performed great military feats, mastered the considerable supply problems they entailed and was victorious in several battles. In all this he had certainly made his plans for conquest sufficiently clear. But neither in 58, when many Gauls believed they could gain their own ends with his help, nor in 57, when most of the Belgae and the Gallic tribes on the north coast submitted to Caesar, could they really see what was afoot. Such a large country, with such proud inhabitants, was not conquered so long as the overwhelming majority of the tribes had not measured themselves against Caesar, so long as they had merely been taken by surprise, so long as they could not believe what was happening to them or still lay in store. All his operations so far had been isolated. Wherever Caesar went he was superior. But he had not been to most parts of the country or only passed rapidly through them. No new political system had been built up.

Of course Caesar may well have believed that the military subjection of Gaul was accomplished. If so he was deceived by the success of his bold incursions, which owed much to the element of surprise. And in any case his success was so impressive that even some German tribes beyond the Rhine wished to submit to Roman rule. In the autumn of 57, however, Caesar had no time to deal with them. He told their envoys to wait until the following summer.

On the basis of Caesar's report, the Senate granted him a *supplicatio* of fifteen days. A *supplicatio* was a feast of prayer and thanksgiving. It was originally a particularly intense way of beseeching or thanking the gods. All the temples were opened. Men and women were called upon to pray and make offerings, the men with wreaths and laurel branches, the women with their hair down. In the course of time the *supplicatio* had become a festival in honour of a victor, and, especially after Sulla's time, the number of days it lasted had been a measure of the success and reputation of the commander.

In the second century he was granted at most five days, as in the case of Marius. In 63 Pompey had been accorded ten. Now Caesar was honoured with fifteen. Cicero later called this a tribute to his *dignitas* – his honour and renown.

The Senate was paying him respect for the conquest of the whole of Gaul. At the same time it was indirectly confirming his command and the legitimacy of his wars. It was honouring him in such a way that the transgressions of 59 were bound to pale. Its decision thus represented a quite extraordinary success for Caesar, however little it meant in material terms. Very quickly, perhaps more quickly than he expected, his hope seemed to be realized – the hope that he could repair all the damage to his reputation and, through great military successes, counter all the criticisms that had been made of him.

The former leading circles in the Senate probably put up some resistance. But, still suffering from the heavy defeat of their anti-Caesarian policy, they probably thought it unwise to pursue it. Numerous senators were personally obliged to the proconsul; his quite extraordinary successes made a strong impression; Crassus no doubt supported him and Pompey did everything to accommodate him.

This must be seen as another triumph of Caesarian diplomacy. As Pompey had recently gained some advantages, Caesar must have made it clear to him that he owed him something in return. Yet it may also be that Pompey was nurturing new plans, which prompted him to agree to a longer *supplicatio* than he himself had been accorded. For Pompey too was anxious to escape the odium of 59 and was once more gaining ground in the Senate.

However, Caesar may have contributed to his own success by the way in which he composed his reports to the Senate. Of these we get some impression from his book *De Bello Gallico*, for although it was not written until 51, when the situation had become much more difficult, we may presume that its account of his problems and achievements does not differ radically from that contained in his earlier reports.

This book, together with another dealing with large parts of the civil war, is of the greatest interest as Caesar's portrayal of himself.

Caesar and the War as Reflected in his Commentaries

PURPOSE AND STYLE · THE SPECIAL
TRUTH OF THE ACCOUNT · THE UNJUST
WAR · THE CRITERION FOR ACTION ·
THE CONCEPT OF HOW EVENTS AROSE ·
THE SUPERIORITY OF THE COMMANDER

CAESAR'S BOOK on the Gallic War was in the tradition of reports by Roman military commanders, but at the same time quite novel in that it was composed in a style that matched the highest literary standards. Though ostensibly a campaign report, it is also a highly idiosyncratic expression of the author's personality.

Such a self-portrait naturally has an apologetic purpose. Hence, Caesar's memoir – as well as the conscious and unconscious wishes that guided it – misrepresents certain matters, passes over others in silence or treats them only cursorily, and gives a somewhat partial account of the whole. This is often hard to check, since for the most part Caesar's report is our only source. Where it is possible to check it, Caesar himself usually provides clues that help in unmasking him. For he leaves many contradictions unresolved, unlike a petty deceiver, who would have been consistent. And he reports many things that today seem discreditable – and probably did at the time. In view of Caesar's evident skill in trimming the facts to his own advantage, it seems all the more remarkable that in many cases he refrained from doing so – even where he was vulnerable on ethical grounds. This does not seem fortuitous. Strasburger speaks of a certain 'immoralism' in Caesar's writings.

Apart from its propagandist tendency, the work has a document-ary purpose. Caesar records his deeds for posterity. For the Roman nobility, and for Caesar more than others, fame was a great spur. He sought to pit himself against transience. And while they had to enlist others to write about them, he could write about himself. Nor did he wish to draw a false picture of himself, since he was certain that he need not fear the judgement of posterity.

Caesar's account gives an impression of total objectivity. He always speaks of himself in the third person, using the first only in his authorial capacity, when admitting to ignorance or proffering a judgement. His language earned the admiration of Cicero, the most competent of his contemporaries, which is all the more noteworthy as Cicero favoured a quite different style. He found Caesar's com-mentaries 'unadorned, straightforward and graceful; any oratorical devices are laid aside like a garment. But, wishing to provide only the material on which others might draw for their historical accounts, he perhaps did the foolish a favour by giving them something on which to practise their hair-waving arts, while deterring the wise from writing at all. For in historical writing nothing is more pleasing than pure, lucid brevity.' Caesar's supporter Aulus Hirtius, who later wrote the eighth book of the *Bellum Gallicum*, refers to this judge-ment and then adds, 'Our admiration is of course even greater than others'; for they know only how well and faultlessly he wrote, while we know with what ease and rapidity.'

The perfection of these reports may lie in their directness, their art in their artlessness, but, as Otto Seel observed, simplicity combines with subtlety of diction, cool detachment with vibrant intensity, elegance with a dryness that does not shun repetition, smooth transitions with abrupt breaks. No Latin author adheres as precisely as Caesar to the rules of grammar. 'And yet, in spite of this, hardly any Latin style is so personal, so charged with individuality.'

Caesar's language is extremely economical. He uses less than thir-teen hundred words, occasional technical terms apart. His vocabulary belongs to the language of everyday speech. As Fränkel put it, 'an ordinary, almost spare language is used to capture extraordinary deeds, whose greatness lies not in any kind of originality, but in an instinctive grasp of what is right, in the intrepidity of total commit-ment, swift execution, and unflagging perseverance'.

What really interests us is Caesar's way of describing events and conditions and at the same time presenting himself. The skilful, yet at the same time artless character of Caesar's narrative argues a degree of stylization, to which of course he subjected himself too. For what we feel to be his greatness presumably has something to do with the fact that he was his own man – that his personality was shaped by his own will and that he found the sphere in which he could realize it. Will and destiny combined in him in a special way, and the former seems to have been the more potent of the two.

He shaped not only himself and his deeds, but also his account of them, in a manner at once so personal and so masterful that this account contains a special truth. To put it briefly: in the *Bellum Gallicum* Caesar presents himself in all innocence as a Roman governor who performs his multifarious tasks in a traditional fashion, conscientiously and circumspectly, as duty requires. He does not appear to be defending himself. Quite the contrary.

Naturally there is no mention of the fact that Caesar, as Sallust writes, 'longed for a major command, an army, and a new war in which his energy could be brilliantly proven'. Nor do we read anything of the principle that Cicero found so laudable, even though it was quite at variance with previous practice – the principle of not merely reacting to attacks and defending the Roman province, but of bringing the whole of Gaul under Roman rule in the interest of a lasting peace. True, Caesar now and then allows us glimpses of a wider context, embracing the whole of Gaul, but he refrains from saying that he ever conceived such a grand design.

Rather, he at first lets it appear as if he proceeded step by step, adopting a fundamentally defensive stance, consonant with the principles of Rome's foreign policy. Allies must be protected and dangerous neighbours opposed. He protected Rome's allies selectively, as his interests required. And in taking preventive measures against the Helvetii he counted on the reader's lack of geographical knowledge, for the territory the Helvetii wished to conquer was nowhere contiguous with the Roman province. At first sight, then, it seems as though he moved from pure defence in isolated cases to the conquest of the whole territory – which, according to Cicero, was how things usually happened in the Roman empire. It might be said that Caesar concealed his intention to conquer. It would be more correct to say that he did not expressly state it.

For he makes no secret of it. Whatever the truth with regard to the Helvetii and Ariovistus, his intentions became clear by the first winter at the latest, when the legions took up quarters in conquered territory. According to his own account, the Belgae recognized this too. Moreover, there was no reason whatever for the conquest of Brittany and Normandy. In 56 his intention becomes quite obvious. In one of his typical sentences, in which the verb is delayed to the end, he writes: 'At about the same time, although summer was almost spent, Caesar, seeing that after the whole of Gaul had been pacified the Morini and Menapii were still under arms and had sent no envoy to talk peace, and believing that he could quickly end this war, dispatched his army there.' It is typical of Caesar's presentation that circumstances are introduced as motives and incorporated into the dynamic of the action, that the syntactic build-up draws the reader into the movement and that the tension is released only when the action begins. Yet this is a stylistic observation. Neither the Morini nor the Menapii had been involved in the fighting. The fact that they were 'still under arms' meant no more than that they were still free and had not yet surrendered to him.

Caesar's account makes it clear that he expected all the Gauls to submit. He gave them orders that they were expected to obey. Every tribe he encountered, with the exception of Rome's long-standing friends, had to submit. They all had to give hostages. If they did, Caesar usually treated them leniently. This was evidence of his clemency. Any prince or tribe who refused to submit was in the wrong and so gave Caesar a pretext for war.

All this was at odds with the Roman principle that only just wars might be waged. And a war was just only if its purpose was to right a wrong. Yet it could hardly be wrong for a foreign power to fail to do what Caesar demanded. And there was a good reason for Rome's defensive policy. After all, the Senate had instructed the governor to help the Haedui 'if this is not detrimental to the interest of Rome'.

It is true that demands of the kind made by Caesar were sometimes addressed to Rome's neighbours, but they were not common – except in the course of a major war – and gave rise at most to minor wars. No one operated on a grand scale outside his own province as Caesar did, demanding universal obedience and submission.

Yet it is not only this demand that Caesar makes clear. More than once he reports that the Gauls wished to be free. On one occasion he states that 'human nature is universally imbued with the desire for liberty and detests servitude.' He understands the pride that caused brave tribes, accustomed to victory, to resist defeat. His description is generally fair and arouses the reader's sympathy for the Gauls – or at least the modern reader's. Yet it is clear that their pride and their desire for liberty were just one more reason for treating them with severity. Caesar proceeded from the premise that they must be subjugated, even if the Senate wished the Gauls to remain free.

As he makes his intention clear without declaring it, he cannot advance any reasons for it. At most he can hint at a few. Occasionally he gives the reader to understand that there was much disorder in Gaul before he intervened. He also speaks of the danger posed by the Helvetii and the Germans, which he dutifully forestalled or con-tained. Yet he does not go beyond hints.

Naturally one must beware of viewing Caesar's desire for conquest with modern eyes. Thoroughly Roman and unused to being challen-ged, he was not plagued by doubts or the need to justify Roman expansion. To this extent he did not differ from his contemporaries. Yet he was not bound by the attitudes that had constantly inhibited such expansion or made it dependent on special circumstances. Above all, even if there was no need to justify oneself for the sake of the peoples involved, it was not self-evident that one might flout the rules enjoined upon a Roman governor.

Hence, what Caesar's book reveals, with little attempt at dissimula-tion, was an enormity even by contemporary standards: one man decided, without authority, to conquer the whole of Gaul, simply because he felt it ought to be conquered, employing an army of eight legions, only four of which were provided by the Roman Senate and people, the other four being raised by himself and supplemented later by two more.

Yet what was he to do? Was he to admit that all this was the outcome of his own arbitrary decision and give his reasons for deeming it right? Would that not have meant severing all his links with the Senate and people? He probably thought it best neither to acknowledge nor to deny his intention, but to imply that it was self-evident – at least after his battles against the Helvetii and Ariovistus, when he found himself more deeply involved in the affairs of Gaul. Anyone who demanded further justification could be indirectly likened to the officers at Vesontio: they had no reason to question the prudence and circumspection of their commander, so why should anyone else doubt his devotion to duty or the propriety of his conduct?

Against any questions and objections Caesar sets himself and his actions. It is through these that he hopes to convince. It is these that are at issue, and ultimately the subject of his book. And by speaking of them in his own way he imposes his own perspective. He never thought to convince his opponents. He addressed himself to those senators and knights who were still undecided, relatively open-minded and impressionable.

He thus defends himself not by justifying his actions, but by rehearsing them. In other words, he adopts an offensive stance. He shows how a responsible, prudent governor must conduct himself: he must not be constrained by petty restrictions, by the need to muddle through, tolerating much, turning a blind eye, and intervening only occasionally; he must not be bound by an attitude that was utterly unimpressive, but in keeping with the current mood. Having no governmental apparatus or sizeable military forces, and therefore unable to achieve much by coercion, Rome usually had to rely on numerous contacts, showing consideration to various parties and adopting a piecemeal approach to problems, though this often went with an excessive degree of carelessness and self-interest. This is what made Caesar so different: he set out to perform his tasks comprehensively and energetically. While seeming to act step by step, as the situation evolved, and to concentrate wholly on the present, he was not content merely to react to events, but took preventive action and never lost sight of the wider context. Aware of every problem and prepared for action whenever it was called for, he set new standards, and by matching up to these standards he was able to demonstrate his superiority.

This he regarded as the proper way to act; to show no consideration, to aim for total success, to behave with generosity and forbearance when necessary, but also with appalling severity and cruelty, as he did in the later campaigns. Yet even the later severity must have seemed to him consistent with his duty. In extreme situations any means was justified. Caesar certainly could not imagine that the way in which he discharged his office would strike any fair-minded Roman as improper. Otherwise he would have been bound to have doubts about himself. His high standards would have been wrong. Time and again he proudly asserts that this or that was intolerable to him and the Roman people and contrary to Roman custom. No compromises are made, no mitigation allowed: the demands of honour are paramount. According to a Greek historian, Caesar once said that this was how the ancestors had acted – boldly, making audacious plans and risking all in their execution. To them fortune meant nothing other than doing what was necessary; inactivity would have been regarded as misfortune.

By performing his duties in this way – which was alarmingly at variance with many well-founded rules, but contrasted agreeably with the negligence and indolence that were prevalent in Rome – Caesar justified himself in a way that could hardly fail to put any would-be critic to shame. Once again, as in so many of his speeches, but this time in a form that has come down to us, he demonstrated his superiority.

In his reports, moreover, he always seems to be fully in control, circumspect and well organized. We repeatedly hear of his arranging for supplies to arrive at the right moment. Nothing disconcerts him; he always knows what is to be done. Admittedly there is much that he cannot foresee, yet he is aware of this and envisages various possibilities. He is therefore cautious, armed against contingency, able to react to any eventuality. Naturally he also has to rely on his junior commanders and his soldiers, whom he praises for doing their duty in exemplary fashion and sometimes fighting battles on their own initiative. Caesar and his soldiers – these are the special assets on Rome's balance sheet. Caesar does not obtrude his own part in military events.

It is certainly not wrong to discern, in his manner of presenting himself, the implication that, because Rome's governors normally acted differently, she needed a buffer against the Germans, her nearest and most dangerous opponents to the north.

The political isolation that forced him into his career of conquest corresponded to his dissatisfaction with the normal Roman tempo. Underlying both was Caesar's exceptional will to assert himself. His dissatisfaction gave an objective content to his determination to conquer. His weakness became his strength.

★   ★   ★

It may be presumed that Caesar's way of describing events – which was no doubt essentially how he understood them – accorded with his conception of how political and military events arise. In an extremely concentrated – and restricted – manner he writes almost solely of the actions of various subjects and ignores the wide inter- mediate area that normally extends between the actors and conditions their actions. Caesar rarely gives an appreciation of the overall situation, of the tasks, the opportunities and the difficulties it entails, before turning to the actions of the subjects. Conditions and situa- tions are usually presented as circumstances determining the action: Caesar sees that such and such is the case and does this or that. Even his descriptions of the landscape are bound up with the action: one follows Caesar's gaze as he surveys the terrain before deciding on the appropriate measures: the landscape is thus drawn into the action. Difficulties are presented as tasks. The less the actors are absorbed in the conditions, the greater they appear. So clearly and sharply are they projected on the screen that they seem to occupy it completely; everything in the background is blurred and unrecognizable. What he depicts is not a total configuration to which many factors contribute, but a limited number of interacting subjects.

Every sentence is trained on a target, an action conditioned by all the foregoing circumstances. There are scarcely any periods of rest. Everything is movement. The immense dynamism of Caesar's rapid, audacious and wide-ranging campaigns is directly mirrored in his narrative. Yet although the action is described baldly, with little plasticity or graphic detail, it is easy to take in. In all essentials the configurations are presented clearly, with the special 'vividness that a game of chess has for the inner eye of an expert and that a clearly appreciated problem or an elegant method has for the mathematician' (Klingner). One is not aware of the observer, only of the doer, proceeding step by step from situation to situation. The opponents too are drawn in Caesar's own likeness; they have reasonable, com- prehensible motives and are credited with the most intelligent inten- tions. The actions of individuals on the opposing side are taken into account and seen to play a large part in determining the events.

Moreover, the régime that Caesar builds up is no more than the sum of interpersonal relations. How persons relate to one another is what counts. There is no talk of institutions, of attempts at persua- sion or reconciliation, of administrative problems, of establishing a system of government. The state of affairs that his conquests were aimed at is described broadly as *imperium in Gallia* ('command in Gaul'). General tendencies – processes at work under the surface, as it were – find no mention. The soldiers march, camps are built, demands are issued, battles fought and conquests made. Caesar gives orders; even security and food supplies are ensured by giving orders to those who are to provide them.

Klingner speaks of a 'ruthlessly simplified approach to things, carried to extreme lengths. Whatever does not pertain to the planning and action of the commander and the politician is excluded.' This accounts for the exceptional clarity and perspicuity of Caesar's account. 'No half-distinct background elements obtrude. We see nothing but the matter Caesar had in hand at any given time.'

Everything is concentrated on action and consists in action; consequently, the fact that Caesar was never entrusted with the task that he mastered so consummately is consigned to the background. Caesar directly involved his readers, like his soldiers, in the accomplishment of an enterprise upon which he himself had resolved.

Only at one point does he break out of the narrow narrative confines. This is in the sixth book, where he gives a comparative ethnology of the Gauls and the Germans. At first sight these chapters seem to have no function. By implication, however, they explain why Caesar broke off his campaign against the Germans without subjugating them: for here one reads that Germany, contrary to current opinion, is quite unlike Gaul. To conquer it would be both difficult and unrewarding. Again Caesar refrains from going beyond implication. Yet should he have said in so many words that he really wanted to conquer Germany too? He neither admits it nor denies it.

A special feature of Caesar's account is the almost total exclusion of emotion. Only the soldiers are allowed to feel fear. Caesar is seemingly immune to it. It has been said that Caesar's commentaries owe their formal assurance to the same strength that produced his actions. There is certainly much truth in this, even if this strength is unlikely to have been as effective in reality as it appears in his account. He cannot have possessed the superhuman superiority that his writings suggest.

Historians familiar with the sources can point to a difficult situation in the civil war for which we have a parallel account, probably based on a report by a member of Caesar's staff. We learn that after suffering a defeat, Caesar spent a sleepless night, tormented by dark thoughts and by the realization that his planning had been wrong. At first he believed the situation hopeless, but after much mental turmoil he finally arrived at a decision. Caesar's own account states: 'Caesar gave up his previous plans, believing that he must change his whole strategy.' This suggests that he was merely adapting himself to a new situation. Similarly, we learn of the doubts and scruples that assailed him before he crossed the Rubicon, but these find no place in his own account.

Yet it runs counter to all human experience to take such a self-presentation, with all its abbreviations, at face value. That such doubts are justified, even with regard to the great figures of history, is amply demonstrated by what we know of various situations in which Frederick the Great or Napoleon found themselves.

We may presume that Caesar enveloped himself in a cloak of outward serenity and superiority. One of his officers tells how on one occasion, in an almost hopeless situation, the soldiers found encouragement 'in the expression of their commander's face, in his freshness and wonderful cheerfulness. For he appeared full of assurance and confidence.' And this was certainly the rule rather than the exception. Caesar's superiority and serenity fascinated simple spirits, but to others they made him inscrutable and sinister, especially as they went with immense concentration. And the outward image he displayed may in large measure have determined his inner attitude. His essentially playful temperament, his wilfulness, and his faith in the fortune conferred by Venus, may have played a part too.

Yet behind this there was doubtless a degree of sensitivity, insecurity, doubt and vacillation; there must have been times when all seemed hopeless and he found himself staring spellbound at disaster. In the seventh book, which describes the great crisis of the Gallic war, he even hints that on occasion he came close to abandoning everything, lest even the old Roman province should fall victim to the Gallic onslaught. Indeed, towards the end of his book he reveals a good deal more of himself and writes with rather more freedom.

There was one question that he could hardly suppress entirely: What was the point of his unremitting activity, his subjugation of Gaul, and perhaps of the sacrifices his soldiers had to make in order to achieve it?

However, it could hardly prevail over the joy he felt at so conspicuously proving his worth: ultimately his strength and the possibilities open to him were equal to the immense task he had set himself. Overcoming all the complexities in his character, he always reverted to action, in which he found concentration and attained an effectiveness that increasingly built up a real world of his own, in which he could enjoy a multitude of opportunities and accomplish great feats, even though they might eventually cut him off from other worlds, especially the one inhabited by his peers, Rome's ruling class, including Pompey. But he would have to wait and see what happened. In 57 the real test still lay ahead, both politically and militarily.

'I often marvel,' wrote Stifter, 'when I come to ponder whether to award the prize to Caesar's deeds or to his writings, how much I vacillate and how impossible I find it to decide. Both are so clear, so powerful, so assured, that we probably have little to compare with them.' And both are presented to us in the commentaries with a naturalness that, on closer inspection, seems positively unreal, yet in a style that suggests the ultimate in objectivity.

Political Successes, Spectacular Campaigns, the First Reverses (56 to the Beginning of 52 BC)

F ROM THE AUTUMN OF 57 Caesar was able, from his base in northern Italy, to devote himself to the internal politics of Rome, which were determined by Pompey's attempt to extend his special position, by the resistance of a widening phalanx of opponents, and by the gradual revival and strengthening of senatorial policy. Caesar's part in all this is at first obscure – until he suddenly brought about a fundamental change in the situation.

Pompey clearly believed that after his success over the recall of Cicero the way was open to him to pursue a more expansive policy. He had triumphed, the Senate was on his side, and his opponents had been weakened. And he presumably felt an even greater urge to exploit the situation, as Caesar, through his unexpectedly great successes, had made substantial headway in terms of achievement, power and popularity. In Pompey's eyes Caesar had been essentially the 'young man' – extraordinarily able, it is true – and this was what he should remain for a long time. But if he seemed to be gaining ground, Pompey should at least improve his own position. He was the first man. This should be clear and made increasingly clearer.

At the time of Cicero's return, there was a grain shortage – no doubt engineered by interested parties, not least by Pompey himself – and Clodius used the consequent unrest to stage great demonstrations of popular anger. Pompey now got the Senate to vote him a *cura annonae*; this was a five-year commission, carrying extraordinary powers, to organize the supply of grain throughout the empire and ship it to Rome. The Senate majority was prepared to agree because of Pompey's new standing, its desire to remove a source of discontent in a difficult situation, and the danger that a much more far-reaching popular law might be forced through if it withheld its approval. The leading senators concentrated on restricting Pompey's influence as far as they could. He now had an opportunity to extend his power in the provinces to Rome itself and to show how important he was. He could impress the Roman people with his achievements and perhaps exert pressure on Rome.

It was probably because of these advantages that Pompey was inclined – or could be prevailed upon – to support such a generous *supplicatio* for Caesar. The decision regarding the *cura annonae* was reached quickly, without reference to Caesar. Caesar had reason to think he had been taken by surprise and must have let this be known. Pompey was no doubt pleased that a conciliatory proposal could be made. Cicero had to introduce it and show himself grateful to Caesar, who had finally agreed to his recall, and above all to Pompey, who had brought it off. He had also had to take the initiative over the question of corn supplies.

In the whole period from September 57 to April 56 Pompey was repeatedly attacked by Clodius and his henchmen. There was lively agitation; the chanting of slogans, jeers and whistles sometimes made it almost impossible for Pompey to be heard – for instance when he wished to speak in court on behalf of one of his supporters. But he was courageous and battle-tried, and not easily cowed. Moreover, he had the support of another gang-leader.

Titus Annius Milo was an energetic, ambitious man, of limited vision, but very determined. With Clodius controlling the streets, Milo realized that it would be useful, indeed necessary, to form a gang of his own. He had bought a number of gladiators and used them to excellent effect. Whereas Clodius used his teams of thugs to give violent expression to the popular anger he whipped up, Milo used violence only as an instrument. While Clodius was motivated by an anarchic temperament and a kind of obscure fury, Milo's approach was more technical. He used violence simply to assert his will. Like so many others at this time he was unscrupulous, intent upon his own advantage, and largely untrammelled by the principles of republican life that most took for granted. Yet like few others he was consistent, looking neither to left nor to right, and concerned only to control the streets through open violence; he wished to make his way as a specialist in the use of violence, as a practitioner of violence. His superbly trained task-force was in many ways superior to that of Clodius, but it had no reserves to draw upon and lacked the support of the broad masses, on which Clodius could always count. They therefore balanced one another out and consolidated one another in their respective roles.

Milo was on Pompey's side. Having contributed decisively to Cicero's return from exile, he now sought to bring Clodius to justice; in this he failed, because Clodius was supported by a broad coalition of friends and relatives, including Crassus and those senators who had been his allies in 58. Then, early in January 56, Clodius was elected aedile and so gained immunity from prosecution. He now tried to have Milo convicted.

A great legal war ensued, against Clodius' helpers on the one hand and against Pompey's supporters on the other. The Pompeians were generally successful in defending themselves and to some extent in prosecuting their opponents. This had much to do with the important role of the knights in the jury courts and the fact that memories of the struggle for Cicero's return were still fresh. Nevertheless, Clodius' most important helper was acquitted, supposedly because of displeasure with Pompey. For all his successes, Pompey's situation generally worsened in the early months of 56.

Early in 56 he had again sought a command, the purpose of which was to restore the king of Egypt to his throne after he had been driven out by his subjects. Pompey wished to take on this task, thereby extending his eastern clientèle and increasing his revenues. A broad opposition emerged. The matter dragged on. The opposition prevailed in the Senate. Once again the majority sided with Pompey's opponents.

Earlier, in December 57, he had tried to advance his cause in the Senate by having one of Caesar's laws brought up for debate by a tribune of the people. This was the *Lex Campana*, under which the Campanian lands were to be broken up. So far only part of them had been settled. This was not to be reversed. But the rest were clearly to be exempt. The Senate had always been anxious to keep these lands in public ownership. Pompey therefore decided to play along. He would not have turned directly against Caesar or disputed the legitimacy of his laws, but he would have removed certain sources of grievance and cautiously distanced himself from his ally. The tribune's speech contained certain barbs directed against Caesar. Possibly the lands still unsettled had been earmarked for his soldiers.

Pompey was not present when the matter came before the Senate. The senators silently took note of this. They were unwilling to consider such a vague offer. In a highly charged session at the beginning of April, Cicero proposed that the Campanian lands should be debated on 15 May. In view of the importance of the matter he urged that as many senators as possible should attend. Everything points to Cicero's being able to count on Pompey's approval. Naturally he was acting in his own interest, as he was keen to effect a reconciliation between Pompey and the Senate majority. But for this it was either too late or too early.

It was about this time that Pompey lost the support of the majority. There were various reasons for this, the main one being the inconsistency of his policy: he not only sought to distance himself from Caesar and reach an accommodation with the Senate, but at the same time he wished to consolidate his special position, which meant making himself increasingly independent of the Senate's authority. He might have achieved something by the former means, only to undo it by the latter. This was the problem that bedevilled his whole career. The aim of winning a special position was incompatible with that of gaining respect and influence in the Senate and the good society of Rome. He may have believed that they were no longer so far apart because the senators recognized that they needed him. But their distrust, fear and disapproval were too strong. Nor was it at all evident that Pompey could help in solving the major internal problems. Against the anarchy on the streets, for instance, he was not only incapable of doing anything, but actually made matters worse. He had lost most of his former popularity with the masses. And against Clodius he could only use Milo, meeting violence with violence. This situation was bound to continue unless regular troops were brought in to restore order. There was no police. No political force was strong enough to prevent the violence. It was thus left to the contending parties, the opposing political forces, and these were now so fierce that no authority could effectively represent the whole. Yet this could scarcely be appreciated at the time. Indeed, in the spring of 56 it was still not understood why the violence should be tolerated at all. It was opposed in abstract terms, as it were, because it conflicted with orderly government. If any hope was placed in Pompey, it was that he would use his authority, in concert with the consul and the Senate, to suppress the violence. The senators were disappointed by his inability to do so. Why, then, did they need him?

And finally, if he made advances to the Senate and distanced himself from Caesar, was this not a sign of his weakness – a weakness to which his diminishing reputation bore witness? Was it then not prudent to oppose him?

It is true that since 59 the Senate majority had been opposed to an excessively rigid stand-off. What the leading *consulares* had recommended had proved unsuccessful. They had been responsible for the defeat at the hands of Caesar and then made things even worse. Yet the pressing demands that Pompey had made after his return from the east had been met, the rivalry between Crassus and Pompey had revived, and a new rivalry between Pompey and Caesar was looming. On what could the members of the triumvirate agree, now that they no longer had any common demands that they urgently wanted to have fulfilled? Where was the basis for a united front?

These questions must have exercised the Senate. Crassus and Clodius collaborated closely with those who advocated a decisive Senate policy. One of the consuls of 56, Gnaeus Lentulus Marcellinus, pursued a clear and energetic line, less extreme than that of the former leading circles, but clearly designed to bring the Senate back to the centre of politics. Given the rivalries between the men who held power, it was possible to hope for a restoration of senatorial leadership. It was a well-tried axiom that the authority of the Senate depended on its decisiveness and unity. And this was not wrong, at least so long as the power of others did not call the supremacy of the house into question; and this no longer seemed to be the case.

In March, when tribunes of the people submitted proposals in Caesar's favour – probably providing mainly for his newly raised legions to be paid from public funds – the consul, by a series of deft moves, ruled out any legislation.

Lucius Domitius Ahenobarbus, one of the most promising *nobiles* and a close ally of Cato, sought the consulate for 55. He could be expected to continue the policy of Marcellinus. He had already announced that he intended to deprive Caesar of his command.

Even before the elections the two Gallic provinces would be available for allocation to the consuls for 55. No intercession was possible against the Senate's decision regarding the provinces. It was thus in the Senate's power to relieve Caesar of his command in 54. This was precisely what the senators wanted. If Pompey brought up the *Lex Campana* for debate they could agree to discuss it and so widen the gap between him and his father-in-law. He may even have welcomed the prospect of putting an end to Caesar's successful campaigns.

During these months, however, Caesar had not only observed Rome's internal politics, but taken a hand in them. The news from Rome must have alarmed him: he could see his achievements threatened and his plans put in jeopardy, while he was several days' journey away, able to act only at a distance. Yet he knew how to make himself felt: he kept abreast of affairs, writing letters, receiving messengers and important guests (Clodius' brother Appius, for instance), planning, issuing instructions, commenting, advising, bribing and demanding. He insinuated and intrigued. It may be presumed that he secretly incited Clodius and Crassus to make life difficult for Pompey, perhaps also to reassure the senators regarding their future collaboration. He may even have had the proposals in his favour introduced in order that they would fail. At all events he wanted to make Pompey susceptible to a new agreement.

It is clear that Caesar wished to have his command extended until he became entitled, ten years after his first consulship, to seek a second. Despite all his successes, his return to internal politics would probably be much easier if he could become consul immediately. For his opponents had recently been gaining strength and might try to arraign him. In the meantime he was planning fresh campaigns: he could conquer Britain and cross the Rhine into Germany. He could use his base in Illyria in order to make war in the Balkans, where there was much mineral wealth and a new power base could be built. During the winter of 57/56 he had paid his first visit to this part of his province, in order, he tells us, to look around.

However, his command could be extended only if he made a new alliance with Pompey and Crassus. He probably realized that this meant risking whatever credit he had gained with the Senate and good Roman society, but he was prepared to take the risk. Perhaps he thought that spectacular new successes would in the long run make up for any temporary loss. At all events, the senators were not going to break him – or impede him. Never again were they to be allowed any say. At least not while they refused to recognize Rome's most valuable and proven citizens and acknowledge their claims. If they now raised their heads again, despite the heavy blows they had been dealt in 59, it would be necessary to inflict further defeats on them. Everything else was bound to fall into place.

In March 56 Caesar met Crassus at Ravenna. Then he travelled to Luca (now Lucca), the point in his province nearest to Rome; here, in mid-April, he met Pompey. Pompey was on his way to Sardinia to organize grain supplies, but was prepared to make a slight detour.

Caesar had already prevailed upon Crassus to make a new alliance. He now remonstrated with Pompey. He painted a lurid picture of Pompey's unfavourable situation: he was weak and had no prospect of coming to terms with the senators, who once more felt strong. He deployed all his charm in order to reconcile Pompey with Crassus. He probably represented the differences between Pompey and himself as mere misunderstandings. In this way he prepared the ground for the proposals he wished to put to him: the three of them should join forces again and forge a new alliance. Caesar's command should be extended, his additional levies legalized, and the number of his officers raised to match the increased size of his army. In return, Pompey and Crassus should be given provincial commands with large armies for a term of five years. In order to put all these measures through, Pompey and Crassus should both become consuls for a second time in 55. Caesar may also have mentioned a second consulship for himself.

Such a compact afforded Caesar the possibility of winning fresh fame and fortune. Crassus would at last have a chance to make great conquests. For presumably it was agreed at Luca that he should be given the province of Syria, from which he could lead a campaign against the Parthians. Pompey, however, could have acquired a large field for military activity only in the Balkans. We may wonder, however, whether at his age he wanted to start a large-scale war, whether Caesar did not lay claim to this region, and – not least – who would then have looked after the interests of the triumvirate in Rome. They ultimately agreed that Pompey should have the two Spanish provinces. It was clearly intended that Pompey should have a military force in Caesar's rear. There was little to do there, and thanks to his *cura annonae* he could always repair to Italy or the outskirts of Rome. On the other hand, Crassus obtained an area where Pompey had many clients. The three thus interlocked their power bases in order to secure themselves against one another. Whereas in 60/59 Pompey and Caesar could have formed a dual alliance, Crassus was now indispensable. Pompey probably thought it best to stay in Rome, where he could go on improving his political position without interference from the other two. And Caesar probably made it clear that he acknowledged his precedence. Pompey was to recognize that he could achieve his ends only in alliance with Caesar.

The triumvirate was thus renewed. It cut deep into Roman politics. A new tier of rule was established: the three were superior to all the others because they had long-term commands and large armies, and two of them were bent on waging wars that could only enhance their power and prestige. The break-up of the homogeneity of the Roman nobility had gone a step further. And for Pompey the old distinction between a provincial command and a position of power in Rome was removed. The provinces were encroaching upon Rome – and becoming more important.

Caesar had brought off a diplomatic masterstroke. He had realized something that his opponents had never dreamt of: he had found a new basis on which the rivalry between Rome's most powerful politicians could be overcome. In 60 the triumvirate had been set up to gain old objectives, but its revival meant a breakthrough to something new. However much the other two were attracted by the prospect of a second consulship and a new command, it was Caesar who now dictated the course of events. Because he wanted to retain his province, they too had to be given provinces. And because he could assert himself only by opposing the Senate majority, they too had to break with the Senate. Crassus probably did not find this difficult, so long as he could at last embark upon a career of conquest. For Pompey, however, it meant the dashing of many of his hopes and the frustration of many of his endeavours.

We do not know whether they considered what effect this ruthless carve-up of large parts of the Roman world would have on the future of the republic. They probably thought mainly of themselves and their opponents and were taken up with considerations that led them, at Caesar's instigation, into quite new dimensions. They wished to re-assert themselves by blatant force.

They agreed that each must involve his friends and allies in their joint policy. Clodius must make up his quarrel with Pompey. Cicero must keep his hands off the *Lex Campana* and openly commit himself to the triumvirate.

Immediately after these negotiations Caesar set off for Gaul, where an uprising had taken place. It had started among the Veneti, a prosperous tribe in southern Brittany that conducted trade between Britain and the continent and maintained a large fleet. The Veneti, who lived in relatively well protected towns, had captured some Roman officers sent to requisition grain and wished to exchange them for the hostages they had been forced to give the Romans. They were joined by many other tribes in northern Gaul (up to the Rhine delta) and by auxiliary troops from Britain.

Caesar sent orders from Italy that ships were to be built on the Loire and rowers and steersmen hired. 'As soon as the season allowed,' writes the dutiful proconsul, he went to take charge of operations. One should remember that the meeting at Luca, according to our calendar, took place not in mid-April, but in the middle or towards the end of March.

Caesar describes the great difficulties posed by the Veneti; he even finds it necessary to explain why he nevertheless opposed them. It was no irresponsible enterprise, no wilful venture that put his soldiers at risk. Rather, he was motivated by a serious consideration: if he yielded to the Veneti, others might follow their example. All the Gauls were fickle, always inclined to rebellion, and imbued with a great love of liberty. It sounds as if Caesar had seriously thought of accepting the wrong done to the Roman officers and making an exception of this one tribe by refraining from subjugating it like the others. In reality, however, these reflections probably served only to emphasize the need to subject the whole of Gaul.

After Caesar had despatched troops to secure the critical regions of Gaul, he himself went to the theatre of war. But he quickly saw that he could accomplish little on land. Nor was he any match for the Veneti at sea, since the Roman ships were designed for Mediterranean conditions. In particular, the Venetic ships were much taller than the Roman, so that it was hardly possible to attack them with missiles or to board them, and they were too stoutly built to be damaged by the ram. Not even by building turrets on the Roman ships, writes Caesar, would it have been possible to match the height of the enemy ships, from which the Roman could have been attacked by missiles.

Yet once again, as so often in these Gallic campaigns, the superior military technique of the Romans proved decisive. Experiences of siege technique were applied to war at sea. Sharp-pointed hooks were attached to long poles on the Roman ships. During battle the Romans sailed close to the enemy, and as soon as the hooks were almost touching the halyards that attached the yards to the masts, the ships were rowed hard ahead, so that the halyards were cut and the yards fell down. Thereupon two or three Roman ships surrounded the enemy ship and the soldiers boarded it. The remaining Venetic ships tried to escape. But the wind dropped, and the Romans were able to storm one ship after another. In this way the whole of the enemy fleet was destroyed.

The battle took place under the eyes of Caesar and his army. From the cliffs they were able to observe precisely what went on. Caesar attaches importance to the fact that this spurred on the soldiers in the ships to display the highest courage.

When the Veneti surrendered, Caesar determined 'to punish them with the utmost severity, in order to make sure that in future the barbarians would be more scrupulous about respecting the rights of deputies. He therefore had the whole of their senate executed and sold the rest of the men as slaves.'

This remark is among the most curious in Caesar's work. Had he forgotten that he had said the 'deputies' in question were in fact officers sent to obtain supplies? Or did he wish to credit them later with the status of deputies? In any case the reason for his severity is threadbare and far-fetched. Presumably he did not care. At the same time the considerations with which he justifies the opening of the war offer a much more plausible explanation for its conclusion: the rest of the Gauls were to be taught a lesson. The revolts that repeatedly flared up among these restless tribes seem to have led Caesar to conclude that they could not be bound merely by treaties. Hence, the cruelty of the measures he took was the outcome of his determination to subjugate the Gauls – which he had made the principle of his governorship.

In the same year Caesar's legate Crassus forced most of the Aquitanian tribes to submit to Rome. Owing to the onset of winter Caesar had to postpone a campaign against the Morini and the Menapii on the North Sea coast.

News of the alliance concluded at Luca soon reached Rome, though probably the details at first remained secret. During the early summer the first proposals in favour of Caesar were forced through. His independent levies were legalized. It is said that many senators voted for Caesar because he had previously sent them rich gifts; after he had filled their coffers, they persuaded the Senate to defray some of his costs from public funds. Yet we should resist the temptation to oversimplify. Many senators probably gritted their teeth and supported the decision because, given the renewed triumvirate, the matter would certainly have been decided by a popular law. This would have meant another discomfiture. The more prominent senators probably absented themselves from the session, and the protests of others went unheeded. Caesar was then voted ten legates. In deciding on the allocation of provinces to the consuls who had yet to be elected, the Senate left him with his two Gallic provinces. Cicero had been obliged to speak for Caesar. In the Senate he had to contend with furious reproaches. How could he support the man who had done so much to engineer his banishment? He was obliged partly to belittle his switch of position and partly to justify it. Matters soon became even worse for him. He had to be reconciled with various of his enemies and defend in court not only Caesar's helper Vatinius, whom he had so vehemently condemned, but even his arch-enemy Gabinius. But Pompey and Caesar mercilessly extracted whatever they could from this soft, vulnerable, respected and eloquent man. And the more they pressed him, the less he could resist.

The elections were due to be held in July, but were postponed because of objections. Pompey and Crassus dared not declare their candidacy in the regular way. This is hard to understand, given their large clientèles and the respect they probably still commanded in spite of everything. A powerful current of opinion must have arisen against them throughout the citizenry. For once it seemed likely that the elections would be highly politicized. The question was whether Rome was prepared to yield to the claims to power made by Caesar and his two allies, claims that were gradually coming to be felt intolerable. Cato missed no opportunity of declaring that the liberty of the commonwealth was at stake, and many believed him. When the elections had once more been deferred because of objections, the consul Marcellinus asked Pompey and Crassus whether they wished to become consuls. Pompey answered: Perhaps yes, perhaps no. When pressed he was prepared to say that for the sake of the honest citizens he certainly did not wish to be consul, but for the sake of those who disturbed the public order he very much wanted to be. Crassus declared, equivocally yet unequivocally, that he would do what was best for the commonwealth. Both probably wanted to be called upon as saviours of the republic. As the deadline for the announcement of candidacies had passed, the consul refused to consider them. This meant that the elections were further deferred.

In November the senators put on mourning. They complained publicly and absented themselves from the games and meetings of the Senate. They were demonstrating their powerlessness or, to put it another way, their inability to act politically: they could do no more than demonstrate and appeal to a public that no longer existed as a real force. When the electoral assembly was finally able to meet in January, the opposition was very strong and the election of the two allies far from assured, despite the large groups of men sent to Rome on leave by Caesar. The other candidates therefore had to be terrorized into giving up their candidature. Only Domitius, with Cato's support, held out till the day of the elections. However, when he went to the Campus Martius – during the night, for reasons of safety – the torchbearer who preceded him was murdered, and hand-to-hand fighting broke out. Several others were wounded before he returned home. Now at last the election could take place. Immediately after it the assembly wanted to elect Cato praetor. Pompey prevented this by claiming to have seen lightning. During the election of the aediles there were fresh clashes, during which Pompey's toga was spattered with blood.

Soon after the election, work went ahead to put through the agreed programme. The consuls were entitled to levy as many troops as they deemed necessary in their provinces. When the law came up for debate, its opponents were given only a limited time to speak; when Cato went beyond the limit an official pulled him down from the platform; when he went on speaking from the floor he was removed from the forum. He returned, however, and began speaking again; he went on speaking even as he was being arrested. The vote was deferred. The following day the forum was occupied. Cato and his friends found their path blocked. Some pushed their way forwards; Cato and a tribune were lifted on to the shoulders of friends and protested loudly, whereupon fresh fighting broke out, in which four men were killed and many injured. One tribune of the people, to avoid being barred from the forum, had spent the night in the Senate, where he was locked in and later violently handled. After the voting a friend brought him out covered with blood and showed him to the bystanders. Crassus personally punched one of the senators in the face.

The wielders of power, intent only upon establishing their own positions, asserted themselves with a violence that matched the ruthlessness they showed in sharing out magistracies and provinces among themselves. Cato and others objected strongly, but could offer no more than token resistance.

Not only was Caesar's command extended, but it was agreed that the Senate should not discuss his provinces until 1 March 50. This meant that the first consuls on whom they could be conferred were those elected for 49. Before this was resolved, Cato had gone to Pompey and begged him to consider that he was saddling himself with Caesar. He might not realize this yet, but when Caesar really began to exert pressure they would both fall upon the city, since he would be unable either to throw him off or to bear his weight. It was an impressive image: two men of gigantic stature towering over the city. But it was to no avail: Pompey was too convinced of his superiority.

Caesar thus gained time and could set about his plans for expansion and conquest in earnest. First, however, he had to repel a surprise attack. Two Germanic tribes, the Usipetes and the Tencteri, had crossed the Rhine near the sea. Caesar tells how they had been exposed to pressure from the powerful Suebi and reached the Rhine only after aimless wanderings. In view of the 'fickleness' of the Gauls, he feared that this incursion might lead to a major war. Some tribes, he writes, had summoned the Germans to help them. He therefore set out with his army earlier than usual, summoned the Gallic leaders to a meeting, informed them of his decision to make war and ordered them to provide horsemen.

When the Germans sent deputies to treat with Caesar, he demanded that they should settle beyond the Rhine in the territory of the Ubii. He also granted them a truce. Meanwhile, if Caesar is to be believed, the German cavalry attacked the Roman. Although the Romans – or rather their Gallic allies who supplied the cavalry – were far superior to the Germans, they were defeated. Believing that a trap had been set for him, Caesar no longer felt bound by his earlier promises. He wanted to defeat the Germans as quickly as possible, as the bulk of their cavalry was absent, searching for corn supplies. Next day a much larger deputation appeared, including all the leaders and elders of the Germans. They wished to justify themselves for the cavalry attack. And it is unlikely that they would all have come together had they not been serious about this. Caesar, however, asserts that this 'proved their already notorious treachery and duplicity'. He 'was glad to have them in his power and ordered that they should be detained'. This was a blatant breach of the rights of deputies, to which Caesar had attached such importance in the case of the Veneti. He quickly ordered an attack on the leaderless Germans. They could scarcely defend themselves and took flight with their women and children. Having reached the confluence of the Meuse and the Rhine they could go no further. Some were massacred on the riverbank; the rest threw themselves in the river and perished. Two whole tribes were wiped out. Only the cavalry survived, together with the deputies, whom Caesar now contemptuously released without maltreating them.

This attack on the Germans may be explained by his anger at the fresh German invasion of territory he had recently conquered, by his annoyance at discovering that his rule was still insecure in the area, or by his determination to put his new plans for conquest into action with all speed. Yet it also indicates that Caesar was waging war with increasing impatience, severity and cruelty. His uncompromising will caused him to act resolutely, but clearly did not bring him the speedy victory he desired. It enabled him to conquer, but not to secure his conquests, at least not politically. He might thus on occasion resort to brutality. It was not to become the rule, but it might have been indicative of Caesar's situation.

May he not have been impelled by the impatience that from time to time assails someone who has put immense effort into attaining a certain goal, thinks he has succeeded, and then presses on with increasing haste, refusing to be held back for fear of losing direction, becoming less and less able to wait and let things take their course, and seeking to force events? If so, such impatience would explain the enormity of Caesar's conduct. Although it would not make it less heinous, it would make it possible to construe it differently – as deriving less from an urge to destroy than from a quickening of Caesar's inner tempo, which broke through the inhibitions that should have restrained him.

Having defeated the two tribes, Caesar decided to cross the Rhine. He states that he wanted to demonstrate Rome's power to the Germans on their own territory, so that they would no longer be tempted to cross into Gaul. Moreover he could as usual claim to have been called upon for aid: the Ubii, who had become friends of Rome, needed help against the Suebi. Finally, he felt obliged to punish the Sugambri for having given refuge to the horsemen of the Usipetes and Tencteri and refused to hand them over. He may have found it insolent that they should turn his arguments against him. He had maintained that the Rhine was the boundary of Roman rule; if he, they said, thought it wrong for the Germans to cross into Gaul, why did he claim any imperial power beyond the Rhine?

The Ubii offered to ferry the Roman army across the Rhine.

Caesar, however, thought this unsafe; above all, he considered it beneath the dignity of the Roman people. What would it look like if divisions of the Roman army were to cross the Rhine in German fishing boats? He was well aware that the breadth and depth of the river, to say nothing of its powerful currents, posed the greatest difficulties. Either they would succeed in building a bridge or it would be better not to cross at all.

Caesar states that he himself devised a way to solve this quite exceptional technical problem, but his account is too brief to be comprehensible. At all events, he found ways of linking the thick beams, sunk in the river by cranes and then rammed home at a slant, in such a way that the stronger the current, the more firmly they held. He then laid transverse beams over them and covered the structure with poles and wattlework. Upstream he had other stakes rammed into the river-bed to protect the bridge 'so that if trunks of trees, or vessels, were launched by the natives to break down the structure, these fenders might lessen the force of such shocks, and prevent them from damaging the bridge'.

After ten days' work the bridge was completed; Caesar does not say how long the planning and the preparation of the tools had taken. Nor do we know the precise location.

Since the Sugambri had withdrawn into the depths of their forests, the Romans could only set fire to their villages and cut down the corn-crop. Caesar granted peace and friendship to some German tribes, at their request, in exchange for hostages. Learning that the Suebi were also hiding in the woods, 'he judged that he had accomplished all the objects for which he had determined to lead the army across the Rhine – to strike terror into the Germans, to take vengeance on the Sugambri, to deliver the Ubii from a state of blockade.' Enough had been done to satisfy both honour and expediency. After eighteen days he returned and destroyed the bridge behind him.

It may be that in 55 Caesar had no plans for a large campaign in Germany; presumably he wanted above all to secure his rear for a quite different enterprise. But to have achieved so little must have been repugnant to his dynamic temperament.

From the Rhine he led his troops straight to the channel coast. For despite the lateness of the season he planned an expedition to Britain, for which preparations had obviously already been made. Again he states that he was concerned about the security of Gaul, because during recent years the Gauls had repeatedly used British auxiliaries against him in the field. Other reports, however, reveal that he suspected great riches on the unknown island. It was a source of tin, and it was reported that gold, silver and iron occurred there too. There were also said to be unusually large pearls. The island might also supply slaves and other goods. In the first place he wanted to investigate the situation, for no one could properly inform him about the land and its people. Obviously his expedition was intended as a prelude to a later and more extensive campaign. For he can hardly have been content simply to penetrate into unknown regions, although that too was important; for in this he was emulating Pompey.

When Caesar wanted to land on the south-east coast of Britain, he met with fierce resistance on the shore with its steep cliffs. The Britons were ready for him. He therefore sailed further and dropped anchor off a flat beach. But the barbarians, as he calls them, had followed and prevented the Romans from disembarking. The water was fairly deep and their weapons were heavy. The Britons were able to fight from dry land or the flat shore. Panic broke out in the Roman fleet. Thereupon Caesar sent the men of war into the enemy's flank. 'The natives, frightened by the shape of the ships, the motion of the oars and the unfamiliar type of the artillery, came to a halt and retired, but only for a little space.' At that moment the eagle-bearer of the well-tried tenth legion loudly beseeched the gods for a happy issue and cried: 'Jump down, comrades, unless you wish to betray your eagle to the enemy; it shall be told that I at any rate did my duty to the republic and my general.' The first men followed him and were followed by others. The impulse that pervaded the Roman ranks was enough to get them into motion. But the fighting was fierce; they could not stand firm or keep rank. The battle made no progress and confusion spread. The enemy attacked on horseback, surrounding one man and striking down another. As the unequal battle dragged on, made more difficult by the water, Caesar had the boats of the warships and scout-vessels manned with soldiers and sent them to the points where his men were hardest pressed. In this way the Romans got the upper hand and finally went over to the attack, now here, now there, on a broad front. They reached land and defeated the enemy. Admittedly they could not pursue them far, because the ships with the cavalry had not been brought across. 'This one thing was lacking to complete the wonted fortune of Caesar,' he remarks in conclusion.

The Britons sued for peace. Caesar reproached them, though it is hard to see for what. True, he had reported that some British tribes had submitted to him, but not that all had done so or that those who had were among the men who had just fought against him. All the same he pardoned them. They were to bring hostages, but this took some time. Then the Roman fleet was partly destroyed and partly damaged by a storm tide. The soldiers were afraid of being cut off from their supplies. It was probably then that the unrest in the Roman army, reported in another source, took place. The Britons decided to resume the fight, calculating that if they now defeated the Romans they would have no more trouble from them for a long time.

Caesar had already guessed what was afoot. He was prepared for everything and took the necessary measures to meet the enemy. He had the fleet restored, taking whatever material he needed from the damaged ships; the rest was quickly brought across the channel. After he had won two victories and set fire to all the farms far and wide, the Britons again sued for peace. Caesar was glad to be able to break off the campaign. He doubled the number of hostages, ordered them to be sent to the continent, and then withdrew to Gaul with his army.

Apart from the fame of having landed with armed force on an island of which little was known and whose very existence had been disputed, he had nothing to show. Only two British tribes sent the promised hostages; the others sent none.

When the Senate discussed Caesar's report in autumn 55, there was for the first time, as far as we know, severe criticism of his way of waging war. Cato declared that sacrifices must be offered to the gods, lest they should punish the soldiers for the crimes perpetrated by their leader against good faith and honesty and so that the citizens might be spared. He proposed handing Caesar over to the Germans whom he had so disgracefully ill-used. This was an old remedy – handing over a general so that Rome would not be visited by divine punishment for breaking a treaty. It had last been applied in 135, when a consul had concluded a treaty with Spanish tribes and the Senate refused to ratify it; he was handed over to the victims of the deceit as the one who had broken the treaty.

Fear of divine punishment still lived on at some level of Roman consciousness. It was bound to be increasingly nourished by the uncertainties of the age: the more insecure life became, the more the gods were to be feared. It was therefore natural for Cato to appeal to religious scruples. He may well have shared them; but his philosophical convictions too could have led him to no other conclusion. Caesar's warfare was a crime – embarking on it in the first place, and then repeatedly attacking Gallic or Germanic tribes. The breach of faith with the Germans was especially heinous.

Yet however serious he was, however much he appealed to the Senate, he can hardly have hoped to succeed: again it was largely a demonstration. The Senate majority again granted Caesar a *supplicatio*, this time lasting twenty days. Caesar himself complained about Cato in a vituperative letter. Cato's reply, in which he levelled fresh reproaches at Caesar, was so serious and convincing that Caesar's friends rued their decision to have his letter read out in the Senate.

At about this time, in the late autumn of 55, Crassus set out for his province. It was no secret that he wished to take the field against the Parthians. Caesar had written and encouraged him to do so. A tribune of the people, however, observed unpropitious omens and forbade him to leave for his province. No war might be started without cause. When Crassus ignored the intercession, the tribune tried to have him arrested; this was prevented only by the intervention of some of his colleagues. The tribune ran out to the city gate and set up a basin of glowing coals; as Crassus passed by he burnt incense, offered a libation, and called down fearful curses on the consul. Yet another demonstration – very impressive but apparently ineffectual. But, in this case the gods were not invoked in vain.

Apart from a few minor reforms to the legal system and moves to restrict electoral malpractice, the consulate of Pompey and Crassus was distinguished only by a great festival. Pompey opened the theatre that he had built on the Campus Martius with the proceeds of the spoils from his eastern war. The complex was not quite finished, but he wanted to inaugurate his consulate with the most magnificent games in living memory.

'The Roman people hates private luxury, but loves public liberality,' Cicero tells us. Roman grandees therefore built no palaces – though they did build themselves fine country villas. On the other hand they set up temples or halls, unless they were content to be remembered by their countrymen for great displays or games. But none had conceived anything so grand and splendid as Pompey's theatre. After all, he had to demonstrate his outstanding stature. He had to surpass everything that had gone before, and this was not easy.

In 58 Sulla's stepson, Marcus Aemilius Scaurus, had broken all records when he was aedile. For his games he built a splendid theatre, the semi-rotunda for the spectators being as usual of wood, and presumably fairly tall; but for this he could resort to existing, re-usable materials. Its novel feature was that the stage wall was built in three large storeys; the lowest was of marble, the central one of glass, then a great luxury, while the uppermost was adorned with gilt panels. Gigantic marble columns (said to number 360) had been transported to Rome for the building. And in accordance with a common custom, which Caesar also observed, Scaurus put on an art exhibition on the stage wall, with Greek statues placed between the columns. Pompey imitated this, and it became a tradition. Costly curtains, paintings and other items were there to be admired. Hosts of animals had been imported to take part in the fights and to illustrate the wonders of nature; among them were five crocodiles and a hippopotamus.

It was not easy to compete with Sulla's stepson, one of those who had inherited the profits of the civil war. When the young Gaius Scribonius Curio later staged great funeral games in honour of his father, he devised something even more startling, though less costly: two wide theatres, also of wood, that could be made to revolve. They were pivoted on a strong post and must have been mounted on rollers. The whole structure was probably hauled round by mules or horses.

For performances the stages faced away from each other in the morning; the theatres were then turned through a hundred and eighty degrees, probably pushed. They then formed an amphitheatre for gladiatorial games. The elder Pliny, to whom we owe this informa-tion, wondered which was the more to be marvelled at – the inventor or the invention, the engineer or his client; that someone was so bold as to think up such a thing, so bold as to undertake it, or so bold as to commission it. But what was most terrifying, he said, was the idea that the spectators dared to take their seats in such a precarious structure. Here, in fact, the people themselves were risking their lives – as the gladiators usually did. The people who had conquered the world hovered on an artificial structure and applauded its own peril. Did people at the time perceive its deeper significance? Everyone gathered there. It must have caused great excitement in Rome and dominated the talk of the town, the curiosity of the citizens, and popular memory for a long time. When the pivots were turned on the last day there was a fresh diversion: it was left as an amphitheatre. The two stages were separated and athletic contests took place, then suddenly the walls were drawn up and the victors of the gladiatorial games met in combat.

Pompey documented his unique personality by building a permanent stone theatre. A similar theatre had been commissioned by the censors in 154 BC, but met with such fierce opposition that it had been pulled down. Such a building was said to be detrimental to morality, a notion that was no doubt suggested by the many popular decisions that the Greeks were accustomed to making in theatres. Pompey countered any such objection by building a temple to Venus Victrix, the goddess of his victories, in the apex of the semi-rotunda designèd as the auditorium. The seats on which the spectators sat formed the temple steps, as it were. He was able to point to certain parallels in a shrine from the Sullan period in Praeneste (Palestrina). This may or may not have carried conviction, but we hear of no opposition to Pompey's building. Work on it probably began in 60 or 59.

Unlike Greek theatres or earlier Roman theatres, Pompey's was completely free-standing. The old wooden structure was transformed into stone, so to speak. An ingenious system of substructures was erected: barrel vaults running round the building and rising towards the centre, and on the outside arcades, interspersed with half-columns that probably supported architraves – three rows one above another, the outer façades presenting a regular and monumental image enlivened by the play of light and shade. The effect must have been not unlike that of the later Marcellus theatre. The stage wall was of the same height as the auditorium, richly articulated and adorned with many statues. The stage itself was probably covered. For performances a curtain could probably be lowered from above. Behind the stage was a large courtyard with gardens, surrounded by colonnades, housing the props and if necessary providing shelter from the rain, but above all serving as an agreeable resort that the citizens found pleasing and imposing. They were richly adorned with pictures and statues, including masterpieces of Greek art. One large room, presumably situated at the axis of the complex, later served as a meeting place for the Senate. It was the scene of the meeting on the Ides of March 44, when Caesar was murdered. In this room stood a statue of Pompey. It may also have contained the group of fourteen statues representing the nations he had conquered in the east. At any rate it is attested that there was a statue of Pompey surrounded by the nations in the complex of the columned hall.

The building was about three hundred and forty metres long and up to a hundred and seventy metres wide, the inner courtyard measuring more than seventeen thousand square metres. The theatre could hold over ten thousand persons. At that time there were hardly any buildings on the Campus Martius. Pompey's great building thus towered up out of the plain, almost as though competing with the Capitol: he was leaving an unmistakable mark on the city. (Today only the foundations survive, but the buildings in the Via di Grotta Pinta partly follow the former outline of the auditorium.)

For the opening Pompey presented theatrical performances that were marked above all by their spectacular staging. One tragedy, for example, featured six hundred mules. At the same time there were five days of animal-baiting in the circus. Five hundred lions and over four hundred panthers perished. The great attraction was reserved for the last day – eighteen elephants fighting against heavily armed men.

The masses marvelled, but in this case they were not amused, for the great beasts aroused their compassion. The trumpeting of some of the wounded elephants was so heart-rending that the horrified spectators pleaded for them not to be killed. This, we are told, displeased Pompey. Cicero remarks that there was something human about the elephants. He asks, 'How can a civilized person take delight in seeing a frail human being torn to pieces by a powerful beast or a magnificent animal run through with a spear?' This was probably an uncommon reaction. For most of the spectators it must have been a horrifyingly splendid spectacle, well suited to a people that ruled the world, and all the more to be relished, the less this people could actually achieve. Here it might fancy itself powerful. What Goethe said about the amphitheatre is precisely apposite: it was 'properly designed to impress the people with the image they had of them-selves, to hoodwink them'. Pompey also displayed one of nature's wonders – the first rhinoceros ever seen in Rome.

Until 55 Caesar had built nothing. Perhaps he first wanted Pompey to complete his theatre. Of course he had long planned to set up great monuments to himself in Rome. In 55/54 he decided what they were to be. He wanted to build a whole forum, a square surrounded by columned halls bordering on the Forum Romanum, rather smaller than Pompey's theatre; yet it was not to stand somewhere on the city's outskirts, but right in the centre. Unlike Pompey, he concen-trated his attention on Rome itself, rather than its periphery. He wanted to be present, in the form of buildings, where the decision-making took place, and at the popular assemblies too. For them he began to build a large covered voting hall of marble on the Campus Martius. It was to be surrounded by a colonnade more than a mile long (1.4 kilometres). This would have exceeded Pompey's theatre in area. In 54 we find that the work had been put in hand and was already in progress. Yet the purchase of the land for the forum had already cost sixty millions. The total cost amounted to a hundred millions.

For all this Gaul had to bleed. That the rivalry between the great men of Rome required such sacrifices was probably of little concern to Caesar. He simply had to win in the building stakes. Suetonius reports that he often destroyed towns more for the booty they could yield than for any offence they were guilty of. All the offerings that had accumulated over the centuries in the shrines of Gaul he turned into money, to be thrown into the Roman power struggle. This was in addition to the normal booty, including the sale of slaves. In this way alone so much gold flowed into Rome that its price fell by twenty-five per cent.

Public building in Rome had long been bound up with politics, though normally it involved political contention between nobles, who were concerned more to outdo their rivals than to win popular support. Interest now concentrated on the rivalry between Pompey and Caesar and on the fact that both of them flaunted their superiority to the Senate. They set out to demonstrate it to all and thereby give permanence to the new power relations.

In the present situation this was particularly significant. For politics involved an exceptional number of ostentatious gestures. Just as the populace translated its discontent into symbolic popular anger and just as the leading senators insisted on proclaiming their impotence, their rights, and the will of the gods, so Pompey and Caesar were determined to flaunt their greatness in aesthetic terms.

It was symptomatic of the curious transitional situation in which Rome now found itself. Those who had long enjoyed authority were no longer strong enough to assert their political claims, and the new wielders of power still lacked the legitimacy they needed if theirs were to be recognized.

To distinguish in this way between the 'no longer' and the 'not yet' is of course a matter of historical judgement – in both senses of the word 'historical': in the first place, we can see things more clearly than those who lived at the time because we know what happened subsequently; in the second, we operate with different categories because we have learnt to think historically. Within a historical perspective every state of affairs is transitional, and doubly so in the critical phases of a system. Even if one is mistaken in the individual case, there is today a great readiness to distinguish between the 'no longer' and the 'not yet'. This was hardly possible in ancient times, and certainly not in the Roman republic. The traditional was regarded as enduringly normal and right, even though at the moment it appeared, for some incomprehensible reason, to be weak and ineffectual. When Cicero complains that *res publica amissa est*, this does not mean that the republic is lost in the sense that it has perished, but in the sense that something that was present is momentarily absent. The republic could be rediscovered, so to speak, and restored to its former strength. In periods of despair – about 55 – he feared that nothing could be done in his lifetime about the superiority of Pompey and his allies. And he was then fifty-one. It never occurred to him that the old republic could never arise again.

Yet even if the perspectives of posterity can be so clearly distinguished from those of contemporaries, the fundamental state of affairs – not just its temporary aspects – must have been perceptible. Seen from the present day, this fundamental state of affairs can be described roughly as follows: power and legitimacy were on opposite sides. And, remarkably, there was no way of changing this directly. Resolutions, laws and authorities could be engineered, but only by force, and any use of force constituted a setback in the quest for a recognized position of power. For Pompey this was all the more irksome, the more powerful he became. Having achieved all he could hope for, he was farther than ever from the secure position of superiority he sought. He could do a great deal, but nothing he did was of any avail, because he could not justify it to the citizenry. What was he to achieve through new laws? There was no question of his openly usurping power. Caesar and Crassus would not have allowed him to, and Pompey himself cannot have wished to. All he could do was to seek approval, somehow, in the Senate and the citizen body, so that he could add authority to the power he possessed. This accorded with Pompey's far from belligerent nature, with his vanity – and also with reason. He would have given much to win such approval.

Yet this was exactly what Cato and his allies were anxious to prevent. On their side were many others who were more or less resigned, ready to make concessions, but unwilling to acknowledge Pompey's superior power as legitimate. Even Cicero, who had allied himself with Pompey, found nothing to say in his favour except that he was powerful and that any resistance to him would prevent the return of peace. And however much large sections of 'good society' were troubled by the growing anarchy, corruption and violence, this was an argument against Pompey and his allies, not in their favour.

This strange situation poses a great problem for comparative historical study. A cursory answer might be that at that time the citizens of Rome were not as dependent on peace and order as they would have been in a specialized society, accustomed to a fully-fledged state apparatus. On the other hand, given their conviction that the traditional order was the only just one, they were certain that true normality was at some stage bound to return. And this certainty was embodied to an astonishing degree in the person of Cato.

The power that this man enjoyed – despite his inability to achieve anything in positive terms – is one of the strangest phenomena in the whole of history. He exercised his power much as Pompey and Caesar exercised theirs, chiefly as an individual, though he had a few notable allies. Yet he lacked the means of coercion that Pompey and Caesar possessed. His only strength was the general conviction that legitimacy lay only with what he deemed legitimate. Traditional authority, the responsibility of the Senate, still had such a strong hold on men's minds that no one could effectively contradict the one person who represented it so courageously, resolutely and censoriously. Whatever the senators resolved in defiance of Cato, it did not alter the general feeling – or even the feeling of the senators themselves – that he was right. Yet this was due not so much to his personality as to the fact that in a curious way he stood for the republic. And where legitimacy was at issue, no one could be against the republic.

In such a situation Pompey could make no headway. However many successes he might score, there was nothing he could do to win legitimacy for the position of First Man to which he laid claim. For there was no cause, no universal demand, that he could embrace in order to establish his own position in Rome. He could speak only for himself, his achievements and his ability, and point to all he had done, and wished to do, for the republic. He did this by building his theatre, a monument to himself and his achievements. Beyond this he could be only indirectly effective.

In other words, the real problem of the Roman republic was not a matter for political contention. There could be no question of legitimating Pompey's special position; what was judged right and proper in Rome was at no one's disposal. One could only demonstrate for or against it, and perhaps hope – and try to ensure – that the unsatisfactory conditions one lived in would soon be at an end.

Roman politics were thus condemned to a leaden paralysis. Cicero expressed this in 53 when he wrote to his young friend Curio, who was about to return to Rome after his quaestorship, 'You will find hardly any field in which to occupy yourself – everything is so paralysed and already almost extinguished.' Everyone held everyone else in check.

Pompey had much to do with this. He made sure that political life was largely blocked by tribunes of the people. The consular elections for 53, for instance, were obstructed for a whole year until finally, in the middle of 53, the Senate instructed Pompey to see that they were held. Everyone did as he pleased, in competition with everyone else. Electoral bribery was so rife that there was a shortage of money; the interest rate rose from four to eight per cent. Two of the candidates promised to give the century that voted first ten millions for their votes. They assured themselves of the consuls' support by promising them four millions if they failed to produce prominent witnesses to testify that non-existent resolutions had been framed for them to be provided with provinces. The courts still functioned, but acquitted most of the accused. Yet they did send one of Pompey's most important friends into exile.

Similarly the normal course of affairs and the formulation of political objectives were gravely impeded, as was the ability of the magistrates to act decisively. Pompey was clearly content for anarchy to reach such a pitch that the Senate would be forced to call him in to restore order. He hoped that by serving the Senate and showing himself trustworthy he could win its esteem.

June 54 saw the first calls for a dictator to be appointed, and the 'rumour of dictatorship' would not die down. There must have been some thought of giving Pompey a special commission, such as had once been given to Sulla, to restore public order.

Cicero too felt that some such authority was needed. In *De re publica* he wrote of the responsible statesman who in times of emergency comes to the aid of the endangered commonwealth as a dictator, or at any rate in an unconventional manner. He lacked Cato's conviction that the main priority was to defend the republic against any superior power, to counter any attack, so to speak, so that it could once more function legitimately. It may be that Cicero, not having taken up a firm position between Cato and Pompey, tried to consider the whole of the republic from a detached, theoretical standpoint. And in his soft, sensitive, nervous way he believed that intervention was urgently needed in order to put things to rights. But Cicero could offer only theoretical concepts. And in speaking of the responsible statesman he was thinking not of Pompey, but of himself. For to him there was nothing more important than philosophical education, and his model was Plato.

The crucial difficulty was that the traditional Roman dictatorship, with its comprehensive powers, had always presupposed that the institutions of the republic were strong enough and society confident enough to preclude the abuse of such powers. This was no longer certain. The very reason that would formerly have made it possible to think of one man as a saviour of the republic was now a reason for ruling out any one man. If he had no power he could achieve nothing. If he was powerful he was distrusted. And Rome was not faced, as it had been in Sulla's day, with the problem of providing a victor in the civil war with the office he sought.

Hence the call for a dictatorship only made for bad blood. Pompey himself – whose 'honest face', according to Sallust, concealed his 'shameless intentions' – publicly denied any suggestion that he sought such a position. But no one believed him. And no one wanted him. He only lost more credit. The *rumor dictatoris* was anathema to good society.

While futile disputes went on in Rome, Caesar continued to demonstrate his military achievements. The conquest of Britain was obviously planned for the following year. The Romans excitedly followed events; many young noblemen joined Caesar, hoping to serve as officers in the invasion. Many merchants too wished to accompany him.

Caesar seemed insatiable; he was clearly not satisfied with the conquest of Gaul. His successes led him to set himself further goals. Rome had never sought security from dangerous rivals through a balance of power, but through their removal. On her borders, however, Rome preferred to play off the various powers against one another. Caesar's policy was not guided by these principles, but by unbounded expansionism.

With a great armada of more than eight hundred ships, some of them owned by Roman merchants, Caesar crossed the channel. Throughout the winter the soldiers in Gaul had collaborated with engineers and workmen in building a new fleet. Rome's soldiers had to possess such skills. Caesar had decided on the shape and dimensions of the ships: they were to be lower, wider and more manoeuvrable. To prevent unrest on the continent, he took the leading Gauls with him, except for the few on whose loyalty he thought he could rely. The expeditionary force comprised five legions and two thousand horse. Three legions and an equal number of horse had to be left behind to guard the ports, to provide logistic support and to monitor events in Gaul.

This time the landing went off without difficulty. The sight of eight hundred ships appearing on the horizon almost simultaneously struck such terror into the Britons that they withdrew from the shore. When the armies met, however, the Britons caused the Romans great hardship through the extreme mobility of their tactics. Their chief weapons were the battle chariots, which at the beginning thundered across the battlefield and from which they first shot at the Romans and then attacked them at various points. The fighters would jump from the moving chariots, then be taken aboard again when necessary and driven to another part of the field. The ponderous Roman cohorts were no match for them. Even the cavalry was confused by the enemy's mobility. In the following night Caesar seems to have hastily devised a different tactic and instructed his officers accordingly. At any rate he succeeded in beating the British by means of greater flexibility and a fierce attack.

From now on the British evaded the Romans; they vanished from the scene, only to reappear somewhere by surprise and fall upon the Romans when a favourable opportunity presented itself. Caesar's army reached the Thames and crossed it. But victory eluded them. There was no booty; a few tribes surrendered because they were at enmity with Cassivellaunus, the leader chosen by the majority, but naturally they had to be treated with special leniency. Much time was lost when a storm seriously depleted the fleet and Caesar had to reinforce it. Unfavourable winds had in any case delayed the Romans' departure for Britain. The army lived in constant fear of being taken unawares and cut off from the continent. It had been able to fend off one attack on the harbour, but feared a second. And in the rear was a restless Gaul, which had just suffered a bad harvest. The commander himself was alarmed by Pompey's plans for a dictatorship. He therefore took advantage of a recent military success and a request for peace from Cassivellaunus in order to break off the campaign. He imposed a tax on the British, as if they were already inhabitants of a Roman province. They certainly never paid it. On about 20 September the Roman expeditionary force returned to Gaul.

Not only had Caesar lost a campaign and a year of valuable time, but a heavy blow had been dealt to his certainty of victory and continued expansion. He may have hoped to reach his goal in the following year, but he may also have decided to turn his attention first to Germany or Illyria. He may have been driven on by impatience, a craving for spectacular successes, and the hope of booty. In fact he had overreached himself. It is not clear whether he already knew this. If he did not, the situation in Gaul soon made it clear.

He now received word that Julia, his only daughter, had died in childbirth. It is clear that he had been very close to her and was deeply affected by her death. We know hardly anything about her. She was probably one of the few people in whom he could confide – perhaps the most important, or indeed the only one. For we hear nothing of Caesar's having a close friend. True, he had political allies, and above all faithful followers and supporters. But whatever details we have about these suggest that they were hardly privy to his concerns and intentions, but were merely given political instructions. Inwardly Caesar was probably always isolated, and even more so after the death of his daughter. Who, after all, was in a position to understand him?

It is reported that Caesar hardly showed his grief. He may have been inwardly shattered, but this must not appear on the surface. After three days he resumed his duties as commander. A later author reports that he conquered his pain as quickly as he was accustomed to doing everything. Caesar's contemporary Cicero did not share this view. And insofar as it implies anything more than outward discipline, it is probably just a cliché.

Pompey wished to bury Julia on one of his estates. But 'the people' – whoever that was – removed her body from the funeral ceremonies and cremated it on the Campus Martius, where her remains were interred as befitted a highly deserving citizen. Caesar responded, contrary to convention, by promising gladiatorial games and a feast for his daughter.

His grief was compounded by a political problem. Caesar's understanding with Pompey depended largely on Julia. They were united in their love for her, and she seems to have removed many conflicts and difficulties. Hence, Caesar's political isolation increased too.

This was particularly unfortunate in view of the political anarchy in Rome and Pompey's efforts to secure a dictatorship. It probably seemed as though the weakness of the Senate and the magistrates, which Caesar had been so keen to foster, could redound to the advantage of his ally and rival. Perhaps Caesar himself was surprised by this sudden turn. We do not know what he hoped for or sought to achieve in this situation. He had scarcely any leverage against Pompey, especially as Crassus was absent. Presumably he decided to demand certain returns for any progress Pompey might make. He also attempted, probably in 53, to interest Pompey in a new matrimonial union. His great-niece Octavia, the sister of the later Augustus, should take over Julia's role. He himself wished to divorce his wife and marry Pompey's daughter. But Pompey refused.

In the autumn of 54 a rebellion erupted in Gaul. The Gauls had become painfully conscious of what had befallen them so surprisingly and what it meant to be no longer free – having to pay taxes, provide troops, and put up with interference in their internal affairs. Above all, Caesar writes, those tribes who excelled all the others in bravery 'could not bear to submit to Roman rule'. They began to understand that they had a common interest in opposing Rome.

Shortage of grain forced Caesar to find winter quarters for the legions in different parts of Gaul. He himself intended to repair to northern Italy as usual. Then the Carnutes murdered the king he had appointed. The Eburones, under their leader Ambiorix, lured the one-and-a-half legions stationed on their territory into a trap and wiped out all but a few men. This was an unprecedented loss for Caesar. On hearing the news, he resolved not to have his hair or his beard cut until he had taken vengeance. Then the Eburones, together with the Nervii and the Atuatuci, stormed the Roman camp in the territory of the Nervii, employing to perfection the siege technique they had learnt from the Romans. Their towers, their sheds to protect attacking soldiers, and their siege-works later earned Caesar's admiration. On launching an assault they sent burning javelins into the huts in the Roman camp; these then went up in flames. Meanwhile they attacked. In heat and thick smoke, and under a hail of missiles, the Romans had to defend themselves against a superior force. Yet they continued to resist tenaciously until Caesar came to their aid; he found not even one in ten unwounded.

As a prelude to his campaign Caesar had outwitted the Gauls, who had observed his approach and advanced with greatly superior forces. He had restricted the proportions of his camp in order to appear even weaker than he was. He had ordered the cavalry to yield in the face of the enemy attack. While fortifying the camp the soldiers had to dash to and fro as though driven by fear. The Gauls believed they had already won and ventured forwards into unfavourable terrain. Caesar then defeated them by ordering his men to rush out of all the gates. In this way the enemy was beaten. For the time being the rebellion collapsed.

But Caesar could not risk leaving Gaul. He wintered near Samarobriva (now Amiens) with three legions. Here the Romans had large arsenals with all the heavy military equipment, as well as the hostages, the archive, and a great store of grain. Throughout the winter he received disturbing reports. He summoned the leading men of all the tribes and managed to persuade many to remain loyal to Rome. But some did not come, and almost all who did were suspicious of Caesar.

Before the winter was over the Treveri rebelled. Their prince, Indutiomarus, convened an 'armed diet'. This meant the opening of hostilities. Every able-bodied man immediately had to appear armed; the last to arrive was tortured to death by the others. The Roman commander, however, whose camp Indutiomarus intended to storm, used the same tactic as Caesar: he lured the enemy on by simulating fear and was able to beat them. The prince was captured and killed, and his head brought proudly into the Roman camp. 'Then Caesar had somewhat more peace in Gaul.'

In the spring of 53, 'expecting a major uprising', Caesar had more troops levied. Two legions were enrolled; a third he borrowed from Pompey. This doubly made up for the troops he had lost. The Gauls were to see how quickly Rome could raise a new army. While winter was still on he made a surprise strike against the Nervii. Before they could assemble, Caesar had taken control of large numbers of men and cattle and laid waste the fields, so that they had to surrender. He similarly took the Senones and Carnutes unawares, then turned against Ambiorix and his allies. At first the Menapii were taken by surprise, and then the Treveri were defeated.

Because the Germans had sent auxiliary troops to the rebels, Caesar crossed the Rhine a second time. However, as the Suebi, whom he intended to attack, had withdrawn to a great distance, he quickly turned back; this time, however, he destroyed only the first two hundred feet of his bridge, on the right bank, leaving the rest standing as a permanent threat. At the end of it he erected a turret four storeys high; he left twelve cohorts (about six thousand men) behind strong fortifications.

Immediately afterwards he settled his score with the Eburones. It all went so fast that he caught them unprepared. Ambiorix fled, and the members of his tribe took refuge throughout the country. Caesar then invited the surrounding tribes to plunder the land. The Eburones, he writes, were 'to be destroyed root and branch for their crime', and he did not want this deed done by his own soldiers.

Thus were the fallen avenged. One campaign followed another; the Gauls were obviously still not sufficiently acquainted with Caesar. Also, they probably could not gather their forces properly, could not supply them as well, and were too naive to understand the full subtlety of Caesar's lightning strikes.

In the autumn Caesar held a diet at which an inquiry was carried out against the Senones and Carnutes. He condemned the leader of their rebellion; he was beaten to death and then beheaded. Having sent the legions to their winter quarters and provided them with supplies, Caesar returned to Italy. It is not clear whether he was really sure that Gaul would remain at peace. But at any rate he had to turn his attention to Rome.

The period of the great successes was over. All that was left to do in Gaul was to secure what had been won, and this was hard enough. He could scarcely hope for further conquests. He had only three more years at his disposal. Meanwhile Crassus had perished after a heavy defeat at Carrhae in northern Mesopotamia (now Haran, on the Turkish-Syrian border). The Parthians had responded to his attack by inflicting an annihilating defeat on his army. The solemn curses uttered as he set out from Rome had been fulfilled.

Caesar had lost his greatest ally. He had always needed Crassus as a counterpoise to Pompey. Because relations between the two were tense, Caesar had power over them both. Had Crassus remained in Rome, Caesar would have been much better able to frustrate Pompey's plan to become dictator; even from the east, however, Crassus could have helped maintain the precarious balance among the three. Now Pompey could exploit any weakness of Caesar's for his own benefit. And the opportunity soon presented itself.

At the beginning of 52 the internal political struggles reached a new pitch. Again no magistrates had been elected. There was not even an *interrex*, because friends of Pompey had prevented an appointment. The election campaigns were this time especially fierce, indeed violent, because Clodius and Milo were candidates, one for the praetorship, the other for the consulship; both made use of their gangs of thugs. Milo's opponents had also taken on mercenary troops.

Finally, on 18 January, Clodius was wounded in an armed clash and then murdered on Milo's orders. Feelings ran high among his large following in the city. Two tribunes of the people had his body, covered in blood, carried to the forum and laid in state on the speaker's platform. They lamented his murder; the crowd, led by one of the best-known of Clodius' mobsters, then took the body into the Senate House, the Curia Hostilia. There they piled up the benches, collected whatever wood was to be found nearby – the podia and benches of the tribunals, the booksellers' tables on the Argiletum, together with the books. Clodius' body was burnt, and the building with it.

It was a beacon fire. The Senate served as a funeral pyre for the popular leader. The populace indulged in a symbolic orgy of destruction, demonstrating at once its power and its impotence. In this long moment of intoxication, made all the more overwhelming as the fire spread to the adjacent Basilica Porcia, it savoured its dominion, which could only be negative.

Because the Senate had been unable to quell the previous disorder, it went up in flames when the violence culminated in the murder of the popular gang-leader. The house from which Rome and the whole world had been ruled was on fire. Rome had been living amid the tension of force and counterforce. Now that the tensions were largely relaxed, the violence was unleashed, and it was virtually inevitable that it should concentrate on the old centre and destroy it.

At the same time the murder of Clodius and the burning of the Senate brought about an abrupt change in Roman politics. There could no longer be any doubt but that the Senate must have recourse to the power and ability of Pompey in order to restore order. This, however, led to a marked strengthening of the Senate régime.

Even Cato thought any government preferable to anarchy. Before the end of January a *senatus consultum ultimum* was passed, charging Pompey with the restoration of public order. As this could not be done at a stroke, he began by levying large numbers of troops: the

safety of Rome was to be secured by military means. Long negotiations ensued. Pompey wanted to be given the dictatorship he had
sought for so long, but the group led by Cato was opposed to it. For
a time Caesar's supporters proposed that the two allies should be
made joint consuls. They can hardly have meant this seriously, since
all Caesar's conquests would have been in jeopardy if he had left Gaul
and returned to Rome. Indeed the Gauls, who were already planning
a revolt, speculated on his doing so. He might have been able to
continue the war as consul. But there was a much more important
consideration: he had to ensure that when his governorship came to
an end he did not become a private citizen, exposed to the risk of
prosecution, but moved directly to a consulship. The right to pursue
his candidacy *in absentia* was thus the compensation that he finally
negotiated for himself. In return he agreed that the highest
magistracy should once more go to Pompey.

The form in which this was to be effected was agreed chiefly with
Cato and his friends. In view of the clash of interests it was decided
that Pompey should be elected sole consul. He was thus accorded
exceptional scope for action – and a great honour – but it fell short of
dictatorial authority. Cato could not bring himself to propose this,
but Bibulus did, whereupon Cato declared that as it had been
proposed he would assent to it. At last, almost exactly two months
after the murder of Clodius, the way was clear for Pompey's election.

He had attained his goal. His special status was at last
acknowledged by the Senate majority. He was resolved to do all in
his power to restore order and the rule of law, and to act wherever
possible with the concurrence of the Senate. He thanked Cato and
decided to make him his adviser. At the same time the senators
naturally hoped that they could eventually draw Pompey so far over
to their side as to be able to count on him to oppose Caesar. They
would then have to see whether they still needed him.

The Senate had already decided to entrust the rebuilding of the
Curia to Sulla's son Faustus. It was to be called the Curia Cornelia,
after Cornelius Sulla. This symbolized the hopes and pretensions of
the house: in opposition to Clodius' supporters, the senators wished
to renew its link with the victor in the civil war who had re-
established a solid senatorial régime. Faustus happened to be married
to Pompey's daughter. The session at which the decision was made
took place in Pompey's theatre. Beforehand, however, the room in

which it was to be held had to be consecrated in a religious ceremony, for the validity of the Senate's decisions depended on the proper venue. The Senate thus yielded to Pompey by meeting in his curia. And he was glad to join it in renewing the link with Sulla. This was a new start.

Despite Pompey's initial concession to Caesar, the accord between Pompey and the Senate greatly added to the governor's troubles. For a long time Pompey continued to attach great importanace to maintaining good relations with Caesar – which strengthened his hand in dealing with the Senate – but he was not necessarily reliant on them any longer.

Caesar's position was extremely difficult, if not desperate. For six years he had devoted immense effort to the waging of war. All the time he had had to keep his eye on Rome's internal politics and try to influence them. He had long since conquered Gaul and initiated further conquests. In Rome he had won significant influence in the face of great difficulties. Now everything he had built up seemed to be collapsing. While he endeavoured to salvage what he could in Rome, a new rebellion had broken out in the whole of Gaul, this time on a large scale.

Since the end of 54 his army had incessantly trekked back and forth. It is true that wherever Caesar went he was still able to beat his opponents. But this was often due solely to the extraordinary speed with which he could mount a surprise attack or to the extreme concentration he brought to the task of outwitting an opponent far superior in numbers, a task in which he relied on his intelligence and the bravery and discipline of his soldiers. He now had the initiative only to a limited extent.

Above all there was no knowing how the struggle would end. Although Caesar was able to set some warning examples, he had to keep within bounds, lest he provoked a general revolt. Time and again he was obliged to show clemency to rebels who surrendered to him, though it was often quite patent that this amounted to no more than a pause in the fighting.

It may be wondered how Caesar could endure these huge burdens. He was now forty-seven and still had to exert himself endlessly. For a long time he had been able to imagine that he had reached his goal in Gaul and was nearing it in Rome. Now it all seemed to be slipping away from him. The Gauls were jubilant, knowing of his problems in Rome and believing that he would no longer find it so easy to turn his attention away from Italy. He may often have been close to despair, able to assert himself only by dint of extreme efforts. At this time he seems to have been embittered and dug himself deeper and deeper into the world he had built up around himself, single-handed, during his governorship – the world of his army and his activity, the world of his cause, and probably also of his thoughts.

Caesar's Gallic World

THE PASSION FOR ACHIEVEMENT

FULFILLED · THE ART OF COMMAND ·

CAESAR AND HIS SOLDIERS ·

DIPLOMACY IN GAUL · ACTION ON THE

GRAND SCALE · 'THE SELF-ABSORBED

GENIUS' · THE HUMAN COST ·

INCOMPREHENSION AND SUSPICION IN

ROME

CAESAR'S LIFE as governor of Gaul was exceptionally intense and fulfilling. He locked himself into a gigantic task that he was nevertheless able to master, in which he could prove himself, grow in stature, and become what he had long intended to be. Although he had to contend with immense problems, he was able to overcome them by adopting a form of action and a way of life that matched, confirmed and enhanced his own view of himself. He created a world that was wholly his, even if not everything in it obeyed him. There were no doubt times when the difficulties and hardships he faced brought him close to despair, when he was tempted to abandon everything and his energy was dissipated in restlessness. All this probably prevented his life from being happy. It may often have been a torment. Time and again he may have found himself staring into the abyss. Yet he went on risking fresh ventures, and the fact that he always won, against great odds, must have made him feel all the more powerful. Indeed, however he perceived it, there was something unique about his ability to develop his personality and live life to the full in a world of his own. Here he had no need to parcel out his rich talents between action and negation, self-confidence and envy, triumph and complaint. He may well have complained and been assailed by envy, and he was certainly often negative; but this was merely the reverse side of achievement, as fatigue is the reverse side of exertion. It did not paralyse him or block his activity; he was not reduced to playing the role of the witty, cynical ex-consul for which, in other circumstances, he would have had to settle. Instead, he was able to realize almost everything that lay within him.

What he had taken on was a self-imposed task, his own cause. It involved both freedom and obligation, for it soon developed its own laws. It afforded Caesar great opportunities for action – some there for the taking, others yet to be won. His success amply rewarded all his efforts and sacrifices, even if at first he hoped that it would come more easily, so that he could go on to even greater achievements. And everything depended almost wholly on him. Much as he relied on others to do his bidding, to work for him, fight for him and devote themselves to his cause, he ultimately owed everything he accomplished to himself, to his art as a commander, a leader of men, and a diplomat, to his untiring energy – and to his good fortune.

It is not easy to say what constituted Caesar's art as a commander. At the very least he had a superb talent for organization and planning, matched with extraordinary alertness and circumspection.

Cicero called him 'terribly active, swift and prudent'. The Greek historian Cassius Dio found him ingenious in discerning what was necessary, interpreting it convincingly, and addressing it with the utmost skill. In his own writings Caesar perhaps somewhat exaggerated this faculty, but to a large extent he must have been able to foresee the host of possibilities inherent in any situation and to arm himself against them. He was well acquainted with the power of chance and sufficiently imaginative to envisage whatever might happen at any time. He planned everything very precisely. It was one of his maxims that not the slightest scope should be allowed to chance. It was a common human failing, he observed, that the surprising inspires in us greater trust – or greater alarm – than the known. He tried to avoid this failing in himself and to exploit it in others. The intelligence that he brought to the waging of war made him far superior to the Gauls, who had great courage, but lacked cunning and were not sufficiently detached from events to appreciate the multitude of possibilities and so avoid being taken in by the first impression that Caesar conveyed to them.

At the same time he must have possessed the courage, fine judgement, resolution, and coolness that Clausewitz identified as the distinctive qualities of the 'martial genius' – the ability to 'hit at once upon a truth that is invisible to the normal mind or becomes visible only after long consideration and reflection'; the resolution that can become 'a habit of mind'; courage not only 'in the face of physical danger, . . . but in the face of responsibility, thus, to some extent, in the face of spiritual danger'; presence of mind, fortitude, force of temperament and, not least, precision in the use of subordinates. In Caesar these qualities were compounded by his inventiveness in devising and exploring new techniques. Finally, he was so observant, so versatile, so quick to learn, that he constantly reviewed the course of events and made fundamental changes in his strategy and tactics. This is attested more than once. He seems to have raised Roman military technique, tactics and strategy to a new plane. And he was obviously able to translate knowledge into practice. For 'the mental reaction, the ever-changing shape of things causes . . . the active man to carry the whole mental apparatus of his knowledge within himself, so that everywhere, at every pulse-beat, he can produce the necessary decision from within himself. Knowledge must therefore be translated into true ability by being perfectly assimilated into his own mind and life' (Clausewitz).

The course of the war suggests that even when Caesar was dealing with single tribes he had to secure his operations on all sides and keep an eye on large parts of Gaul. There is no doubt that in his military planning he showed himself an outstanding general, able to make extensive military dispositions and deploy vast resources. This is confirmed in every case for which we have fairly precise information – in his large-scale operations against Ambiorix in 53 or the comprehensive strategy he later employed during the rebellion of Vercingetorix. A vital role was played by his famous 'celerity', the surprise attack, the *blitzkrieg*, which enabled him to attain his objectives throughout a wide theatre of operations.

Warfare of this kind must in the long run have posed unusual difficulties. Owing to the Gauls' love of liberty, Caesar had to reckon with facing brave and dangerous enemies in many places at once. It was often impossible to beat them – if they broke off the fighting early and sued for peace – for Caesar was usually in no position to refuse them peace terms. Yet there was no guarantee that they would not strike again at the first opportunity. True, he always took hostages. It is curious that he never says what he did with them. Often enough he would have had reason to execute them. On occasion he may even have done so, but if he did it brought him little benefit.

Such difficulties made Caesar's war different from those of, say, Alexander or Pompey, in which a king was usually the focus of resistance; when he had been roundly beaten, his kingdom could be subjugated. It differed also from many other wars fought by the Romans, in which one power was pitted against another. Caesar had to contend with numerous powers, a turbulent country and, it must be repeated, valiant opponents. Given such circumstances, the way in which Caesar conquered Gaul – with superiority, boldness, swiftness, energy and perseverance – makes him one of the great commanders of all time. It was not for nothing that he earned the boundless admiration of Frederick the Great, Napoleon and many others.

Another quality that contributed to Caesar's achievements was his brilliance as a leader of men. His army was a superb military machine, his soldiers brave, tenacious and experienced. In an emergency they could do whatever was required without orders. But they could not do this from the start: they had to learn. Some things Caesar taught them himself by drilling them, discussing events and the nature of the enemy with them, and continually instructing them in new tactics and new techniques. Yet he must also have imbued them with his own will, in particular by encouraging them, and more generally by arousing in them a fierce soldierly pride, a passion for the task in hand. This is attested by many instances of admirable devotion and loyalty.

Caesar made heavy demands on his army. To quote Clausewitz again: by pursuing an 'activity with extreme effort, the fighting man comes to know his strengths. The more accustomed a commander is to making demands on his soldiers, the surer he is of having them fulfilled. The soldier is as proud of overcoming hardship as he is of surmounting danger.' Caesar shared this pride with his soldiers; hence, the heavy fighting, the long marches, the endless digging that was involved in setting up camp and laying sieges, were not simply laborious: they engendered a consciousness of ability that enhanced the honour of the army.

To quote Clausewitz for the last time: 'Of all the glorious feelings that fill the human breast in the heat of battle, none is so potent or constant as the thirst for fame and honour, to which the German language does scant justice by reducing it to ambition and a craving for renown, which are two unworthy subsidiary notions . . . According to their origin these feelings must surely be reckoned among the noblest in human nature, and in war they are the true breath of life that endows the monstrous body with a soul.' All other motives are important at times, but 'unlike ambition, they do not make the individual martial act the property of the leader' – and, we might add, of the army. Since Caesar's soldiers made the martial achievement their own, it became the yardstick by which they wished to be measured. Whatever value men attach to achievement and success was sought by the soldiers in the profession of arms.

Since Caesar lived with his army, at least throughout the summer, he knew not only the senior officers, but the subalterns and non-commissioned officers too. He was continually on the move inspecting his troops. Despite his far from robust constitution, he shared in all their dangers, exertions and privations. He is even said to have tried to overcome his indifferent health by long marches, frugal rations, and exposure to the elements, and by making harsh demands on his physique. He marched at the head of his army in the blistering heat, and during the fighting he would, if necessary, take up a position in the front ranks; if his soldiers fell back he would bring them to a halt by stopping them individually and turning them back to face the enemy. For himself he claimed no privileges, but only higher obligations, and of course superior insight. Yet the soldiers were allowed to share this insight. He did not simply give them orders; he also knew how to convince them. He informed himself about the situation and, if he deemed it important, told the men of his reflections. For all the assurance with which he conducted himself and issued his commands, he treated them as comrades. They felt that he knew them and that they could rely on him.

Caesar's leadership radiated strength and confidence. The men knew that he would expect nothing unnecessary of them. By his own example, by the compelling and convincing manner in which he demanded the highest achievements, he seems to have inspired them to do everything to prove their worth to him. Caesar reports how one of his legates encouraged his men: 'Show the same bravery under our leadership as you have so often shown your commander. Imagine that he is here, watching us all.' And in an emergency, when things became difficult, Caesar's confidence was sufficiently strong to re-store morale. His 'vigour and wonderful cheerfulness' communicated themselves to the soldiers. They too must have believed in Caesar's fortune.

Caesar addressed his soldiers with sovereign assurance. He never curried favour with them; the more reliant he was on them, the less he let them see it. At least this is what we read in his account; moreover, the circumstances of his campaigns and the successes he achieved suggest that this picture is not wholly false. Once, when the soldiers were suffering grievously from hunger, he offered to call off the enterprise if it was proving too hard for them. Subtly and shamelessly he was challenging their pride, and they reacted to his offer with indignation. Later, during the civil war, Pompey saw the bread made from plants on which Caesar's men subsisted and ob-served that he seemed to be dealing with wild animals. In delicate situations Caesar would exploit the natural competitiveness and *esprit de corps* of the individual units by appealing to the most tried and tested in order to inspire the others. He probably never forgot to honour the brave. He was generous with rewards and promotions. He is also said to have given his soldiers weapons with silver and gold mountings, for he was concerned about their appearance. He also wished to ensure that they looked after their weapons and did not lose them during the fighting.

He attached great importance to regular supplies and provided for his men to the best of his ability. When grain was plentiful he distributed it without regard to the standard rations. He was both strict and indulgent. In dealing with mutinies and desertions he was merciless. When close to the enemy he insisted on absolute discipline and constant preparedness. He did not, for instance, announce in advance when hostilities would commence or the army would move off, but issued his orders without warning. He often set his soldiers on the march without reason, preferably during rain or on rest days. Yet he was also prepared to turn a blind eye. 'Sometimes, after a great battle, he would release the soldiers from all duties and allow them to stroll about and devote themselves to whatever pleasure they chose; he boasted that his soldiers could fight just as well when perfumed' (Suetonius). He doubled their pay and saw that they had a large share in the booty. Sometimes he gave every man a slave from the booty. This increased their common interest in war and victory, but it was by no means the only bond between Caesar and his army.

'These legions', he is reported to have said, 'can tear down the sky.' At the time he was in southern Spain, 'where the pillars of Hercules marked the end of the world and where the way to the sky, which rested on the shoulders of Atlas, was not far off' (Sattler).

Apart from strategy and leadership, Caesar had to attend to countless political and diplomatic affairs in order to ensure his rule over Gaul and safeguard his military operations. He relied partly on trusted individuals, whom he appointed as kings or promoted in other ways, and partly on the aristocracy as a whole, which he strengthened against individual magnates. He also exploited the hegemony of certain tribes, especially the Haedui, and consolidated their position in order to gain their allegiance and use them to control other tribes. His system rested essentially on personal relations with individuals or small groups. This was consistent with his view of politics, which he saw chiefly in terms of personal connections. It was necessary to distinguish between friends and enemies and to court various powerful, or potentially powerful, figures. Such connections would then supplement the power of the legions; a fine network was stretched over a country that measured more than nine hundred kilometres from north to south. The history of the Gallic campaigns shows that it contributed greatly to Caesar's success, provided that the general mood did not turn against him in large parts of Gaul. When this happened, anti-Roman sentiment took hold of most of his friends, unless they fled or were eliminated. This system of personal connections sufficed for a time to enable him to assert his will, but not to ensure permanent recognition of Roman rule among the Gallic tribes. This would in any case have been difficult, even if Caesar had been more adept at setting up solid institutional links with the Gauls.

Such a system required a high degree of vigilance. Caesar needed a reliable intelligence service to keep him abreast of developments, for not everyone could be trusted. Everywhere he had to mediate, conciliate, explain and give instructions, fostering loyalty by rewarding service with service, and reacting to disloyalty without necessarily being able to punish it. In this sphere too he must have operated with energy and flair.

In this extraordinarily active life, the demonic force that supposedly lay behind Caesar's serenity could be directed outwards. It has been surmised that while campaigning he became so used to giving orders that he could no longer carry conviction in Rome. However, it was not as simple as that: he became accustomed to a world that allowed him – indeed obliged him – to develop his potential without constraint. Apart from giving orders, he had to plan, acquaint himself with foreign conditions, make dispositions, provide for his soldiers, cultivate alliances, react to new situations, and attend to much else besides.

In all these activities Caesar could operate on a grand scale and was free to determine and dictate what happened – without being inhibited, without having to immerse himself in dreary, time-consuming negotiations or seek others' approval for his actions.

True, he constantly had to apply himself to winning the hearts of his soldiers: this was after all how he led them. True, he had to engage in many negotiations in order to gain the allegiance of the Gallic nobility and assure himself, by adroit diplomacy, of the loyalty of individual tribes. And even if he did not need the approval of his subordinates, he may often have found it necessary to convince them that his instructions made sense.

Yet this was all part of his task, and it must have fascinated him. Relations with the soldiers were uncomplicated, and those with his subordinates must on the whole have gone according to his wishes. Negotiating with the Gauls gave him the chance to exercise his diplomatic skills. Even if these and other activities were means to an end – the conquest and securing of Gaul and, indirectly, the advancement of Caesar's inordinately ambitious career – they also accorded with the ideal of the highly active life he had chosen: he controlled everything that happened, at least on the Roman side. He set the rhythm and tempo, in spite of many extraneous demands and periodic initiatives launched by the enemy. On the whole the relation between means and ends must have been such that, while the means might sometimes impose heavy burdens, these were not so heavy as to jeopardize the ends. Caesar was not invariably successful, but his plans were seldom totally frustrated; and when they were he quickly saw the need for a fresh approach. He probably never tired of the effort. He was after all confronted with all the gravity of a major war and intent upon subjugating a country far larger than had been conquered by any previous Roman general. He had ample scope to prove himself – or to fail. It all depended on him. It was an immense challenge.

The strength to meet it may have derived partly from the pleasure and pride he took in his many achievements, but it was also a function of the challenge itself. Caesar's activity during these years was at any rate motivated not solely from within, but also by external circumstances, to which his inner energies responded and with which they interlocked.

An additional motive was the need to gain influence in the internal politics of Rome. Caesar found himself facing extraordinary demands, and he seems to have been equal to all of them. He radiated an immense dynamism. From Cicero we learn that he was hardly accessible even to members of his own staff – not because he was arrogant, but because he was so much in demand.

A man who could achieve so much stood head and shoulders above all the others. Only Pompey approached him in stature. We may presume that his endeavours really were inspired, as he would have us believe, by a consciousness of serving his native city. He saw himself as Rome's representative, always jealous of her honour. Rome had to appreciate this.

In view of his notable conquests and his involvement in internal politics, the magnitude of Caesar's accomplishments far exceeded anything expected of even the most outstanding and responsible Roman generals and governors. He led his life at an exceptional level of intensity because he had to accomplish much more than anyone else if he was to win recognition in Rome, and because he was fighting a war that he had started deliberately. After all, he saw his strenuous expansionist activity as proof his own worth. The exceptionally high standards of achievement that he had set himself not only contrasted with the general ineffectuality prevalent in Rome, but were the basis of the claim to distinction with which he confronted his peers.

There was nothing in Rome's internal politics to compare with Caesar's activity. They centred upon trivialities; one had to contend with the ineffable reservations and obstructions of one's peers and could do nothing important without first convincing the Senate or the popular assembly – a Senate that could now be convinced of hardly anything, and a popular assembly that could receive proposals only from a magistrate. There was scarcely any scope for action, scarcely any serious decisions to be made. Any significant enterprises were stifled before they could be initiated. Futility and tedium prevailed. It was virtually impossible to do anything for Rome. Most of the time was spent jockeying for positions that were quite meaningless to those not directly involved. Infinite effort and ingenuity went into accomplishing very little. It was doubtful whether the ends justified the means. One could either join in the game or suffer from it. There was nothing here of the grandeur that was Rome – nothing that could impress posterity.

It was thus not only his role as commander, but the challenge to his whole personality, that Caesar enjoyed in the fresh air of Gaul. Here he could breathe freely and develop his potential to the full; here his life took on a higher meaning. He now had the chance for achievement that he had craved.

It was the consummation of the self-confidence and self-absorption that Strasburger attributes to Caesar: 'He obviously believed in a natural identity between what was desirable for him and what was

desirable for the world, so that, while conflicts between persons
might arise, there could be none between his political (and military)
planning and the moral law.' In this context Strasburger speaks of the
'unparalleled immorality . . . with which Caesar not only acted, but
recorded his actions'. For he 'wrote about these terrible events with
an elation that is tolerable only if one credits him with the higher
innocence of utter demonic possession'.

Yet the enormity of these events was regarded at the time as
concomitant with war. The fact that Caesar had started the war was
reprehensible according to the standards of the age. He may have
been troubled, even haunted, by the thought that this war, started for
the sake of his own fame and perhaps also for the glory of Rome,
claimed hundreds of thousands of victims – to say nothing of all the
devastation. Who can say that he was not profoundly affected by the
questions, reproaches and accusations, voiced or unvoiced, that were
addressed to him, and by the untold sufferings of a brave and noble
people? After all, he had not only daily, but nightly, dealings with
them; his soldiers sang songs about his amorous liaisons, and one
Lingonian later boasted of being Caesar's offspring. All we can say is
that he did not record his feelings. At most they can be discerned in
the care he took to justify his initial involvement in Gallic affairs. In
any case, such matters had no place in a history of the war. And if
they troubled him, he seems to have come to terms with them
privately. He was inclined to be reticent and unforthcoming about his
motives. He may have consoled himself by reflecting that he could
not foresee the extent and duration of the war. The thought of the
consequences of the civil war that tormented him at the Rubicon
might have forced itself upon him as a result of his experience in
Gaul. He tried to be generous whenever he could. The atrocities he
ordered were exceptional and can probably be explained by Roman
military practice. But above all he himself became a prisoner of the
war, as week succeeded week and month succeeded month: he can
have had little time for detached reflection. According to the Roman
view, a war once begun could not be abandoned until victory was
won. Rome's honour was at stake. And Caesar had to prove himself.
The way in which he did this is reflected in the elation of his account.

In proving himself, Caesar cultivated aspects of conduct that found
little favour in contemporary Rome and neglected others that the
Romans would have deemed necessary. By the standards of the

Roman oligarchy he acted in a highhanded and inconsiderate manner – not as a normal military commander, but as an autocrat, not as a plenipotentiary of the Roman republic, but as the master of a whole world.

However well he served Rome's interests – or rather however efficiently he performed the task he had set himself – his ego always occupied centre stage. He raised himself above all the customary ties and connections in a way that the Roman aristocracy found deeply suspicious.

At this time he is reported to have said more than once that it would be harder to demote him, the first citizen, from first to second place than from the second to the last: in other words, he was now so intent upon occupying the first place that, were he to lose it, everything would at once cease to be important to him.

Since he conducted his governorship for many years with such intensity, the life he led in Gaul gradually became his true element and the Gallic theatre of war his world. It was a large and self-contained sphere of activity, in which his way of conducting himself became more and more second nature. Surrounded by admiration, dependency, officiousness and respect – and supported by fortune – he did not always need to impose his will: he could let it impose itself, and at times show great generosity. He probably seldom met with opposition, but when he did he would not tolerate it. We may cite the case of Catullus, who criticized him in a satirical poem, saying that Mamurra, Caesar's adjutant, now possessed what had once belonged to Gaul and far-off Britain. Mamurra lived in great style, having built himself a palace in Rome that was long famous for having cost so much. Mamurra had also stolen the poet's mistress. Catullus therefore asked Caesar, 'Was it for this, great captain, that you visited the farthest isle in the west?' He called Caesar a libertine, a man without shame, a glutton, an inveterate gambler. He also alleged an erotic relationship with Mamurra: 'Schooled in one bed, two lechers equally lustful, companions and rivals with the girls.' The proconsul complained to the poet's father, a prominent citizen of Verona, which lay in Caesar's province. The poet had to apologize. Having done so, he was at once restored to Caesar's favour and invited to dinner the same day; and Caesar maintained his old bonds of hospitality with the poet's father.

The aura of habitual success that formed around Caesar shielded him from the questions, doubts and presumptions of his peers, from their claims and expectations. He could no longer orient himself to anything but the bald ethic of achievement to which he was dedicated. This was the source of his greatness and his limitations.

Caesar had never been properly understood in Rome. Things now became worse. At the beginning of the civil war Cicero called him a *téras*, a wonderful, frightening, monstrous and inscrutable phenomenon of a higher order – so alien had he become to his peers. Yet it was in such alienation that he had found his freedom.

There was admittedly something very Roman about Caesar's relentless activity. Idleness always gave the Romans a bad conscience. When Caesar described his goal as 'leading the way with deeds' (*operibus anteire*), this was merely a Roman version of the old Homeric ideal to which Cicero also subscribed: 'to be the best and excel others'. Cicero even justified his own philosophical writings by claiming that the cultivation of Greek philosophy in Rome was an act of public service. Caesar took a similar view when in 54, on the way from northern Italy to Gaul, he wrote his book *De analogia*, a plea for pure and precise Latin.

But Caesar's activity was unquestionably excessive and verged on the unscrupulous. For the achievement ethic had always existed side by side with that of discipline, class solidarity and equality, which Caesar had completely abandoned.

In Caesar's other world, the world of Rome, he succeeded in winning many hearts; he bestowed and received favours, formulated wishes and demands, and met with compliance. But he convinced no one: he was not trusted and not needed. He could find no foothold.

Caesar certainly possessed extreme charm. Cicero praises his fine, noble, cultured manner (*humanitas*) and his obliging nature. When Cicero asked him to take one of his friends on to his staff, Caesar replied that he would make him king of Gaul and that Cicero should immediately send him another. Whereupon Cicero recommended a lawyer, Trebatius Testa, and received the witty response that this suited Caesar very well, for not one of his senior officers could draw up a bond (an equivocal allusion to the parlous state of their finances). He dedicated his *De analogia* to Cicero, praising him as the father and master of Latin prose style and stating that he had done a great service to the fame and honour of the Roman people. What we hear from Cicero, many of whose writings have been preserved, must have been experienced by others too: Caesar not only wrote countless letters to Rome, but couched them in a highly personal, elegant and witty style. He had a penchant for understatement. When making a large loan to Cicero, he remarked that he would gladly help him insofar as his indigence allowed. When Cicero wrote, in his letter of recommendation for Trebatius, that he was placing him in Caesar's 'victorious and reliable hands', he added, 'Let me lay it on thick, although you disapprove.'

By tempering his dynamism with the gaiety and pleasantry of his conversation, Caesar made social intercourse easier. Yet one gathers from Cicero's letters how difficult and laborious it was to respond to the great *imperator* with elegant banter. With his dynamism he had long since erected barriers around him and was not easily approachable.

Keen observers now feared – not for the first time – that Caesar's aim was tyranny. 'Of course,' Cicero is said to have commented, 'when I see the meticulous care with which he tends his hair and the way he uses only one finger to scratch himself, it again seems to me impossible that this man could contemplate such a crime as the destruction of the Roman republic.' It is not known when Cicero said this, but it may have been about this time; Cicero had just seen Caesar at Ravenna.

However, quite apart from the public image he presented, we must have serious reservations about the suspicions entertained by his opponents. The fact that he acted highhandedly and with extreme effectiveness in Gaul did not necessarily mean that he sought autocracy in Rome.

In Rome tyranny was a frequent reproach, but never a role. Whether the leading senators, the chief men in the republic, ever understood Caesar's situation is no clearer than whether Caesar knew – or cared – why they found him so alien. Presumably what they thought they knew of him and what he thought he knew of them was enough to account for their mutual animosity. His gaiety, charm and urbanity seemed to betoken a superiority that made them all the more distrustful. It may be that their apprehensions were no less to blame for pushing him to the margin of the republic than his behaviour was for arousing such apprehensions. Here one could cite Thomas

Mann's remark about Frederick the Great: 'In any case, perhaps his intentions were honest – and he merely deluded himself about how dangerous he was? The man who was a mystery to all – perhaps he was also a mystery to himself?'

The consequences of Caesar's increasing isolation were patent: despite all his services to Rome, he was losing ground in the city. He could of course blame this on changes in internal politics. All the same, it may be assumed that Caesar, to whom activity had probably always been a compensation for a degree of isolation, became all the more restless as he saw his success diminishing. And he measured success against his increased expectations. He now suffered reverses both in Gaul and in Rome. With regard to Rome, however, one wonders whether his diminishing success was not the result of his increasing restlessness and dynamism. May not his growing estrangement from Roman society have been due to the very means by which he sought to approach it, indeed to force himself on it?

The Crisis of Caesar's Governorship
(49 BC)

For all Caesar's intense involvement in warfare and politics during the first six years of the Gallic war, the real test of his ability still lay ahead. The wide scope for action that he had enjoyed – or won for himself – in both Gaul and Rome had been due to a particular power structure, the main feature of which was his opponents' inability to make their power felt.

The Gallic tribes had not yet accustomed themselves to a policy embracing the whole of the country. Caesar's superiority was due largely to the fact that he had broken down the internal boundaries and was able to operate on a grand scale. The Gauls still thought in terms of individual tribal lands and their immediate or wider environment and thus allowed themselves to be played off against one another, hardly aware of the real power they could muster.

The general stalemate in Rome had given Caesar substantial opportunities for intervention. He could exploit the weakness of the Senate and the weakness of Pompey. Here too he had been able, through the agreement reached at Luca, to widen the field of politics by involving the provinces to a greater extent. Cato and his friends had not succeeded in mobilizing the forces they had at their disposal in the citizen body, while Pompey, wagering on anarchy, could only act negatively.

From the beginning of 52 events in Rome began to move to Caesar's disadvantage. Pompey had begun to collaborate with Caesar's opponents; there was a danger that the Senate régime would gain strength and that sooner or later Caesar would face a united front. The opportunities for pushing through demands and the prospects of favourable conditions for his return were disappearing.

The revolt in Gaul, in which the tribes for the first time acted in concert, had soon taken hold of half the country. It began among the Carnutes, who had killed all the Roman citizens in the town of Cenabum. The news spread like wildfire and reached the Arverni that very evening. They acted on it by electing a young nobleman, Vercingetorix, as king.

Vercingetorix was a handsome, imposing man of about thirty, highly gifted and ambitious. He detested the Romans, but had observed them closely in order to study their tactics and work out ways of defeating them. He seems to have campaigned zealously for the revolt; he travelled the country tirelessly, inspiring his compatriots and convincing them that Gaul must be free again. They must all join forces. He represented the national cause. He devised the strategy for the rebellion, and when it came he was the obvious leader.

He was joined by many tribes – the Senones, Parisii, Pictones, Cadurci, Turoni, Aulerci, Lemovices and Andes, and all those living by the Atlantic seaboard. Vercingetorix was given supreme command. He decided what contingents were to be supplied by each tribe and instructed them to have a certain quantity of weapons ready by a certain date. He regarded the cavalry as the most important part of the army.

Everything was expertly prepared. Caesar writes that in the exercise of his command Vercingetorix 'added the utmost care to the utmost severity'. Among other things he laid down draconian penalties. Major offences were to be punished by torture of all kinds, followed by burning; those guilty of minor offences were to have their ears cut off or one eye put out, after which they were to be sent home, so that all might understand the gravity of the war. A great army gathered.

Vercingetorix persuaded the Bituriges to join him. Others won over the Ruteni, the Nitiobroges and the Gabali. A Gallic army invaded the Roman province of Transalpina, heading for Narbo (Narbonne).

This was the stage things had reached before Caesar, having obtained what he wanted in his negotiations with Pompey, could leave Italy and cross the Alps. Reports from Gaul must have reached him, but he received them with iron calm. Meanwhile even the roads by which he could have joined his legions had been blocked. Hardly any news got through. The fate of the army was unknown.

The negotiations had been very difficult, especially as Pompey was no longer willing to contravene the law. Caesar's candidacy *in absentia* was to be approved by a tribunician law. Cicero had to travel to Ravenna to act as go-between and promise Caesar that he would do everything to prevent his friend, the tribune Caelius, from using his veto.

Caesar's governorship had two more years to run. Would this suffice for the final pacification of Gaul? Was it too long to hold up the imminent alliance between Pompey and the Senate? What now followed was the real test of Caesar's strength and confidence.

When he entered Provence it was still winter – the beginning of February by our calendar. Caesar had brought fresh recruits from Italy and now raised several thousand men in the province – paying no heed to whether they were Roman citizens or not – and hurriedly secured Gallia Transalpina. Then he had to see that he reached his legions.

He found the solution in two bold forced marches. At first he approached the Cevennes from the territory of the Helvii and crossed them, in spite of deep snow. The soldiers are said to have had to clear a way through snow two metres high. To the enemy's surprise Caesar was now in the territory of the Arverni and had it laid waste by his cavalry. Vercingetorix had to return home in great haste. Yet this had all been a diversionary manoeuvre, for while his troops at first remained there, Caesar secretly left with a small following and made for Vienna (now Vienne), in the far north of the old province, again crossing the mountains. He had previously ordered his cavalry there; he found them rested and now pressed on with them in an uninterrupted march to the borders of the Lingones, where two of his legions were stationed. On arrival – and before the Arverni knew of it – he assembled his whole army there.

Vercingetorix, however, marched into the territory of the Boii in order to lay siege to a town that was tributary to the Haedui. The Haedui were still loyal to Rome. If Caesar could not rescue the town there was a danger that the whole of Gaul would turn against him. What was to be thought of a power that could not protect its friends? Despite serious supply problems he could not avoid this hazardous enterprise. On the way he took Cenabum and set fire to the town after his soldiers had plundered it. Vercingetorix raised the siege and moved against Caesar. At Noviodunum, south of Cenabum, there was a cavalry engagement. For the first time we hear of Caesar using German horsemen. Vercingetorix was beaten. Caesar could now turn towards Avaricum (now Bourges) in order to conquer it and bring the Bituriges back under his control.

The Gallic leader now changed his tactics. He began setting fire to farms, villages and even towns that could not be defended within a wide radius of the Roman army. This was the same procedure that had brought Cassivellaunus success in Britain. The Haedui, who were supposed to provide the Romans with supplies, delivered virtually none. The Romans ran out of corn and had to fetch cattle from more distant villages. (Meat was not a normal part of military rations; the soldiers lived mainly on grain, which was eaten in the form of bread, or as a stew, with the addition of oil, vegetables and herbs.) Caesar now gave his soldiers the option of raising the siege of Avaricum if they were too tormented with hunger. They proudly rejected the offer, saying that they had never departed from any place without completing a task. 'Never a word was heard,' their general records, 'that was unworthy of the dignity of the Roman people and of their previous victories.' The siege-work was extremely arduous, especially as the Gauls had by now learnt to parry the practice of the Romans. Twenty-five days were spent building a ramp, a hundred and ten metres wide and twenty-seven metres high. It was banked up over wooden gangways and supported by beams. The Gauls undermined it and set the wood on fire. The Romans, however, managed to prevent the enemy from escaping. Finally, when a violent storm came up and the Gauls withdrew from the wall, the Romans were able to conquer the town. As usual in extreme cases, they were not sufficiently disciplined. The Romans indulged in a terrible massacre. All the inhabitants they could seize were killed, including old men, women and children. Caesar states that eight hundred survived, out of a population of forty thousand. Perhaps, after their immense hardships, he had to give his men free rein. It may be presumed that by now Caesar and his men were desperately weak and that their weakness was transformed into atrocious violence. The violence was probably also meant to have a deterrent effect.

But the opposite happened. Vercingetorix, who had initially been in favour of abandoning and destroying the town, skilfully roused the fighting spirit in his troops; whereas a leader's authority is usually diminished by defeat, his daily increased. Caesar reports this himself, and his words reflect a degree of respect for this considerable foe. Vercingetorix raised fresh troops to make up for his losses and tried to persuade other tribes to defect from the Romans.

Caesar then turned towards Gergovia (Gergovie), seven kilometres south of Clermont-Ferrand and the main fortress of the Arverni. The town stood on a high hill and was difficult to approach. As it appeared impossible to storm, the Romans besieged it. Vercingetorix had pitched his camp near the town; as he controlled all the high ground, he was unassailable and would have the advantage in any hostilities.

The Romans' difficulties were compounded by mutinies among the Haeduan auxiliaries who were marching towards Gergovia. When Caesar moved against them with part of his army, Vercingetorix attacked the rest and almost defeated them. It was becoming increasingly clear that Caesar had to reckon with the defection of the remaining tribes. His only course seemed to be to join up with the other part of his army, led by Labienus, which had been sent to conquer the region around Paris, probably in order to prevent the spread of the rebellion among the Belgae. Yet Caesar's departure must on no account look like flight.

He therefore attempted a surprise raid on the high ground held by Vercingetorix. Three camps were taken, but in the flush of battle the soldiers disobeyed Caesar's orders and pursued the enemy to the town walls, where they were met by fresh Gallic troops and driven down the hill exhausted, suffering heavy losses. Forty-six centurions fell – almost one in seven. Of the men, however, only seven hundred were killed. It was nevertheless one of Caesar's most grievous defeats.

He tried to save what could be saved. He called the army together and severely reprimanded the soldiers for their indiscipline and presumption in believing they knew better than their commander, but finally praised them for their valour and spoke words of encouragement. For two days he had the army ranged in battle order. Vercingetorix dared not leave the town. At last Caesar could withdraw without too much disgrace and move to the territory of the Haedui.

There too the whole country was by now in disarray; even Rome's friends had joined the seemingly victorious cause. Above all, the Romans had lost Noviodunum (now Nevers); this was a severe blow, as Caesar had deposited all the Gallic hostages there, together with his stocks of grain, the state chest, most of the baggage, and the fresh horses purchased in Spain and Italy. With the help of the hostages, the Haedui now put pressure on the tribes that still remained loyal to Caesar. They took them to the fortress of Bibracte (now Autun). They destroyed whatever grain they could not take away with them. Everywhere Caesar's army was cut off from its supplies. About the same time the Bellovaci defected, and Labienus was forced to withdraw southwards to Agedincum (now Sens).

Caesar appeared to be no longer in control of the war; the whole of Gaul seemed lost. He had to ask himself whether he should not turn round and fight his way back to the old Roman province, lest this too should be lost. From his account it seems as though only a few of his more timid associates thought this unavoidable. In his lapidary way he continues: 'What spoke against such a course were the shame and disgrace of the thing, the barrier of the Cevennes and the difficulty of the roads, and more especially his pressing anxiety for Labienus and the legions which he had sent with him on a separate mission.'

Here he presents himself to the reader with his usual superiority, stating only the reasons that finally weighed with him. Yet he cannot have seen the matter so clearly from the start. In view of the recent defeat, the rapid disintegration of his rule in Gaul, the imminent collapse of all he had worked for, and the acute danger to his army and his province, Caesar would have had to be made of stone had he not at first been devastated and plunged into a mood of hopelessness and despair, alternating with wild resolution, before he knew what he had to do. He forced a crossing of the Loire and marched northwards, so that he was soon able to join up with Labienus.

When the Gauls called a council in Bibracte, it was attended by all the tribes except the Remi and the Lingones, who were still loyal to Caesar, and the Treveri, who were fighting the Germans. Vercingetorix was confirmed in his supreme command. He ordered new levies. A scorched earth policy was to be pursued everywhere. Finally he ordered various tribes to invade Gallia Transalpina at three points.

Vercingetorix had exceedingly large forces at his disposal and was greatly superior to the Romans, especially in cavalry. As Caesar could get no more supplies from Italy and the province, he called upon the Germans who had surrendered to him to supply sizeable contingents of horse and light infantry, who were accustomed to fighting together. He then moved eastwards to the borders of the Sequani (north of the Jura) in order to bring help to the province. On the march he encountered Vercingetorix. A fight ensued in which the Romans were at first hard pressed – the Gauls even took Caesar's short sword – but the impetuous German horsemen finally secured the victory. The Gauls had believed that the Romans were about to quit their country and merely wanted to inflict a last heavy blow on them. All their horsemen had sworn a solemn oath to ride twice through the Roman columns. This made their defeat all the more painful. Vercingetorix therefore withdrew temporarily to Alesia, supposedly with eighty thousand foot and numerous horse.

This was to be the site of the decisive encounter. Not only was it the most difficult battle the Romans had faced: it was preceded by the heaviest labour. For it was here that the biggest earth-moving operations of the whole war took place. Alesia stood on the summit of a hill, almost impregnable and impossible to take by storm. Caesar therefore set about building a siegewall. At the same time he had to anticipate the possible arrival of a relieving army and was therefore obliged to secure his rear with forces far inferior to those of the Gauls.

Soon after being surrounded, Vercingetorix had sent the cavalry out of the town at night with orders to urge the tribes to call all their able-bodied men to arms. The town's supply of corn would last for thirty days, or rather longer if it was strictly rationed. He dared not risk an open battle with the Romans.

The wall built by the Romans extended for fourteen kilometres, with twenty-three forts as strongpoints. Since Caesar's army was too small to occupy these fortifications in the proper strength, he sought to secure the approaches by additional devices. Towards the town he first had a deep trench dug, almost seven metres wide and with perpendicular sides. Barely four hundred metres away from it two other trenches were dug, each five metres wide and five deep. The first of these was filled with water diverted from a river nearby. Behind them he had a ramp and palisade built, four metres in height, with breastwork and battlements; all round the works he placed turrets at intervals of twenty-seven metres.

During these works the Gauls made repeated sorties, and the Romans, being occupied with obtaining grain and timber, had difficulty in resisting them. Caesar therefore devised further obstacles. He had tree-trunks and large branches sunk into the ground; the tops of the branches were sharpened and partly interwoven. In front of them pits were dug; in these sharpened stakes, burnt at the tips, were sunk, so that they protruded only four fingers' breadth out of the earth. They were firmly stamped in and covered with twigs and brushwood to conceal the trap. In front of them were logs with iron hooks attached. Similar devices were placed on the other side. All the men had to go out and collect thirty days' supply of corn and fodder.

In this position Caesar hoped to be able to resist the Gauls. He knew that they must relieve Vercingetorix, and this was where he wished to meet them, relying on all these fortification works as the only means to counter their numerical superiority.

Meanwhile the Gauls had called a council and decided what contingents each tribe must supply. They all agreed that they must make every effort to relieve Alesia. There was no opposition, even from the friends of Rome. According to Caesar, two hundred and fifty thousand foot and eight thousand horse were assembled. Full of confidence, they marched on the town. There the might of the whole of Gaul was to show itself and be victorious.

It had taken them more than thirty days to arm, and in the town the food was running out. The inhabitants considered what was to be done. Critognatus, an Arvernan noble, proposed that they should revert to an ancient practice and 'keep themselves alive with the bodies of those who, because of age, seemed no longer suitable for war'. He failed to persuade the others, but they nevertheless resolved to send away everyone who could not fight – the old, the sick, the women and the children. The preservation of the eighty thousand men who were to fight in the final battle for the freedom of Gaul took precedence over all other considerations. Those who were expelled approached the Roman lines in a long procession and begged to be taken in, if only as slaves. Caesar, however, decided that they should return to Alesia, so that the enemy's food would run out all the sooner. He hoped to end the siege before the relieving army arrived. He showed no more humanity and no less determination to win than the Gauls in Alesia. It is true that he was fighting to preserve his rule, while they wished to regain their liberty. But this did not alter the fact that military considerations were paramount, and foremost among these was the safety of his men. His decision was harsh, but we know too little to judge whether it was contemptible. As they were not allowed back in the town, the men, women and children camped miserably under the walls, where most of them probably died.

Shortly afterwards the huge Gallic levy arrived and halted about one and a half kilometres from the Roman entrenchments. On the very next day they advanced to do battle. At the same time Vercingetorix led his men out of the town and began to fill up the first Roman trench with hurdles and earth in preparation for the sally.

Caesar describes the battle in detail, with his usual objectivity, but underneath his account one senses all the tension of an extremely varied series of encounters in which the issue was decided only at the last moment. He 'had disposed the whole army on both sides of the entrenchments in such a way that when the time came each man would know and keep his proper station'. The first cavalry engagement took place before everyone's eyes. Caesar describes how the Gauls 'sought to inspire their countrymen with shouts and yells from every side' and how both Romans and Gauls fought with extreme bravery 'because no deed, of honour or dishonour, could escape notice'. The Gauls used their well-tried tactics, grouping lightly armed men and horsemen in such a way that in each group the former could protect their own horsemen and fend off attacks by the Roman. At first the Gauls gained the upper hand. After long and bitter fighting Caesar's Germans attacked again, just before sunset, concentrating on one point; they succeeded in breaking through and the battle turned in the Romans' favour.

Next day there was calm, but at midnight the Gauls attacked again on both sides. From outside came loud battle cries, while inside Vercingetorix had the trumpet sounded as a signal to attack. The Romans defended themselves with heavy stones, sharpened stakes, burnt at the tips, and bullets cast from slings. At daybreak, having fallen into the traps set by the Romans, the Gauls broke off the attack.

Only now did they study the situation more closely. They realized that the Roman fortifications had a weak point in the north, where a hill extended so far that it had not been included in the Roman works. The Roman camp was half-way up the hill. At night the Gauls sent a contingent of picked men, said to number sixty thousand, to this point. At midday, having rested, they began to attack the Romans from above, while Vercingetorix attacked from the other side. Battle cries rang out on both sides, causing alarm among the Roman soldiers, who had to defend themselves in front and heard danger approaching from the rear. At the same time new attacks began at other points.

Both sides fought with the utmost ferocity, knowing that the battle must now be decided. The Gauls must break through the Roman fortifications, for another failure would mean defeat. The Romans must hold them back in order to win the first round. Being superior in numbers, the Gauls could send in fresh reserves to relieve their exhausted soldiers. Large numbers of them were employed in filling in the trenches and making all the complicated Roman devices unusable. Others moved up in close formation under their shields towards the Roman wall. The Romans' strength began to flag, and they were running out of weapons.

Caesar had sent six cohorts to the aid of the most hard-pressed detachment on the hill, ordering them to break out down the hill if the worst came to the worst. He then raced through the fortifications, shouting encouragement to his men everywhere and telling them that the fruit of all previous engagements depended upon that day and hour. He was seized by a desperate resolve to make the last supreme effort and assert himself against superior forces.

But the enemy was too strong. From higher up the hill they attacked the Romans with a hail of missiles. The Romans abandoned their turrets, and the Gauls began to tear down the rampart and breastwork with grappling-hooks. Whereupon Caesar withdrew men from other points in order to throw them against the enemy. It was not enough. He quickly formed new cohorts and sent them into the fray. Finally he himself hurriedly led the last reserve to the centre of the fighting. The battle began afresh. The Gauls launched a new attack. Caesar writes that they recognized him 'by the colour of his cloak, which he usually wore in battle as a distinguishing mark'. The formulation is striking: if he had been wearing the usual commander's cloak, the *paludamentum*, he would have had to say so. Obviously he wore a special garment even in the field. And its colour must have indicated to the Gauls – and the troops he was leading against them – how serious things were becoming when the battle was renewed with such ferocity.

Again battle-cries resounded from both sides. The Roman cavalry, which Caesar had sent from a relatively quiet point, now appeared in the enemy's rear. At last the fortunes of battle changed. The Gauls in the relieving army, seized with sudden terror, broke off the attack and took flight. Seeing this, Vercingetorix's men withdrew. The Romans were by now too exhausted to pursue the fleeing enemy. Only the cavalry was able to catch up with the rearguard and kill a large number. The Gallic army broke up. The battle was lost. The country had not regained its liberty.

In Alesia, Vercingetorix declared that they must bow to fortune. He told his men to kill him or hand him over alive, whichever they pleased. They sent word to Caesar. He ordered them to surrender their arms and bring the leaders before him. He sat on the fortifications outside his camp as the men approached in a long procession. They laid down their arms in front of him. Vercingetorix came in shining armour; even in defeat this courageous and handsome man impressed the Romans. He threw himself down before Caesar, stretching out his hands in front of him in token of surrender. Many onlookers were moved with compassion, but Caesar coldly reproached him for breaking the old bond of friendship and had him laid in chains. He had to wait in prison for six years before he could be displayed to the Romans in triumph, after which he was mercilessly executed in the Mamertinum, next to the forum. Caesar pardoned the Haedui and Arverni, however, returned their prisoners, and restored them to favour. He needed them to bolster his rule over Gaul. Of the remaining prisoners he gave one man to each of his soldiers. The others he probably sold for his own profit.

Few battles, says Plutarch, have been fought with such outstanding bravery and such a wealth of technical invention or 'martial genius'. Caesar had indeed won a great victory. The war was not over. But much of central Gaul was once again in Roman hands, and the links with the old province and Rome were secured.

At the end of 52, as had happened more than once before, the Senate granted Caesar a twenty-day *supplicatio*. Apparently many wished to document that the Gallic war was finally at an end: a successor to Caesar could now be appointed.

While Caesar was cut off in Gaul, certain events in Rome had greatly weakened his position. Although the law allowing him to conduct his candidacy *in absentia* had gone through, Pompey had brought in several bills intended to curb various abuses. Though useful reforms, they were detrimental to Caesar's rights. It was decreed, for instance, that in future no candidacy might be pursued *in absentia*. We do not know why Caesar's friends raised no objection. Perhaps they were unaccustomed to acting on their own initiative, as they normally received instructions from Gaul. They noticed the omission, if at all, somewhat late in the day. When the couriers finally reached Caesar, the law was already cast in bronze and deposited in the archive. Pompey went and made a correction, but it was doubtful whether this would have furnished grounds for a legal claim. Even worse, another law, designed to restrict electoral bribery, laid down that in future five years must elapse between a consulship and a governorship; meanwhile, magistrates from earlier years were to be sent to the provinces. This meant that Caesar could be replaced immediately after 1 March 50. His command would then have run its term.

He could now be replaced by a successor in good time and brought to court before he could seek the protection of a new consulate. Caesar could of course defend himself politically; Pompey had not yet finally gone over to the opposite camp. Moreover, decisions of the Senate could be blocked by tribunician veto. But this would not be easy, as order had now returned to Rome. Milo had been condemned and banished. On this occasion Pompey had had the forum guarded by soldiers, so menacing in their demeanour that Cicero dared not make a vehement plea for the defence. Numerous other trials followed, and many ended in convictions. In 52 Pompey's Spanish command was extended.

The following winter Caesar could not leave Gaul. There was a threat of further risings. Several tribes armed for war; they had decided to operate separately again, believing that they could defeat the Roman legions one by one, as Caesar had distributed them widely across the country. It was a harsh winter, and the soldiers suffered greatly from the almost unendurable cold. However, Caesar, having pitched his winter camp at Bibracte, first had to march westwards into the territory of the Bituriges, then, after a short pause, north-westwards against the Carnutes, and finally northwards against the Bellovaci. He clearly could not entrust these operations to subordinates. To spare the soldiers undue hardship, he took different legions with him each time. He also promised to reward them with large shares in the booty. The war against the Bellovaci was especially protracted. They had won support from other tribes throughout a wide area and brought in German auxiliaries. Caesar met with varying success. We learn that in Rome it was again hoped that he would be defeated. But in the spring or early summer of the following year he was once more victorious.

Yet still the country was not pacified. The unrest continued and the Gauls went on re-arming. Caesar's legates had to move against various tribes, though they met with little opposition. The Gauls calculated that since this would be Caesar's last summer as governor they had only to bide their time, in the expectation that they would have less trouble from his successor. Here and there Caesar struck all the harder, taking more and more hostages and executing the leaders of the rebellious tribes. After the fall of Uxellodunum, a town belonging to the Cadurci, into which a large number of rebels had withdrawn, Caesar ordered that all who had borne arms should have both hands cut off. He spared their lives, 'so that the penalty for their wickedness should be all the more obvious'. Aulus Hirtius, who wrote the eighth book of the Gallic War, which describes its last two years, observes that Caesar knew that he was famous for his clemency and had no need to fear that such actions would be ascribed to natural cruelty. He did not see how he could prevent further rebellion without recourse to such deterrent measures. A fine explanation!

It seems that Caesar already attached great importance to his reputation for clemency. We hear of defeated enemies appealing to his *clementia* and *humanitas*. On the whole he probably showed consideration to his defeated enemies whenever possible. He was too controlled to give vent to anger and too attentive to be indifferent to abuses. He may even have been motivated by an awareness that the war stemmed from his ambition for conquest and wished to act as mercifully as the exigencies of war allowed. In any case Caesar took pleasure in showing generosity, not only to his soldiers, but to everyone under his authority. It was a mark of his superiority; it did not preclude punctiliousness and severity, but embraced them. He was certainly not naturally cruel. Occasional cruelties were the outcome of his high pretensions, his growing impatience, and his boundless determination to assert himself.

He was also well aware that if he was to hold Gaul he had to secure the allegiance of many of its leaders. If he wished to win over important figures and tribes, he must not treat their supporters badly. Having dealt a severe blow on one occasion, he had to make up for it on another.

Hence, in 51 and 50 Caesar was at pains to mitigate the impression of his victory at Alesia and the force and circumspection of his military operations by a show of mildness and friendship. The whole of Gaul, writes Aulius Hirtius, might re-open hostilities at any time. Knowing that his governorship was soon to end, Caesar had to do everything in his power to prevent renewed fighting, lest he should leave a theatre of war behind him on returning to Rome. 'He therefore showed the tribes every possible honour, conferred substantial rewards on their leaders, and imposed no new burdens on the country, so that he might demonstrate to Gaul, exhausted by so many defeats, the benefits of submission and so preserve the peace without difficulty' (Hirtius).

The reckoning of the war was terrible, even if the figures are greatly exaggerated. Caesar is said to have captured eight hundred towns, subjugated three hundred peoples and defeated three million armed men, a third of whom were killed and another third imprisoned or taken into slavery. The country had lost countless lives and untold wealth. Gaul could still defend itself, but if peace returned it might be persuaded, for the time being, to accept the new situation. This was what Caesar had brought it to. The battles of 51 remained as echoes of the great uprising of Vercingetorix.

Caesar thus finally crowned his great military achievements by a political one; he established peace over a wide area, thanks to an enormous expenditure of intellectual energy and a willingness to compromise.

In the autumn of 51 he had visited Aquitaine for the first time. Having inspected it and finally subjugated it, he left for Gallia Transalpina in order to hold court, to reward those who had kept faith with him and Rome throughout the long war and to cement the bonds between them. Finally he returned to the Belgic area and wintered at Nemetocenna (now Arras).

During this time, according to Hirtius, 'his sole aim was to maintain friendly relations between the tribes and Rome, to forestall any hopes of armed conflict and to give no occasion for it.' In 50 Gaul remained quiet. The country was so well secured that its tribes did not even exploit the civil war by falling upon Caesar's rear or rising up against Rome, which was now exposed to attack from outside. Caesar really had conquered the country.

Yet even if Caesar had solved one of his great problems – by dint of immense effort and sacrifice, and at the cost of great losses, not least among his own men – the other loomed all the larger, the more pressing it became: how was he to return to internal politics? A powerful group in Rome was intent upon stripping him of his governorship and then, as Cato repeatedly declared and even swore, to bring a prosecution against him. It was obviously intended that his trial, like Milo's, should take place under military guard.

Marcus Claudius Marcellus, one of the consuls of 51, declared that, since the Gallic war was over, a successor should at once be sent to relieve Caesar. For a long time the Senate evaded his proposals, especially as Pompey felt bound by his promise not to allow any discussion of Caesar's provinces until 1 March 50. On 29 September 51 it was nevertheless decided to put the subject on the agenda for 1 March; it was the consuls' task to do everything possible to see that it was resolved. Finally, a report was to be presented to the Senate about the soldiers in Caesar's army who had completed their period of service or had other grounds for release. This was no doubt seen as a way to undermine Caesar's standing with the army. Four tribunes of the people had interceded against these resolutions. But the will of the Senate was plain, and the new consuls were likely to support it. Pompey had stated publicly that if a veto was entered against the resolutions that were due to be taken in March, this should be regarded as disobedience on Caesar's part. Asked what would happen if Caesar wished to seek the consulship during his command, he replied, 'What happens if my son beats me with a stick?' This clearly meant that such a thing was impossible, but if it happened all the same it was so bad that it could not be tolerated.

Marcellus, moreover, affronted Caesar by having a citizen of Novum Comum whipped. Novum Comum (now Como) was a colony founded by Caesar himself. Marcellus wished to demonstrate that the man was not a Roman citizen, even though Caesar, by an improper interpretation of the legal regulations, had conferred civil rights on him and others. He told the man to show Caesar his stripes.

The prime mover behind this policy was Cato, and its chief advocates three members of the family of the Claudii Marcelli – two brothers and a cousin, each of whom succeeded in becoming consul in 51, 50, and 49. They and their allies were convinced that Caesar's return to internal politics must be prevented at all costs.

This could of course mean civil war. Cicero records that the second consul of 51, a lawyer named Servius Sulpicius Rufus, repeatedly recalled the civil war of the eighties and predicted that a new one would be even worse. He maintained that anything for which there was a precedent could easily be regarded as legitimate; and one always added a little of one's own. He insisted that nothing must be allowed to happen that could lead to a civil war. The possibility of a war recurs elsewhere in the reflections of good observers of the contemporary political scene.

It thus became clear what was expected of Caesar. Sulpicius assumed that he would use armed force to resist being relieved of his governorship, which was properly a matter for the Senate to decide. Cato also feared that if he were again consul in Rome he would act arbitrarily and illegally, as he had done before, and that the result would be autocracy. These senators saw the alternative as 'Caesar or the republic'. The one precluded the other. If the old and universally acknowledged order, which rested on the will of the whole citizen body, was to continue in existence, Caesar must be removed.

They had been able to deal with Pompey, however long and strenuously they had opposed him. They knew that he fundamentally respected the old order. This was after all why his policy had been so vacillating and hypocritical. His power and pretensions were so great that Cato and his friends thought him dangerous. But by nature he was a conformist who ultimately fitted into the framework of the aristocracy. Hence, when there was nothing else for it, they could do a deal with him. Presumably they already knew that once they had recognized his power it could no longer harm them; there was no longer anything to prevent his respecting the Senate.

Caesar, however, was disrespectful and alien; he did not fit into the republic. His cold, unconcealed ruthlessness, his scorn of the traditional institutions and their representatives, his self-centredness, his total lack of scruple in starting the war in Gaul, his attacks on other nations and, not least, the insufferable superiority he displayed in his dealings with everyone – all this was enough to put him beyond the pale of the aristocracy and make him seem a sinister figure, quite apart from the demonic element in his nature. He could not be expected to comply and, if necessary, bow to the judgement of his peers.

All Roman actions were conditioned by the fact that Roman society ultimately agreed about its order; this unanimity was so strong as to stamp itself, as it were, on the thinking of its members. They did not desire what they had no business to desire. This applied to Pompey too. The individual might be permitted a certain latitude and certain liberties; he might on occasion kick against the pricks. Yet there were limits, and these limits, for all their elasticity, became all the firmer the more they were stretched. Everyone knew this. The citizens lived in the closest mental contact with one another; this constituted the common reality.

Yet Caesar, as Cato sensed, did not share this reality; he inhabited a different reality. While Pompey always remained attached to the aristocracy, however much it tried to keep him at arm's length, Caesar was detached from it. In a sense he always had been, but he made it clear in 59; since then he had become his own master, and there was probably good reason to think that he would be quite unscrupulous in asserting himself in Rome.

It is admittedly not clear whether Cato and his allies really knew what they were letting themselves in for by their implacable opposition to Caesar. Did they really think they could attain their ends without a civil war? And if not, were they aware of what that might mean? Insofar as it is possible to judge from their later thinking, they seem to have been unable to estimate the dangers they were conjuring up. They certainly relied on Pompey, and overrated his power and potential. They probably underrated Caesar, however much they feared him, because they hated him. Their conviction that the old republic was the only true order may have made it impossible for them to imagine that its opponent could be so strong. They were more impressed by Caesar's alien quality than by his ability, more by the evil he represented than by the power he commanded.

It was an open question, on the one hand, whether Pompey really wished to side fully with Cato, and, on the other, what line the majority of the Senate – and the citizens – would take. The future depended above all on how Caesar would defend himself against his opponents and whether he could ultimately assert himself against them. None of this became wholly clear in 51. Caesar's plan was to seek a second consulship – probably that of 48. Then, with the help of his veterans, he intended to bring in a land law. The Senate would either agree to it or once more be heavily defeated. Yet this presupposed that Caesar could pursue his candidature without giving up his governorship. It was his only way of avoiding prosecution.

Given this situation, he had three aims. First, he tried to ensure that the magistrates for 50, and especially the consuls and tribunes of the people, should include men who would be forceful and decisive in representing his cause. Secondly, he attempted to win over the Senate majority. Thirdly, he did all he could to consolidate his power in Gaul, so that if the worst came to the worst he would be prepared for a civil war.

In 51 he had already moved one legion to Gallia Cisalpina, ostensibly to secure the local colonies against surprise attack. In the early summer of 50 he travelled throughout the province and was received with great honour. 'There was nothing they did not devise to adorn the gates, the streets and all places lying along Caesar's route. The whole population came out with their children to meet him; everywhere sacrificial animals were slaughtered, and in temples and market-places were divans covered with carpets for public banquets' (Hirtius). The pomp and enthusiasm with which both rich and poor greeted Caesar were intended to honour not only the successful commander and governor, but also the man who had long sought to obtain full citizenship for the inhabitants of the province. Caesar is unlikely to have spoken – at any rate publicly – of the possibility of war. Yet he no doubt intimated that no one could safeguard the interests of the inhabitants – or of Rome – as well as he. And he is certain to have spoken of the injustice his opponents in Rome wished to do him.

He then returned hurriedly to Belgic territory, reviewed his army and assured himself of the loyalty and devotion of the soldiers and his friends in the region. He had new weapons and military equipment made and raised fresh troops in order to bring the army up to strength.

In the autumn he sent the legions into winter quarters with the Belgae and Haedui, the bravest and most respected of the Gauls, so that the Romans could keep them under supervision. Only then did he return across the Alps to Gallia Cisalpina and set up his headquarters at Ravenna.

Towards the Senate he showed himself markedly loyal. 'For he believed that his cause would be easy to support if there could be free expression of opinion in the Senate.' This, it seems, was the nub of the matter. If the senators were against him, this could only mean that they were under pressure from a small but powerful clique of opponents. Just as they could not imagine what forces Caesar might bring to bear, he could not imagine that the Senate majority itself was ill-disposed to him. Before 58 he had occasionally almost won them over; since then he had distinguished himself as no other commander before him and made the most strenuous efforts on behalf of the city. Since ancient times the Romans had set the highest value on the fame of the military leader. Caesar could thus claim recognition and high honour. That he was denied them by his opponents could be explained only by their obduracy, partiality and self-seeking. He presumably found it inconceivable that the Senate majority would willingly agree to his being relieved of his command, summoned to Rome and put on trial. How did his opponents stand? What had they done while he was conquering half the world? Moreover, many of the senators had received gifts, loans and other favours from him.

Yet however unjust he might be to Cato and his allies, his assessment of the Senate majority was not wholly wrong. His achievements probably impressed many of the senators, and they cannot have found it easy to condemn the conqueror of Gaul. After all, they had often conferred high honours on him. Yet for this very reason there were many who did not relish the prospect of his return, and they were by no means in favour of his being elected consul without laying down his command. Their reluctance to pass a resolution to replace him was no doubt due largely to their fear of him. To this extent their judgement was much the same as Cato's. The difference was that they did not want to risk a war. Caesar was not to be trifled with and must therefore be allowed to have his way.

In 50 Caesar's cause was championed above all by the tribune Gaius Scribonius Curio and the consul Lucius Aemilius Paullus. He had gained their support by paying them sums running into millions. We hear nothing, however, of his having influenced the elections in their favour. Paullus had rebuilt a large basilica by the forum and thereby incurred large debts. Curio's debts were due mainly to his splendid theatre-building and the games he had sponsored. Yet no price was too high for Caesar to pay, though he himself, for all his huge wealth, is said to have reached the limit of his resources. Curio needed ten millions and Paullus thirty-six. The annual tribute from Gaul amounted to forty millions, though the proceeds of the booty probably far exceeded this sum. Pompey is reported to have said more than once that Caesar would have to start a civil war because he would otherwise be unable either to complete his building works or fulfil the expectations attendant upon his return.

Gaius Scribonius Curio was a 'brilliant good-for-nothing' (*ingeniosissime nequam*), a highly gifted, fiery and exceptionally charming young man of about thirty-four who had so far proved somewhat unreliable. He belonged to the plebeian nobility; his father was one of the most distinguished *consulares*, a relatively independent politician, and the two often collaborated.

Curio had always been pre-eminent among Rome's gilded youth. In 61 he was the leader of the 'downy-bearded youths' who championed Clodius with ebullient zeal in the popular assembly to prevent his being prosecuted over the Bona Dea scandal. In 59 he was the one person to attack Caesar loudly and insolently during the terror that followed his opponents' withdrawal from politics. He was probably unsurpassed in his extravagances, devoted to wine and women, generous in all things, lavish even with what did not belong to him; he set the tone among his contemporaries, the best known among them being the witty Marcus Caelius and Mark Antony, who was destined for subsequent fame. They all emulated the model apparently set by the elder generation, though they lacked the scruples that ultimately kept their elders – with the exception of Caesar – within bounds.

Cicero praises his rhetorical skills and tells us that they were due not to training, but to nature. He spoke with freedom and facility, at times pointedly and always thoughtfully. Everything seems to have come easily to him. It is impossible to decide whether he acted out of courage or levity. As late as 51, his friend Caelius remarked that he never did anything with deliberation. He possessed exuberance and *joie de vivre* and took such delight in the freedom he enjoyed that he behaved with the utmost nonchalance, shunning conformity; he did not want to be part of anything, but insisted on being his own man. He wanted to be something quite special, not as others would have him be, but as he chose to be. After all, everything came his way. He did not have to exert himself, and he was certainly not assiduous, nor was he easily impressed. He was guided by impulse, delighting in his levity and versatility, taking things as he pleased, and committing himself to nothing.

Eduard Meyer remarks: 'Like Caesar, he combined a total disregard for the principles of political morality, a supreme and ostentatious nonchalance of bearing, with shrewd political judgement and a well-founded and ennobling consciousness of his capacity for achievement; he was far closer to Caesar than to Clodius, for instance, whose heir he seems otherwise to have been and whose widow, Fulvia, he married.' Mommsen spoke of Curio's 'charming candour' and discerned in him a spark of the Caesarian spirit. True, he had not grown up as an outsider, and he did not have Caesar's energy and single-mindedness – at least not until 50 BC. Caelius once said of himself that what he loved most was to be concerned about nothing; this applied to Curio too for much of his life. This meant that for a long time he conserved his rich potential undiminished. It accounts for his brilliance and the qualities that many found so enviable, whatever misgivings they might arouse in more earnest spirits. Caesar seems to have had a special liking for him. He describes him very sympathetically in his book on the civil war. And during the civil war he entrusted him with the most important tasks, in spite of or even because of his levity – or because he knew what lay behind it.

Curio had not at first intended to become a tribune of the people and sought the office only when a vacancy arose. Nor was he intent on siding with Caesar. Like his friend Caelius, he could just as well have represented the Senate's cause. At that time neither young people generally nor the most interesting and liveliest spirits among them felt drawn to Caesar. But they wanted to attract attention; Curio also needed money, and only Caesar could provide it.

Caesar, however, at first seems to have wanted nothing to do with him. Perhaps he still bore him a grudge for opposing him in 59, perhaps he underrated him, perhaps he found him too expensive, or perhaps he had already reached a covert agreement with him. At all events, Curio at first did everything in his power to oppose Caesar and did not go over to his side until the end of February 50. From then on he lacked neither strength and imagination nor constancy and enthusiasm. All his energies hardened in the new alliance.

When the Senate discussed Caesar's provinces on 1 March and the consul Marcellus advocated the immediate sending of successors,

Curio prevented any decision. The consul Paullus helped to drag out the deliberations. We do not know the precise course they took. But the tactics of the Caesarians are clear: Curio was not unconditionally opposed to replacing the governor, but he insisted that if Caesar laid down his command Pompey must do the same. The republic would then be free again. He made himself the advocate of the traditional order; it must be restored. He was not troubled by the fact that Caesar's command had run its term, while Pompey's was due to last several more years, and that to shorten Pompey's would be to dishonour him. What was the purpose of Pompey's being governor of the Spanish provinces if he was still in Italy? If certain senators felt threatened by Caesar's legions, Curio said, Pompey's were equally dangerous. Would it not be better if no one could rely on a large army in order to exert pressure, if the commonwealth were again ruled as it had been formerly, by the Senate and people? By agitating for peace, Curio was appealing to the dearest wishes of all the citizens, to weaknesses that were all the greater as the citizen body could scarcely defend itself militarily any longer. Curio therefore proposed that Caesar and Pompey should both lay down their commands.

He thus defended Caesar's cause in a highly imaginative way, by taking the offensive. To prevent Caesar's replacement he shifted the focus: the issue was no longer Caesar, but the republic. The changes of 55 – and indeed of 59 – should be reversed. Pompey was thus forced on to the defensive. True, he probably now had to rely more than before on Caesar's opponents. But Curio, like Caesar, was chiefly concerned with the Senate majority. It is said to have been ready to accept the proposal, but the opponents prevented a vote. (Admittedly, we have this information only from a Caesarian source, but it may still be correct.) In principle most of the senators were on the side of Pompey. And they can have been in little doubt that, if both leaders laid down their commands, Caesar would be politically far superior, thanks to his tactical skills and the strong support of his veterans; he might even succeed in drawing Pompey back on to his side.

What Curio championed, the freedom of the Senate and the people, was thus Caesar's strength. Yet however weak the republic seemed to Curio and Caesar, the senators were equally weak. They must have told themselves that if Curio's proposal were accepted it would at least be possible to avoid civil war. We do not know how many senators attended the meeting in question and what effect Curio's rhetoric had; perhaps he took the Senate by surprise. Pompey was very indignant. In order to be more of a match for Curio he is said to have taken fresh lessons in rhetoric in his old age. He saw that he had to make some concession: Caesar should retain his command for the summer and be relieved only in mid-November. Many senators were agreeable to this. But it did not solve Caesar's problem. Curio therefore entered his veto.

The Senate did not have to put up with this. It could use the device known as 'negotiation with the tribune', which meant putting strong pressure on him and, in extreme cases, threatening to suspend him from office. There were precedents for this. However, in view of the independence of the tribunate, it could not necessarily be maintained that such a suspension was lawful. And in any case this procedure could succeed only if the Senate resolutely supported it. Since Curio was not easily impressed, they had to be resolved on extreme measures. But resolution was precisely what the Senate lacked. It therefore rejected the proposal to negotiate with the tribune. Cicero's perceptive friend Caelius, in a letter, drew the conclusion: 'They have decided in effect that he must be admitted to the election without surrendering his army and his provinces.' It is true that he had observed a short while earlier: 'If they use every means to put pressure on Curio, Caesar will defend the intercessor; if they recoil from this, as is likely, Caesar will stay as long as he wishes.' In short, even the strongest pressures were unavailing if there was an army in readiness to punish any violation of the hallowed rights of the tribunes of the people. What Caesar had simply brushed aside when it impeded him as consul he was now prepared to defend with his army. And since everyone knew this, the Senate majority avoided a test of strength.

Roman politics were again paralysed. There was a smog of lethargy and anxious tension in the air. In April came news that the Parthians were threatening war – in response to Crassus' invasion. Pompey proposed a motion in the Senate that two legions should be sent to the east, one taken from Caesar's army and one from his own. The Senate agreed. Pompey then announced that he would make available the legion he had lent Caesar in 53. Caesar would thus lose two legions, which in any case he seems to have replaced by new levies. He gave every soldier a handsome gift of money. Young Appius Claudius, who took over the troops in Gaul on the Senate's instructions, reported that the proconsul was extremely unpopular with his soldiers; they complained about their endless hardships, the long and repeated wars, and demanded to be released. If only Pompey were to show himself, he said, they would defect to him. He had probably heard what he wanted to hear; the soldiers may well have complained and cursed, but he failed to understand that the soldiers were devoted to Caesar, in spite of – indeed because of – the hardships they had endured, and that they would in any case willingly obey his commands if they remained under his authority.

Pompey had just recovered from a serious illness. He had seen how first Naples – where he had been staying – and then one Italian city after another held thanksgiving festivals for his recovery. On his way back to Rome he was ceremoniously received everywhere, greeted with flowers and escorted by torchlight. He had the impression that he enjoyed universal favour and felt strong *vis-à-vis* Caesar. In answer to anxious questions as to how he would defend himself if it came to a civil war, he replied that he had only to stamp his foot, and horsemen and foot soldiers would spring out of the ground. Yet while the prosperous strata of society, those whom Cicero called the 'good' and who probably regarded themselves as such, might demonstrate massively in Pompey's support, they were far from willing to take up arms for him or the republic.

From his sickbed Pompey had sent a letter to the Senate, declaring his willingness to lay down his command early. How different from Caesar, who refused to lay down his even after it had run its term! Pompey admittedly did not say when he intended to do so; Curio pointed this out and complained about it.

Otherwise nothing happened for a long time. Yet the more things dragged on, the more determined the consul and his allies became to cut the knot. The showdown between Caesar and the group led by Cato could be postponed for a while, but not prevented.

In the middle of 50 Marcus Caelius was already speaking of an impending civil war. To Cicero he added: 'It will not escape you that in internal disputes, so long as they are conducted by civil and not military means, one must be on the more honourable side, but when it comes to war one must be on the stronger. Whatever is safest must then be considered best.' And he soon saw that Caesar was the stronger. Caelius realized that what counted was Caesar's army. On the opposing side he saw neither strength nor determination.

Caesar was of the same opinion. He banked on the weakness of Pompey and the Senate. He therefore offered to stand down only if Pompey did the same. We hear of no further offers of compromise. Caesar seems to have pursued a hard line.

At the end of November there was uproar when the censor Appius Claudius, the uncle of the young man just mentioned, wanted to expel Curio from the Senate. His colleague Piso, Caesar's father-in-law, and the consul Paullus objected, and the censor had to give way, but he expressed himself in such an offensive manner that the tribune of the people sprang towards him and tore his toga. The consul Marcellus therefore made a report on Curio, demanding that he should be reprimanded or even suspended from office. Curio merely declared that he had sought to do what was best for Rome and submitted to the judgement of the house. The majority of the senators decided in his favour.

Shortly afterwards Marcellus, hoping to force a decision, made an inflammatory speech attacking Caesar and demanding that he should be declared an enemy if he did not give up his Gallic command forthwith. It was early December, and Marcellus was nearing the end of his consulship. Curio then again called upon the Senate to vote on his own proposal for the termination of both governorships. This time he succeeded. Marcellus seems to have demanded only that the parts of the proposal should be voted on separately. The details are not known, except that a majority was in favour of relieving Caesar of his command and against relieving Pompey of his. And the proposal to relieve them both was carried by three hundred and seventy votes to twenty-two.

The first vote was of academic interest; it was the second that counted. If they had to choose between Caesar and Pompey they preferred Pompey. Above all, however, they were in favour of peace and this meant, since Caesar seemed determined to do anything, that they were in favour of Caesar. Strictly speaking, they were right, if they were against a civil war, for that could only make things worse. But in all likelihood they voted for it less because it was right than because they were cowardly – less because they were thinking politically than because they were weak. Had they been resolute and really approached the problem politically, they could perhaps have induced Caesar to make concessions before it was too late. As it was, he could only despise them. And on the other hand Marcellus became increasingly troubled. Hence the Senate, in its desire for peace, promoted war.

When Curio left the Senate, he was loudly applauded by the crowd of bystanders. As the senators dispersed, Marcellus shouted to them that they had now appointed Caesar himself as their master.

He was greatly alarmed. He therefore seized on rumours that had been circulating for weeks, according to which Caesar was bringing several legions across the Alps. 'Now,' he said, 'the danger is imminent.' Curio rejected this assertion, which was in fact incorrect, and even now the senators could not bring themselves to declare Caesar an enemy. Instead they put on mourning.

Marcellus, however, declared that if the Senate would not do its duty he, as consul, must act on his own initiative. He went with the consuls designate for 49 and a number of other senators to visit Pompey, who was staying on the outskirts of the city. He carried a sword in front of him; handing it to Pompey, he called upon him to protect the city. At the same time he conferred on him the command of the two legions, which had not yet set off for the Parthian war, and empowered him to make further levies. Marcellus was acting out of his general responsibility as a citizen and specifically as consul, without instructions or special powers. He saw it as his duty. It was a *fait accompli*, designed to commit the Senate majority, and Pompey himself, to removing Caesar from his command and being ready for war if Caesar did not yield.

Pompey's attitude, as usual, was equivocal. The removal of Caesar would not have served his turn, since the Senate would then have had less need of him. On the other hand, even he was bound to fear Caesar's return to internal politics. No doubt there was still the possibility of a fresh alliance with Caesar. But this would have shifted the balance of power strongly to Pompey's disadvantage. He would have had to fight, he would have become dependent on Caesar, and above all he would certainly have lost the generally respected position that was so important to him. He was therefore probably inclined to come out against Caesar. Moreover, Curio restricted his room for manœuvre by depicting him as the real obstacle to an understanding with Caesar. The controversy had now been narrowed down to the alternative 'Pompey or Caesar'; it was already thought that unless one of them was sent to make war against the Parthians there would inevitably be a violent clash between the two.

Until now Pompey had avoided committing himself. He temporized, and was certainly reluctant to risk a war. 'I believe,' writes Montesquieu, 'that what harmed Pompey most was the shame he felt on reflecting that he had lacked foresight when he helped Caesar to power. He accustomed himself to this thought as late as possible; he failed to go on to the defensive because he did not want to have to admit that he had put himself in danger. He assured the Senate that Caesar would not dare to begin a war. And because he had already said it so often he went on repeating it.'

Even now Pompey could not make up his mind, but he did not turn down the request; and the impression was bound to arise that he had accepted it. Caesar's envoy, arriving in Rome on 6 December, no doubt in order to negotiate with him about the impending crisis, was so outraged that he at once hurried back to Ravenna. At last Pompey started to arm. He took command of the two legions. On 10 December Curio's tribuneship ended, and he immediately joined Caesar. He is said to have advised him to go to war. Among the new tribunes of the people, Mark Antony took over Curio's role as Caesar's representative. On 21 December he made a fiery speech attacking Pompey. If Caesar's supporters were already adopting this tone, Pompey wondered, how would Caesar himself act? He convinced himself that war was inevitable. Antony, however, issued an edict prohibiting compliance with Pompey's call to arms.

With Cato and his supporters, Pompey now bent all his efforts to persuading the Senate to decide against Caesar on 1 January 49. Supporters, including many old soldiers, were brought to Rome; troops were ordered into the suburbs; the atmosphere was highly charged, intimidating to some, encouraging to others. Before the session, pressure was brought to bear on many senators.

On 26 December Caesar had sent Curio with a letter addressed to the Senate. On the morning of 1 January 49 it was delivered to the consuls at the entrance to the temple of Juppiter. Two tribunes of the people insisted that Lucius Cornelius Lentulus Crus, who was to conduct the session, should read it out. In it Caesar once more listed all his services to the city and referred to the fact that the people had rewarded him by granting him the right to seek office *in absentia*. He was prepared to lay down his command. But only if Pompey laid down his. Otherwise he would retain his legions, for he did not wish to hand himself over to his opponents. Indeed, he was prepared to free the commonwealth from the rule of the clique that was clearly depriving it of its liberty. Caesar must have calculated that he would finally be able to intimidate the feeble Senate.

The consul, however, permitted no discussion, but began his report as soon as Caesar's letter had been read. He declared himself ready to obey the Senate, but only if the senators made a clear decision, without letting themselves be influenced by Caesar. Otherwise he himself would seek Caesar's favour. In fact he could have done so, for he was deeply in debt and Caesar had offered him financial support. Whereupon Pompey's father-in-law, Quintus Caecilius Metellus Scipio, rose to make a similar declaration on Pompey's behalf. If the Senate still hesitated it would in future seek Pompey's help in vain. They appealed to the anti-Caesarian opinion of the Senate majority; if the Senate could not bring itself to deprive the proconsul of his governorship, they would draw the obvious conclusion and seek an agreement with him.

Some of the senators spoke against this, including Marcellus, the consul of the previous year, who insisted that the levies must continue, so that the Senate could decide freely, protected by a Pompeian army. It is not clear whether he thought the outcome of the deliberations too uncertain or whether he had by now realized how difficult it would be at present to oppose Caesar with military force.

For Caesar had in all about eleven legions, many of which he could bring to Italy within a few weeks. Pompey, on the other hand, had seven legions in Spain, which might be able to attack Caesar's rear, but not for some time. In addition he had two legions in Italy, but these had fought under Caesar and were therefore not wholly reliable. He could not make large-scale levies until he was empowered to do so. And if he did, there was a danger that Caesar would march on Rome. Of course no one knew the real balance of forces, especially as Pompey, for tactical reasons, acted as though he were sure of victory.

The Senate accepted a proposal by Metellus Scipio that Caesar should dismiss his army by a certain date, obviously 1 July; if he refused, this was to be taken as a hostile act. The tribunes of the people interceded against this. The Senate was in favour of going into mourning. The tribunes interceded again. Nevertheless the senators left the meeting, went home, and returned in mourning. In the evening Pompey summoned them all to visit him at his suburban residence, where he praised those who had shown resolution and criticized those who had vacillated. The city filled with his soldiers, old and new.

Some senators, including Caesar's father-in-law Piso, said they wished to go to Ravenna and treat further with Caesar. Although they sought leave of absence for only eight days, their request was refused. In any case it was not necessary, as conciliatory approaches had already come from Caesar.

Curio had not only brought his letter, but been given instructions as to how he was to act if the Senate refused to be intimidated. Cicero acted as go-between. It turned out that Caesar was willing to hand over the province of Gallia Transalpina and eight legions. He dropped the demand for Pompey to lay down his governorship, but expected him to depart for Spain. Cicero finally won a further concession: Caesar would even give up Gallia Cisalpina and two further legions, retaining only Illyricum and one legion. This he insisted on. In return for these concessions his privilege of seeking office *in absentia* was to be recognized.

If this offer had been taken up, Caesar would have virtually forfeited the possibility of waging a civil war. At all events it would have been difficult for him to mobilize those legions that had passed under a new command. The Senate would admittedly have had to jump over its shadow and promise Caesar's soldiers proper provision, so that they would no longer be dependent on him. In one crucial point Caesar remained adamant: he continued to insist on becoming consul without entirely surrendering his command, so that he could not be arraigned.

Pompey was prepared to consider Caesar's offer. However much he feared Caesar's renewed consulship, civil war appeared to him worse. Caesar may have promised to collaborate with him in future and even to show consideration when in office. But the determined opponents around Cato insisted that he must pursue his candidacy as a private citizen, and it was on this that all attempts at mediation foundered. Cato declared that one should rather seek death than allow a citizen to dictate conditions to the republic.

On 7 January the Senate met again and passed the *senatus consultum ultimum*: 'That the consuls, praetors, tribunes of the people and he who stands as proconsul outside the city shall see that no harm comes to the commonwealth.' Against this there could be no intercession. On the contrary, any future intercession could be suppressed on the basis of this resolution. The two Caesarian tribunes, seeing themselves threatened, left the city with Curio. The course of events from now on was predictable. Pompey immediately began levying troops, ordered the delivery of weapons and collected money from every possible source. In the night of 10/11 January, as soon as he had been informed of the news, Caesar crossed the Rubicon.

As Cicero later remarked, he did not want war, but simply did not fear it. It was his opponents who had wanted it.

This was a curious situation: the man whom the majority of the citizens regarded with antipathy, distrust and fear was less intent upon war, far more decisively and seriously in favour of peace, than those who opposed him and had the sympathies of this very majority on their side.

The demands made by both sides related to the existence of Caesar and the republic. Both were thoroughly justified, but the one precluded the other.

On Cato's side was the work of centuries, one of the greatest heritages of world history: he and his followers were in duty bound to preserve the republic they had inherited from their forefathers. If Caesar saw exclusively self-seeking motives at work in them – and Cicero now and then concurred in his view – he was not entirely wrong. When are such motives ever absent? They identified themselves with the commonwealth, but this did not alter the fact that they felt a duty to it. And they knew that they must triumph now or never, if they wanted to thwart Caesar's unscrupulous and destructive power or indeed an autocracy. Yet this meant risking a civil war, the outcome of which could not be foreseen, especially as it might end with Pompey ruling Rome instead of Caesar.

On the other hand, Caesar could not simply surrender to his

opponents after all he had achieved. It was all very well for Pompey to send him word that he himself was acting solely in the interest of the republic and that Caesar should bow to it too. Who, after all, was the republic at that time, if not Pompey and his allies? True, Caesar had once declared himself willing 'to take anything upon himself for the sake of the commonwealth'. There were some things, then, that the republic could require even of him. But could they include risking his very existence? After everything he had built up, it was indeed understandable and – from Caesar's highly subjective viewpoint – justifiable that he should refuse to take this sacrifice upon himself. Yet this meant that he had to begin the civil war for his own sake.

On the one side, then, there was a great gulf between the necessary and the practical and, on the other side, no less a gulf between the necessary and the permissible. The opposing parties were forced to do what they should never have done.

Moreover, those caught between these extremes were too weak. And it was ultimately this weakness that expressed itself in their indecisiveness. They were for Pompey, yet they acted for Caesar by being in favour of peace. Wishing to stand above the parties, they fell between two stools. Not wishing to take up a position on the side where they belonged, they succeeded in making this side all the more inflexible.

The efforts at mediation at the beginning of January were in any case bound to be made more difficult because the consul Lentulus and his allies could not rely on the Senate majority. A majority whose inclination towards peace stemmed primarily from weakness perhaps made peace negotiations impossible and promoted war, while at the same time weakening the prospects of the side that most of them supported. And because they supported it they added significantly to the authority of Cato and his friends. To this extent they strengthened his cause – yet in effect they helped the cause of his opponents. It cannot be ruled out that many of them were motivated partly by the realization that it was impossible to win a civil war, whoever emerged as victor. The trouble was that their fear of war exceeded their ability to prevent it. And their weakness certainly involved a modicum of political masochism. Otherwise it would have been hard to endure.

This was the climax of the astonishing process by which the

republic moved towards its dissolution without anyone's desiring it – because it was now impossible to act in any way but the wrong way. In Caesar and his opponents two different realities confronted each other. This was a situation that should never have been allowed to develop.

One may of course wonder whether there might not have been a way out – at the beginning of 49 or at the end of 50, or even earlier. And one must not underrate the human capacity for mastering even the most difficult situations. Yet the likelihood is that the situation left no alternative, especially after the end of 50. The laws of action were determined far more by the positions within the overall configuration than by the free will of the actors. This too was a feature of the situation. It therefore seems to us paradoxical, contrary to all the expectations we have of the human capacity for action. Yet we all know that such situations arise from time to time.

How this particular situation arose is made outwardly clear by the course of events. Moreover, when we realize that in the Roman aristocracy it was possible for an outstanding outsider to conquer an entire country in defiance of the Senate and win the devotion of his soldiers, some of the structural background to these events becomes comprehensible. Yet this is to understand only half of the historical process. Even in politics much is decided not by the actors, but through them. The total effect of their interaction always far exceeds what they settle between themselves. One must therefore take a more detached view and study the process of the crisis as such if one wishes to understand its seemingly paradoxical outcome, the impasse of 50/49, and the personality and position of the man who started the civil war for extremely personal reasons – a personality and a position that become increasingly puzzling, the more we have to explain by reference to their particular character.

Montesquieu believed that we were dealing here with a more general pattern. 'If Caesar and Pompey had thought like Cato,' he wrote, 'others would have thought like Caesar and Pompey.' The roles were ready to be filled, as it were, and to play them was not only a matter of personal guilt, but at the same time a recognition of the structure of the age.

12

The Process of the Crisis without Alternative, Caesar's Right to the Civil War, his Greatness

THE EFFECTS OF ACTIONS DEPEND ON
THE WIDER CONFIGURATIONS ·
CORRUPTION · THE IMPETUS OF THE
MAJOR CONFLICTS · SENATORIAL
RESPONSIBILITY AND WEAKNESS: THE
LIMITED CAPACITY OF THE REPUBLIC ·
THE TENDENCY TO ONE-SIDEDNESS ·
CAESAR'S INSENSITIVITY TO POLITICAL
INSTITUTIONS · 'HERE I DEPART
FROM THE BASIS OF LAW' · GREATNESS
AND LACK OF INHIBITION

HOW IS IT POSSIBLE for an order to collapse when all who have a share in it regard it as the proper order? To put it more precisely: how is it possible for it to be destroyed by those who have a share in it, in the absence of any extraneous influence – to

349

be destroyed when no one wishes to attack it, to be annihilated when no one repudiates it?

Such an effect can be produced only by unintended side-effects of action. Whatever is done by those involved – in pursuit of great ambitions or petty interests – certain impulses generated by their actions are bound to feed into the wider context of the process, which none of them desires but which must nevertheless have resulted from their actions, since nothing else can have given rise to it. If this happens, however, the configurations must be of such a kind as to have generated these side-effects. For any effects that arise from an action – beyond its immediate success or failure – are independent of the action itself. Something as innocent as the procreation of a child was of relatively little consequence for the demography of eighteenth-century Europe, but in a modern developing country it may be part of a profound change. If a judge is bribed, this may be either an exceptional occurrence of trifling significance or part of a process of growing corruption; it may even, in a highly explosive situation, be the crucial action that ignites the powder-keg of public indignation. Within processes of change, the cumulative effect of banal events can thus be far more pregnant with consequences than illustrious political deeds, whose effect does not extend beyond the history of events. Everything depends on the configurations within which actions take place.

Human beings always do much the same things: they try to secure and enjoy their lives, attend to their duties and their interests, make the best of their opportunities, engage in administration and politics, contend with their opponents and seek to distinguish themselves. However, when the Romans of the late republic did these things they hastened the dissolution of their order, whereas their predecessors in the classical republic, acting no differently, had simply demonstrated its durability. For the configurations had changed. Defending the traditional now had the same effect as reform; attempts to restore order and lawful procedures actually generated a stronger impulse to dissolution than adherence to humdrum routine. Livy put it succinctly when he said, *Nec vitia nostra nec remedia pati possumus* ('We can bear neither our shortcomings nor the remedies for them'). By acting against one another, the Romans produced side-effects that all tended in the same direction.

To begin with the simplest example, we can observe various vicious circles (or rather vicious spirals, for the momentum was increasing). The exploitation of the provinces led to widespread corruption. This was in many ways an incentive both to further corruption and to efforts to reduce it by setting up special courts and increasing the penalties. Yet these measures led to further exploitation. For now the judges too had to be bribed.

As early as 70 Cicero observed maliciously that he could foresee a time when the provincials would ask for the repeal of the laws intended for their protection. For then every governor or member of his staff would take away only as much as he considered sufficient for himself and his children. One is reminded of the well-known case of those laws which in the long term tend to harm those whose situation they seek to improve in the short term. When Tacitus remarked *corruptissima re publica plurimae leges* ('When the republic is at its most corrupt the laws are most numerous'), he was indicating the connection between increasing abuses, increasing legislation, and the increasing ineffectuality of the laws. Above all, the sum of the infringements was matched by a slackening of society's judgement as a whole, on which the preservation of the traditional order had once rested.

Prices rose. Whoever sought election incurred more and more expense. For him it was merely a matter of being elected and pursuing a virtually predestined career. But it had the side-effect of contributing to the increasing corruption, growing expectations and worsening exploitation. The process admittedly did not follow a straight course; in 52, for instance, morals again improved.

Nor need growing corruption, as history shows, be fatal for a political system. But in Rome it was part of a wider pattern. Within the oligarchy, for instance, corruption was rife because it was linked with the popular enthusiasm for ever more splendid theatres and games, and because the ambition of individuals and families was increasingly freed from the solidarity that had previously existed among the nobles.

Syme speaks of the 'great truth' that 'corruption can be a guarantee of political freedom'. He is probably right in principle, though corruption must not become the motive force behind a process of change. In Rome, however, a lively competition was unleashed, which finally gave Caesar the opportunity to buy important helpers. This in turn was ultimately part of the inner transformation of the aristocracy, which produced a narrowing of senatorial standards and at the same time led certain outstanding personalities to ignore them. As no one was allowed to gain even a toe-hold from which he could reach a position of power and raise himself above the basic oligarchic equality, almost all stirrings of individuality, imagination and independence, almost all attempts to react to novel situations by novel means, were viewed with suspicion and distrust; mediocrity, rigidity and narrowness were at a premium. As a consequence there were frequent failures and a repeated need for someone different to step in, namely Pompey; he then gained an unusual ascendancy because no one else was able to act unconventionally and, in particular, to operate with the new professional armies.

Yet the interplay between the bulk of the senators and the individuals who confronted them as outsiders, was a phenomenon of long standing. It had occurred more than once since the time of the Gracchi. The underlying cause was that Rome was faced with political tasks of such magnitude that they could not be performed within the framework of traditional senatorial solidarity.

Behind this lay the impossibility of addressing the many weighty problems that arose directly or indirectly from Rome's imperial power by employing the institutional, intellectual and moral resources of a 'communal state'. Ultimately it was the contradiction between the forms of the communal state and the exigencies of a world empire that led to the collapse of the republic. Montesquieu recognised this long ago. Here lay the true cause of the moral decline to which the Romans ascribed its downfall, and of its inability to bring so many abuses under control.

Yet while the many unsolved problems spawned a variety of political forces, these were not of such a kind as to attempt in any way to oppose the existing order. On the contrary, all the sources of opposition operated within this order and ultimately remained isolated.

They affected the traditional order only indirectly, by working to the benefit of individual nobles who had somehow come to occupy 'outsider positions'. For the great antagonists of the Senate majority all represented demands that resulted directly or indirectly from the unsolved problems of empire. They embodied a concern with the practical problems of the commonwealth.

★ ★ ★

The decline of the republic drew its real impetus, willy-nilly, from the conflicts between Pompey and the Senate during the late Sixties and the Fifties. These started with Pompey's demands arising from the need to put an end to the depredations of the pirates and the difficulties caused by Mithridates. He was concerned with specific tasks. The Senate's chief concern, however, was with the power that was bound to accrue to Pompey and its fear that he might no longer abide by the rules of oligarchic equality. Hence it opposed him with its veto and its authority, and as a result the republic suffered real damage. In the first place, its institutions were impaired and, in the second, Pompey was obliged to seek more than just a generally acknowledged position of privilege. He now proceeded to build up an independent position in opposition to the Senate, and this produced an enduring rift between him and the leading senators. In the most important political conflict of the age, then, the Senate no longer stood above the parties, as it had nearly always done in the past. Rome was thus deprived of the supreme authority that could ultimately effect a compromise or assert its will. Instead, the Senate suffered repeated defeats, until Pompey finally resorted to the promotion of anarchy in order to force the Senate to accept him. The context within which a process operates not only involves the subsidiary consequences of actions, but also produces 'consequential actions'.

By defending the republic, the Senate – or, to be more precise, its leading members – gave Pompey cause to attack it. By the year 52, when they were finally prepared to do a deal with him, a new and much more incisive conflict was unavoidable.

The real motive force behind the process of decline is becoming transparent. It consists in the fact that, on the basis of defending the republic, certain forces came to oppose each other so intensely and so extensively as to do profound damage to the republican order. The different forces had become so disproportionate as to fuel the process of disintegration.

In earlier times the hand of fate would doubtless have been discerned in a situation in which everyone destroyed what they wished to preserve. It is obviously a case of the kind of involvement in history that can lead conservatives to engineer revolutions and make reformers effectively into enemies of change – that can make a lover of peace into an agent of war, the forces of evil into a power for good. In such a complicated situation the impulses of opponents who are resolved to fight one another to the death may work in identical directions. From one angle they may appear as opponents, from another as partners, if the historically most significant outcome of their contention is the destruction of a community's institutions.

There was a time when such experiences would have been summed up by the apophthegm 'Man proposes, God disposes' – and in relation to some periods, of which the Roman republic is admittedly not one, the past-tense version 'Man mused, God was amused' seems to give it added point. Yet this truth need not always produce a result that is so paradoxical, so much at odds with normal expectations. The result depends rather on the circumstances. Yet if an order fails to contain all the possible tensions, but lets them develop into a powerful conflict with itself, this then finds expression in the consequences – and naturally in some way in the motives – of the actions taken within this order.

But it was probably rare, if not unique in world history, for events to take such a striking course as they did in the Roman republic. The republican order eventually became an issue not between, but beneath the parties; it was no longer the subject, but the object, of controversy. There were no conflicts about the order, but about the price of the order. Superficially the dispute was about numerous individual questions, but in the background lay only the positions of Pompey and Caesar within the commonwealth. The real effect of the dispute consisted in the accumulation of its side-effects, which contributed to the progressive weakening of the traditional order. Controversy yielded to change. There was in fact a notable discrepancy between what was at issue and what was changing. Accordingly Roman society had little control over its affairs; it produced no one who opposed the inherited order, but only outsiders – no innovators who could have pointed it in new directions, but only men of a new and different type. It was not by instituting changes, but by trying to demonstrate their abilities in new ways, by taking up unprecedented positions, that they contributed to the momentum of the process. But the fact that the process took such a strong hold on the republic as a whole was due in at least equal measure to the defenders of tradition.

This situation could arise only because of the universal conviction that the inherited order was the only just one. Had this not been so, the Senate's many failures and its conspicuous weakness would inevitably have raised serious doubts about the viability of its régime.

As there was no alternative to the existing order, the under-privileged, those suffering hardship and deprivation – the potential rebels, in other words – had no chance to unite and produce an intellectual and political counter-force. No new ideas emerged about the governance of Rome, let alone any prospect of linking such ideas with various interests in a purposeful endeavour to create something new. Instead, there was general satisfaction with the old order on the part of the powerful or potentially powerful, and impotence on the part of the dissatisfied. The satisfaction of the former was largely due to their realization that, if necessary, they could assert themselves against the Senate. It was the very weakness of the Senate that made its régime so popular. It was at this stage quite impossible to take a detached view of the existing order and see it as merely one option among several.

It is not clear whether the wider ranks of affluent Roman society clung to the existing order because they were aware of the liberty it afforded them or because they felt existing conditions to be part of a powerful social identity. It may not even have occurred to them to consider the function that the order should fulfil. In any case, they knew of nothing different: the alternative was either an unjust order or none at all.

The unanimity of all the 'good' (that is to say, the wealthy) in their attachment to the Senate was so robust that Cicero could define good politics as politics that pleased the good.

If the inherited order was the only legitimate one, it followed that the leading senators must do everything in their power to preserve or restore it. It was this that gave such force to Cato's position. It was not only Cato's character that led him to resist so tenaciously, but the expectations that society had of the leading senators – and of the rest of the Senate, whose views they took into account when deciding what attitude to adopt. The authoritative role of the leading senators had always found its legitimation in promoting the interests of the Senate and the republic as a whole.

Now that the republic was in such danger, they had to go on representing the interest of the whole. The Senate majority may have been inclined to resignation. And with hindsight we may say that this was the best way to preserve the republic for as long as possible. At all events, any attempt to manage the Senate régime more strictly had always aroused dormant forces and made matters worse. Yet with the best will in the world no one could have known this at the time. Otherwise they would have had to despair of their order. Hence, the less the old institutions worked, the more firmly they clung to them. And this meant that the more the republic needed Pompey – in the end even to ensure public safety – the more resolutely the Senate had to oppose him – until the anarchy became intolerable.

For senatorial responsibility was incompatible with a weak Senate: hence the narrowing of senatorial norms. Hence also the fear that made every danger seem far greater than it was – the fear that Pompey might acquire excessive power. But when one is weak, the dangers not only appear greater, but actually are greater. This led to the struggle that seemed all the more necessary, the more hopeless it became. Finally, the attempt to destroy Caesar's political existence formed part of the logic of this mixture of senatorial weakness and senatorial responsibility.

One should certainly not belittle the influence exercised by individual personalities, especially such men as Cato, Pompey and Caesar; nor should one underrate the role played by chance in individual situations. But in the positions they took up lay a certain pressure that caused them to act as they did, forcing them, almost fatefully, into a configuration in which their actions, through their side-effects, propelled the process of republican decline.

The political capacity of the commonwealth was insufficient to make the configurations themselves – the source of actions that had far-reaching consequences – into an object of political action. Everyone was too much involved in them. There was no opposition that could supply a starting point for a widening of the system. Not enough power could be assembled in one place in order to create institutions, of whatever kind, that would allow the totality of actions and processes to be brought back into conformity with some order, so that whatever the politicians did would once more be part of an order and not of a process of disintegration. There were thus no 'reserves of reform', no ways of enlarging liberty and political participation. Democracy was practicable only in communal states; world empires required aristocratic or monarchic rule. The Roman aristocracy was no longer equal to the task, and no one wanted a monarchy. For no one was prepared to give up his existing liberty or the order of which it was part. Everyone was too republican – except perhaps the urban poor, who did not count and in any case knew nothing better to do than to acclaim great men.

Society could no longer cope with its problems within the framework of traditional liberty, but it was unwilling to renounce this liberty. This was the essence of the crisis without alternative. Yet in the absence of any possible alternative, the existing order was bound to be destroyed – in the struggle to defend it. It was simply worn down, not because anyone wished to destroy it, not even despite the fact that everyone wished to save it – but precisely because of this fact. The only possible exception was Caesar, whose purpose may have been different.

An immense one-sidedness took hold of the various forces. Because Cato concentrated on defending tradition and Pompey was interested in urgent practical problems, a contradiction arose between concern for the constitution and concern for the tasks that needed to be performed. In the prevailing political conditions the one precluded the other.

This tendency to one-sidedness may account for the abundance of imposing figures who people the period. Crassus embodied the ultimate ideal of wealth, Lucullus that of luxury (which was constantly practised, but never acknowledged). Cato realized the ancient ideal of gravity and consistency, but with un-Roman rigidity, Caesar that of achievement, but with un-Roman freedom. The younger generation – Clodius, Curio, Caelius, Antonius, together with their ladies – was one-sided in its rich, but ruthless cultivation of its opportunities, its devotion to pleasure, its quest for distinction without commitment, its reluctance to be harnessed to the duty of constant care and attentiveness. Others were typical of the age in that they represented an unprecedented gamut of possibilities. Cicero combined Roman reality with Greek theory, devotion to recognized standards with an unusual degree of opportunism, firm advocacy of the republican cause with conspicuous disdain for its advocates. What united them all was that they belonged to the most powerful aristocracy the world has ever known. In them the old constraints of Roman discipline gave way to high pretensions. The crisis posed a challenge, but it did not shake their faith in tradition; yet at the same time they were receptive to a wealth of new possibilities deriving from the recent discovery of Greek culture.

What characterized Pompey was the fact that he had absorbed all the contradictions inherent in the notion of a 'communal state' ruling a world empire. He combined the achievement and pretension of a man who had solved numerous imperial problems and could think in Mediterranean terms, yet showed a becoming respect for the Senate and good society. This is what accounted for all his difficulties and his political ineptitude. He was constantly forced to disavow himself and his friends in order to preserve the semblance of total respect for convention; he could do this only by enveloping himself in a fog and constantly saying what he did not mean. With Pompey vacillation and inconsistency seem to have become second nature.

For Caesar, by contrast, the contradictions of the contemporary Roman world lay between himself and the Senate. He adopted a position that was based quite one-sidedly on achievement and ignored respect and discipline. He saw no goodness in the 'good'. It is not at all clear whether he could envisage anything better than the existing order. It is not obvious that he had any new ideas or inspired anything new – only that he thought and acted in a new way.

His was a bright light on the dark Roman horizon. Not being implicated in the complaisances and entanglements of Roman society, he retained, as it were, the purity, freedom and innocence of the consistent outsider.

At the same time he had the remarkable clarity of judgement that intelligent outsiders sometimes possess. Political intelligence is after all – apart from its tactical aspects – essentially a function of the position one occupies. It is easier to form a judgement if one is at the centre of a group and guided by accepted axioms than if one is exposed to the whole complexity of political possibilities. Similarly, it is easier to form a judgement from outside, because it is easier to see through things. Yet the clarity that arises from detachment may be deceptive. Seeing through things can all too easily mean seeing right through them, and this is another way of missing the reality.

Caesar was insensitive to political institutions and the complex ways in which they operate. In Sulla's day he had already come to see the senators mainly as Sullans – not as representatives of the whole commonwealth, but as the heirs of the winning party in the civil war. He had no respect for them then, and they could not win his respect later. The whole system was flawed and inefficient. Since his year as consul, if not before, Caesar had been unable to see Rome's institutions as autonomous entities. They failed to measure up to his high standards. He could see them only as instruments in the interplay of forces. His cold gaze passed through everything that Roman society still believed in, lived by, valued and defended. He had no feeling for the power of institutions to guarantee law and security, but only for what he found useful or troublesome about them.

Accordingly, he could employ the most contrived and preposterous arguments with regard to the principles of the order. Not only when he suddenly saw the tribunes' right of veto threatened, but also in connection with the *senatus consultum ultimum*: at the beginning of 49, for instance, he complained to his opponents that this resolution had previously been resorted to only to deal with dangerous legislative proposals, violence and popular unrest. None of these had occurred. As though his own insubordination did not go far beyond all these!

Thus what struck him most about the Senate was the fact that it was controlled by his opponents. It hardly seems to have occurred to him that it was responsible for the commonwealth. He could impute only selfish motives to his opponents and was apparently unable to appreciate that the Senate was also very much concerned with preserving the inherited republic and the senatorial régime. Hence, when he began the civil war, he held that the body politic as a whole had no part in it. He was probably quite unaware that he was completely brushing it aside, along with all the institutions in which it was involved.

In Caesar's eyes no one existed but himself and his opponents. It was all an interpersonal game. He classified people as supporters, opponents or neutrals. The scene was cleared of any suprapersonal elements. Or if any were left, they were merely props behind which one could take cover or with which one could fight. Politics amounted to no more than a fight for his rights. This was the angle from which Caesar viewed the various positions and from which even the Senate and the citizens were assigned their roles: as those who had to grant him his rights – and would have done so had his opponents not used cunning and force to prevent them. Under normal circumstances his extraordinary services to Rome, his claim to honour (*dignitas*) would have received due recognition. If the Senate and the popular assembly were only free he would have nothing to fear. He therefore stated that he wanted to 'free himself and the Roman people from the domination of this small clique' – meaning his opponents.

It is true that Pompey's soldiers restricted the freedom of the commonwealth to such an extent that Caesar could not assert himself in it. It is true that the Senate majority was weak enough to let Pompey have his way. It could thus be only the 'intrigues of his opponents' that forced him to invade Italy.

Whoever saw things in this way was no longer living in the same reality as the society to which he belonged. Yet there was nothing pathological in the way he had cut himself off from it and built up a partial reality of his own – or if there was, it was part of a disease that had taken hold of the whole commonwealth.

This becomes evident when we realize that his opponents too inhabited only a partial reality. They acted in accordance with old and generally recognized convictions, but the precondition for these convictions – the ambience they presupposed – was no longer present. Hence, they too were one-sided in insisting on their absolute authority. They acted as much in the interest of the inherited order as against it. For although it was established that the Senate was the supreme organ of the republic, it was never envisaged that any impossible or highly dangerous enterprise should be taken in hand to secure its régime. On the contrary, the traditional order had been grounded in reality and always derived its efficacy from the sense of reality possessed by its supporters. As the Senate had more than once acknowledged the reality of the conquest of Gaul, the policy adopted by Cato and his allies was at once justified and unjustified.

They failed to see the absurdity of prosecuting a man who had done so much for Rome during the past nine years. This was no less unrealistic than Caesar's failure to appreciate how alive the republic still was in the minds of the citizens and to understand that what he perceived in the background was merely the truth as seen by an outsider, not the absolute truth. It was a truth that, once perceived, empowered one to act, but not one that would have carried conviction with Roman society.

The epic poet Lucan (AD 39–65) makes Caesar say at the Rubicon,

'Here I leave the basis of law, dishonoured as it is' (*temerata iura relinquo*). Whether or not he said it or thought it, it was true that the common basis of law, which had united the whole commonwealth and embraced both Caesar and his opponents, no longer existed. The preconditions for lawful action were no longer intact, because both sides had moved too far apart. The majority of the citizens, caught in between, no longer had the strength to bridge the gap.

If both sides had abandoned the common ground of reality, if the common reality was fragmenting, so that both sides were right in their own terms, this gave rise to a new, fractured reality. What was peculiar about this reality was that so much power could be marshalled against those with responsibility, yet without raising any doubt about their régime. Only this made it possible for them to feel so secure and carry the Senate with them. Another curious feature was that this power was not linked to a cause, but only to a personal claim – though admittedly it was a claim that had already almost objectified itself in the immense efforts of Caesar and his army in Gaul.

Strasburger wrote that when contemplating Caesar we have a sense of being transported back into the world of Achilleus or Coriolanus, into an archaic age in which men denied the claims of the commonalty and even made war on their own city for personal reasons. For Caesar could not be seen as a precursor of the emperors Augustus, Trajan or Hadrian. Yet in every sense he was a child of his age – admittedly an age that one can comprehend not from a so-called stage of development, but only from its structure: Caesar could emerge as he did only from the crisis without alternative.

The fruitful tension that had shaped the Roman republic from the beginning was released. The archaic aristocratic quest for honour and fame had never been broken or brought to an end, but channelled by a powerful solidarity into the service of the commonwealth. And Rome's aristocracy had been remarkably successful bringing its weight to bear against the ambitions of individuals. Yet as soon as the tension was relaxed, the striving for *dignitas* regained some of its autonomy and force, especially in outsiders. Sulla and Catiline may be cited as examples. In Sulla's case too the preservation of the honour he felt he was owed on the basis of his achievements was a major cause of the civil war, though not the only one. The Roman aristocrat was never a tool of the republic, but only a partner in it. In the absence of an alternative, which in other circumstances might have furnished a cause, there was nothing left for a strong-willed and powerful outsider like Caesar to do but to exaggerate and pervert the ancient claim to *dignitas*. This was not a matter of personality, but of the contradictory nature of contemporary Rome, which determined the conditions under which he developed and was able to create a world of his own. He had to answer for what he did, but what he *was* went far beyond personal responsibility. There was an element of tragedy in the way he and his world came into conflict with the republic.

The enormity of his decision to start a civil war is not thereby lessened, but it becomes more understandable. Yet unless it is pathological, it must be an expression of the greatness of Caesar's personality. Throughout his career, and especially in taking this decision, he displayed not only extraordinary ability and strength of mind, but such staying power, steadfastness, such a wealth of inner resources, that here, if anywhere, the term 'greatness' may be used to capture something essential. Yet we must not lose sight of the fact that in greatness, as in all things human – and perhaps to an even greater extent than in all others – there is a certain ambivalence. It comprehends both the admirable and the abominable. Seel put this very well when he wrote, 'What he was did not exceed what a man can be, but in being what he was he achieved success and a special kind of effectiveness; with so wide a range, and with so far to fall, he displayed an individuality of manner and style that raises him above the amorphous plurality and anonymity of what had gone before.' Seel forbears to say whether this can be called 'greatness', and merely adds, 'It is certainly not "edifying", yet it has such human force that little else in world history is so instructive or teaches us so much.' I am inclined to think that this very ambivalence justifies the use of the term 'greatness'. We should not deny Caesar's greatness, but we must not see it as exclusively good and wholesome.

The identification of such qualities always involves a degree of generalization. Any preponderance of great qualities is perceived as greatness. This is, so to speak, a result of the way perception is pulled in a certain direction – towards the old or the new, towards youth or age, towards good or evil – and then confers the bonus of generalization on the one side or the other. Yet does not something similar occur wherever personality and environment interact? As a concrete example we may cite what took place in Gaul between Caesar, his officers and his soldiers – and his defeated enemies. What made Caesar superior as a commander and a leader of men then appeared as superiority pure and simple – without being denoted by any term that would correspond to our concept of 'greatness'. There were later parallels during the victories in the civil war, this time involving a wider circle of observers; some were impressed by what struck them as brilliance, others more by what they found terrible. Yet all concurred in discerning something extraordinary in Caesar and crediting him with an immense personal capacity that he must have recognized and cultivated in himself.

Yet there is nothing to suggest that Caesar was in some way the executive arm of a higher authority such as the 'spirit of the age'. Nor did he control what happened, at least not until the civil war began; he was merely part of it. In fact he was very much caught up in the process of reciprocal one-sidedness, which worked through him and with which his career and the extraordinary development of his personality were essentially bound up.

At best one may say with Jacob Burckhardt, 'Great men are necessary for our life, in order that the movement of world history can free itself sporadically, by fits and starts, from obsolete ways of living and inconsequential talk.'

This is not to say that Caesar was an exception in his day and did not want to preserve the old republic. It may well be that he too could think of nothing better than to preserve it, but that the preservation of the republic was low on his list of priorities. It was, after all, not entirely fortuitous that the ultimate form of the alternative was 'Caesar or the republic'.

13

The Civil War (49–46 BC)

IF CAESAR WAS TROUBLED, before crossing the Rubicon, by the thought of the misfortune the civil war could bring upon everyone, this shows him to have been painfully aware of the human cost of his rebellion against Rome – of the possible casualties, the suffering, the destruction.

Yet he could not take his fellow citizens seriously as citizens. This is attested by his remark, 'What befits a decent man and a peaceable, honest citizen more than that he should remain aloof from civil disputes?' In his opinion, the fate of the *res publica*, even in the hour of its greatest need, was no concern of theirs.

This is of a piece with his self-justification, according to which the republic had come under the control of a small clique and had to be freed from it. Only then could legality return and the republic once more become the common cause of all, in the hands of the Senate and the popular assembly. If no other means was available, this liberation must be brought about by Caesar's army.

As so often happens, however, those who were to be liberated showed no interest, and Caesar soon ceased to make much use of this thesis. He constantly referred above all to the injustice against which he had to defend himself.

This was the sole reason for the war. There is no point in quibbling. Caesar's claim was not a programme, but a plea for his personal right, for the honour he was owed on the basis of his achievements. The issues were, on the one hand, Caesar's claim to *dignitas* and, on the other, the defence of the republic. Accordingly, Caesar maintained that his quarrel was only with his 'opponents' whom he consistently called *inimici*. They, by contrast, saw him as the enemy (*hostis*), who had attacked the commonwealth from without. He claimed to be involved in 'civil disputes', but to them it was 'civil war'. Cicero remarked that the war was so civil that it arose not from disputes among the citizens, but from the temerity of one depraved citizen. It was a conflict not of citizen against citizen, but between two generals and the armies loyal to them. This was the reality of the war, insofar as it was conducted by military means. To this extent the bulk of the citizens behaved precisely as Caesar expected: they took no part in the war. And even Cicero could excuse this by saying that ultimately Pompey was fighting against Caesar and that both were fighting for autocracy. There was of course a difference: as Cassius Dio put it, Pompey merely wanted to be second to none, while Caesar was intent on being the first among all.

Behind all this, admittedly, lay politics, the continuations of which by other means constituted the civil war. Here matters were more complicated. The crossing of the Rubicon set the seal on Caesar's failure to regain his political credit despite nine years of effort. His opponents' victory in the Senate had been close, yet Caesar would have gained no credit even if they had failed, but only a reluctant recognition of his power and his achievements. He was aware that he had lost, but probably not of how little he had lost. In any case it may not have mattered to him.

Caesar failed because of the republic, but the republic also failed because of Caesar, as soon became evident. Yet this, like any other considerations regarding the dual reality of the time, belongs to the historian's perspective. Contemporaries were unaware of them. To them Caesar was simply an insubordinate governor who was leading his troops against the city and from whom they had nothing good to expect.

By starting the civil war, Caesar became utterly isolated. Everyone's fears seemed to be confirmed. His own father-in-law condemned the launching of the civil war as a crime. To Cicero it was an act of madness. There was universal outrage and hostility. And however much of this was a momentary mood that might quickly subside, Caesar could not easily escape the odium of having plunged Rome into a bloody civil war. In a letter written in January 49, Cicero wondered whether they were really dealing with a Roman commander or with Hannibal. 'This insane, miserable fellow has never had the least inkling of the good! Yet he claims that he is doing everything for the sake of his honour! But where does honour reside, if not in honourable conduct? And can it be honourable to hold on to an army without the approval of the Senate or to occupy towns inhabited by our citizens in order to gain easier access to the city of our fathers?' indeed, how can it be permissible to begin a civil war for the sake of one's own honour?

Caesar might be victorious; there was no doubt of this, despite his opponents' superiority in numbers. Whether he could win over the Roman citizenry, or whether his victories would make this all the more unlikely, is another matter.

But we do not know whether he pursued this aim, or even whether he was aware that he should pursue it. There is more to suggest that he thought only of the simple equation of achievement and rank, rather than the laborious work of talking, listening, answering and persuading. If so, he must at the same time have wanted to win himself the position of first man in Rome, in other words a principate. *Princepes* properly existed in Rome only in the plural; these were the *consulares*, Rome's leading citizens. Pompey, however, had become *princeps*, in the singular; he was the first man, with a substantial lead over all the others. And clearly Caesar too had for some time wanted to win himself a principate. True, at the start of the civil war he gave it out that he wanted nothing more than to live without fear under Pompey's principate. But he had also declared several times that it would be easier to demote him from first to second place than from the second to the last. In any case he could never have avoided rivalry with Pompey. Whether such a principate was compatible with the traditional order is another matter. And another matter still is whether Caesar gave this a moment's thought. Since institutions were not important to him, he presumably had nothing against the inherited order so long as it was possible to win recognition within it and assert himself if necessary. And since he was imbued with the 'genius of self-confidence and self-absorption' – which meant, according to Strasburger, that there was a 'natural identity between what was desirable for him and what was desirable for the world' – it may be that no other questions troubled him. At all events he wanted the opportunity and the power to put through whatever was necessary. Whether he also thought of making himself an autocrat is far from clear.

In any case this was not his reason for starting the civil war. He wished to take by force the rights that the Senate had ultimately refused to grant him. And so he had to campaign for four-and-a-half years to obtain the *dignitas* to which he believed he had established a claim in the preceding nine. After that everything would be fundamentally different.

The Italian Campaign and the Policy of Clemency

THE INVASION OF ITALY, ROME
ABANDONED · PEACE NEGOTIATIONS ·
THE 'PARDON OF CORFINIUM' ·
CLEMENCY AS A CONSEQUENCE OF
THE PERSONAL CAUSE, BOTH KIND
AND MASTERFUL · THE SWING OF
OPINION IN ITALY · POMPEY
ESCAPES · CAESAR IN ROME · 'IN
FUTURE EVERYTHING WILL
BE DECIDED BY ME'

F ROM ARIMINUM Caesar sent a few cohorts along the coast towards Ancona and others across the Apennines towards Arretium (now Arezzo). He quickly had several towns occupied within a radius of a good hundred kilometres, in order to show the extent of his resolve and to secure the starting positions for the coming campaign. He still did not know whether Pompey would hold Rome or withdraw to Greece. Depending on what Pompey decided, Caesar would have to march on Rome through Tuscany or move along the Adriatic towards Brindisi. Before doing either, however, he clearly wanted to await the arrival of further forces from Gaul. Besides, he hoped that fresh negotiations would be successful.

Pompey had privately sent deputies to tell him that he, Pompey, was not acting out of personal enmity, but put the commonwealth above everything. Caesar should do the same. Caesar, however, used the opportunity to put new proposals to Pompey: he was ready to relinquish his command immediately, even if Pompey retained his; he would seek the consulship in Rome as a private citizen. He thus bowed to the Senate's resolutions on the termination of his governorship. All Caesar required was that Pompey should depart for Spain and that all armies in Italy should be disbanded.

This would probably have meant the suspension of the *senatus consultum ultimum* and again produced a power vacuum in which Caesar could count on winning his case in the courts, should it ever come to a trial. He probably had little doubt that he would be elected consul. Finally Caesar asked for a meeting with Pompey in order to settle all their disputes and seal their agreements with oaths.

It is curious that Caesar should only now have put forward these proposals, which, for all their modesty, substantially safeguarded his interests. Had he, at the end of December, underestimated his opponents' determination after all? Did he only now realize what a war could mean and how long it might last? At all events, he once more did his best to maintain the peace.

Pompey was inclined to consider all these proposals; he seems to have turned down only the suggestion of a meeting. The distrust still prevailing between him and his senatorial allies did not allow it. They were afraid of a new understanding between Pompey and Caesar. Moreover, even Cato and most of his allies were prepared to accept Caesar's proposals. They stipulated only that he should withdraw to his province of Gallia Cisalpina, so that the Senate could deliberate and arrive at a decision freely and without pressure.

Pompey, most of the magistrates, and many senators had left Rome on 17 and 18 January on receiving news of Caesar's rapid advance. Pompey had declared himself unable to defend the city. When Caesar's proposals arrived at his headquarters on 23 January he was already in Campania, to the north of Capua.

★　　　★　　　★

The evacuation of Rome sent shock waves throughout Italy. That the magistrates and the Senate should retreat before Caesar did not speak well for him. Above all there was mounting fear. And as nothing could be done against Caesar, anger and disappointment were directed against Pompey. People felt deceived. As the well-informed knew, he had long toyed with the plan of giving up not only Rome, but the whole of Italy, in order to meet Caesar in a large-scale strategic operation from both the east and the west of the Mediterranean area simultaneously, using his superior naval power. Whereas Caesar had at least eleven legions, Pompey commanded only seven in Spain and two in Italy, though he hoped to levy further legions. In the east, however he could call upon his clients to supply huge contingents.

Pompey's calculations were based on his experiences in the civil war of the eighties. How much stronger did he need to be than Sulla, whose victory had also been initiated in the east? Such considerations, however, were beyond the comprehension of the senators, to whom Rome was the world and strategy a closed book. They saw only that the city was being abandoned and that they were militarily more and more dependent on Pompey, and they still recalled his confident assertions about his own strength and Caesar's weakness. He had wanted to reassure them, and was actually convinced of his long-term superiority. Moreover, he had probably thought it unlikely that Caesar would start a war with only one legion. For all his other legions were still in winter quarters in Gaul; two were actually in Narbonne and its environs, as a shield against Pompey's Spanish troops.

With the headlong abandonment of Rome – during which the consuls even forgot to make the necessary sacrifices to the gods and to take the public funds with them – the military situation suddenly became clear to all. This was not at all what Cato's senatorial supporters had intended. Their illusions were rudely shattered. They were now keener on peace than on destroying Caesar, which had become an uncertain prospect and could be achieved only at huge cost. Pompey, for his part, was well enough versed in warfare, but at the age of fifty-six he found it more difficult than before to make bold decisions.

Pompey, Cato and their friends, however, decided that it was not for them, but for the Senate, to decide on the disbandment of the armies and the ending of the levies. According to time-honoured principle they could make practical concessions, but the republican order, the responsibility of the Senate, must be upheld. In other words, they wanted to ignore the defeat they had already suffered and pretend that they were voluntarily conceding to Caesar what he had won from them by defiance. However, the Senate could only pass its resolutions in Rome, in the proper buildings.

Pompey therefore replied to Caesar's proposals by saying that his great deeds undoubtedly entitled him to a triumph and a second consulship. If he ceded his command to the successors appointed by the Senate he, Pompey, would be prepared to go to Spain. First, however, Caesar should withdraw from the Roman territory he had occupied. The Senate would then meet in Rome and decide upon the disbandment of the armies in Italy.

This would have met all Caesar's practical demands. He was refused nothing; everything was generously conceded. The occasion for civil war was removed, the struggle no longer necessary. How serious Pompey and his allies were in their response to Caesar is demonstrated by the fact that they at once made it public. 'He would be mad to refuse, especially as his demands are in themselves shameless,' wrote Cicero.

All the same Caesar did refuse. The reason he gives in his book 'On the Civil War' is not convincing. He complains that Pompey set no date for his departure for Spain and, above all, that he, Caesar, was required to withdraw while his opponents went on arming. In particular, however, he was profoundly disappointed that Pompey was unwilling to meet him.

The excuse regarding the timing of Pompey's departure for Spain seems somewhat tenuous: this could surely have been negotiated. And if the Pompeians continued to arm they were presumably acting no differently from Caesar: no agreement had been reached, and Caesar was bound to be militarily superior once he was joined by his battle-tried legions. It was of course hard for both sides to agree simultaneously to abandon their strategic positions, given the distrust that existed not only between Caesar and his opponents, but between Pompey and his allies. For presumably the precise conditions could have been negotiated only between Pompey and Caesar; and that was not possible. Caesar does not seem to have seen this. But that only shows how he misjudged the situation in Rome.

Caesar's misrepresenation reflects the political difficulty he found himself in through refusing to agree to terms. He could scarcely admit that he did not trust the Senate's decision, as he had always maintained that it was fundamentally well disposed to him. He now appeared clearly as the violator of the peace.

He had probably received Pompey's reply by the end of January. Shortly before that, his most important subordinate, Titus Labienus, had defected to Pompey, presumably because he had always been a follower of his. Labienus is said to have told Pompey that he was underestimating his own strength. Thereupon, according to Cicero, Pompey's strategy became more decisive. Caesar now resumed his advance, as the twelfth legion had almost reached him.

In the early days of February he occupied the whole of Picenum, a region where Pompey had many clients. There was no resistance. The towns readily opened their gates – not because they favoured Caesar, but because his military superiority was evident. Insofar as the hurriedly levied Pompeian troops could not be withdrawn in time they went over to Caesar.

Only when he reached Corfinium, a town in the Abruzzi, did he encounter serious resistance. Here a large army division of more than thirty cohorts (eighteen thousand men) led by Caesar's old enemy Domitius Ahenobarbus, was clearly determined to fight. Domitius hoped to cut off Caesar and annihilate him with his two legions (twelve thousand men plus cavalry). He therefore sent urgent appeals to Pompey, who was in northern Apuleia, to come to his aid immediately.

Pompey, however, appealed to Domitius with equal urgency (he could not order him, as he did not have supreme command) to join him, as there was no prospect of a successful encounter with Caesar in the Abruzzi. Obviously Domitius had no conception of the military might confronting him – of the difference between Caesar's legions and his own troops, who were newly recruited, scarcely trained, and inexperienced.

By mid-February Caesar had surrounded Corfinium. On receiving word that Pompey was not coming, Domitius prepared to flee – or so Caesar tells us. His soldiers thereupon offered to surrender. Next morning Caesar had all the senators, senators' sons and knights brought out of the town and led in front of him; for the first time in nine years he saw his enemy Domitius. There was another *consularis* among the prisoners. They feared that Caesar would put at least the more prominent of them to the sword. He, however, claims to have protected them from being attacked and jeered at by his men. He complained that some of them had ill repaid his favours, but then he dismissed them. He even handed back the military fund that had been surrendered by the officials of Corfinium, 'in order not to appear more moderate with regard to human life than with regard to money, although it was certainly public money, which Pompey had allocated for the soldiers' pay'.

This was the famous 'Pardon of Corfinium', Caesar's first act of clemency in the civil war. It had been widely feared that he would follow the example of Sulla and order large massacres, considering what kind of men he had in his following – all the 'infamous youth'. Even if Caesar wanted to avoid it, he would have to yield to them in this. Together with the anxiety over a general remission of debt and the recall of those in exile, this produced a complex of dismal and depressing expectations. It was rumoured also that Caesar found it difficult to restrain his anger and would lose control at the slightest provocation. He had been abroad for nine years and no one really knew him; bad reports had been put about, and by opening the civil war he seemed to confirm the worst.

Against this he set his policy of clemency. Early in March he formulated it in a letter to the two men in charge of his Roman office. He was determined, he said, to act with the utmost leniency. 'Let us try in this way, if we can, to win back public opinion and gain a lasting victory. For all others have incurred hatred through their cruelty and failed to maintain their victory for long, with the exception of Sulla, and I do not wish to emulate him. Let this be a new way of gaining victory; let us secure ourselves through mercy and magnanimity!'

The letter was obviously intended for circulation. The passage just quoted is extremely interesting in that it is linked with the information that Caesar is making a fresh attempt at conciliation with Pompey. Pompey, he writes, should be friends with him rather than with those who have always been the bitterest enemies of them both, and whose intrigues have brought the republic to its present pass.

Thus, even as he seeks and hopes for conciliation, Caesar speaks not only of wanting to prepare or facilitate an understanding through generosity, but of achieving a lasting victory. This underlines his determination to win. We may presume that his conciliatory gesture included Pompey's political existence and honour. He was not concerned with anyone else; all his offers of peace were addressed to Pompey. Only once, in the middle of 48, did he turn to Pompey's father-in-law, Metellus Scipio. But as he expresses it here, Pompey's honour was a function of Caesar's victory, dependent on Caesar and guaranteed by him. The formulation was not very skilful for an open letter, though Cicero called it 'reasonable for all its unreason'. Yet this was how Caesar saw things. His clemency was a means to victory. Conciliation and victory, generosity and triumph, were obviously meant to produce the same result.

Curio remarked at the time that Caesar showed such leniency only because he thought it popular: fundamentally he was cruel, and his cruelty would assert itself once it became clear that he could not immediately attain his ends through leniency. Curio's testimony is unreliable; we have no reason to think that Caesar was cruel. It is well known, however, that he could show great displeasure and anger; these arose from impatience and a determination to assert his will. And after everything his opponents had said of him and done to him he had ample cause to be angry with them. How would Domitius have acted if the roles had been reversed and Caesar had fallen into his hands? Caesar must often have found it hard to release opponents who were likely to return at once to the enemy camp. He could at least have held them prisoner. Yet he almost always showed mercy, even towards some whom he captured a second time; there was hardly one whom he executed, and few whom he banished.

Cicero later praised him for having triumphed over his victory and thereby become almost godlike. At any rate the psychological achievement is considerable. Montesquieu found Caesar's clemency insulting and remarked, 'One saw that he did not forgive, but merely disdained to punish.' Frederick the Great, commenting on Montesquieu's observation, remarked, 'This reflection is exaggerated! If all human actions are judged with such severity there is no longer any room for a heroic deed.'

★   ★   ★

All the same, there were some things that facilitated Caesar's clemency, that encouraged it and presumably placed it in a wider context. It was by no means just a tactic, but in various ways a consequence and an expression of Caesar's greatness and, seen as such, equivocal.

In the first place it was a consequence that Caesar drew from his cause. No one who wages war only for his own sake, in order to fend off injustice, can very well resort to murder. To the first *consularis* who begged mercy from Caesar, he declared that he had not left his province in order to perpetrate injustice, but in order to defend himself against the indignities done to him by his opponents and to liberate himself and the Roman people. These acts of mercy were thus consistent with his decision to go to war for a highly personal cause. Both are expressions of the same greatness – and the same enormity.

For his clemency was prompted by an immense consciousness of superiority. 'Just as he was determined to take the lead in deeds' (in other words, to be the first man), 'so too he wanted to be foremost in justice and reasonableness', Caesar declared to the Senate, according to one source. When the first beneficiaries of his clemency returned to the enemy camp, he wrote to Cicero that he would not let this deflect him, 'for nothing pleases me more than that I should remain true to myself and they to themselves.' Did he not wish to win them over? Or was he disappointed at being unable to do so at a stroke? At all events, he wanted not only to excel them in magnanimty, but to overcome them with magnanimity. He was so resolved upon this new mode of conquest that he was for the most part able to refrain from any cruelties that his anger might urge on him. Through the exercise of clemency Caesar showed himself superior.

It seems to have communicated itself partly to his soldiers. When one of the Pompeians, having captured a Caesarian ship, regarded the crew as booty, but released the quaestor, the latter is said to have stabbed himself in the breast with his sword, for 'Caesar's soldiers were accustomed to giving, but not to receiving it'.

There was in Caesar's clemency, whether consciously or unconsciously, a monarchic trait. Sulla's murders, while spreading terror, also involved respect. They were committed among equals, even if from the position of the victor. To this extent they were in accordance with ancient aristocratic thinking in terms of friend and foe, if one wanted to practise it in true archaic fashion.

Clemency had until now been an instrument of foreign policy: it was the mercy shown by the victorious city of Rome to its defeated enemies, and as such Caesar had often practised it in Gaul. It implied the pardoning of an offence, the waiving of the right to punish the offender. This is perhaps why Caesar diplomatically avoided the term *clementia* in his writings and chose to speak of compassion (*misericordia*), generosity (*liberalitas*) and leniency (*lenitas*). *Clementia*, unlike these, had a connotation of mercy.

The claim it implies is clear not least from the displeasure that Caesar is said to have shown on learning of the suicide of his chief opponent, Cato: 'I envy you this death, for you envied me the chance to save you.' Many old opponents found Caesar's mercy offensive. They were hurt and insulted by it. Through it Caesar won a second victory, as it were, this time over their honour, by making them a gift of their lives and political positions. The consciousness of having accepted his mercy was extremely painful. Yet almost all survivors begged for it; and the reproaches they had to make to themselves for doing so were turned against him.

All this anger derived ultimately from the defeat that Caesar's opponents had to endure and from the launching of the civil war. This was the real offence. Caesar could do nothing to change it. Clemency could only mitigate its consequences; it could not bring reconciliation. Its efficacy was therefore limited.

Yet much depended on the manner in which Caesar exercised his clemency. He seems to have done little to disguise the enormous sense of superiority that lay behind it. It was thus both philanthropic and autocratic, an expression of the greatness that he had acquired over the years. And it probably became effective and obvious at Corfinium. A letter written by Caesar not long afterwards seems to confirm this.

In a letter written at the time Caelius asked Cicero, 'Have you ever read or heard of anyone fiercer in attack and more moderate in victory?'

The release of Caesar's captured opponents apparently soon led to a change of mood in Italy. When it was seen that there were no murders or confiscations and that Caesar behaved 'honourably, moderately and wisely', earlier apprehensions evaporated all the more readily as Caesar had such great successes. Of this 'insidious clemency' Cicero wrote, 'Whatever evil he refrains from arouses the same gratitude as if he had prevented another from inflicting it.'

The whole of Italy began to come to terms with him. We know that inwardly the 'good' remained on the side of Pompey and the Senate, but they did not want to lose anything in Italy; as far as possible they wished to go on living as before. And what were they to do? They did not want to take up arms, and they could not influence events. Whatever they may have thought, they were powerless, since they refused to defend themselves. It was clear from their behaviour that the citizens' republic had outlived itself. They might of course hope that Pompey would return victorious, but meanwhile Caesar was the stronger of the two. One always has to be on one's guard against the stronger party, especially in a civil war.

Moreover, Pompey and some of his prominent allies, in their impotent rage, severely threatened all who did not join them; there was even talk of proscriptions. Cruelty was the mark of the champions of the old republic. Cicero appositely characterized the situation when he asked, 'Is there anything more pitiful than that the one should sue for approval in an altogether disreputable cause, while the other, representing the best cause in the world, should be bent on giving offence – that the one should be considered the saviour of his enemies, the other the betrayer of his friends?'

It was a curious situation. It may be explained by saying that the Pompeians demanded sacrifices of the citizens because they were fighting for the cause of the republic. Yet the almost unbridgeable gap between what they demanded and what the citizens were prepared to offer shows that they might still have the cause of the republic on their side, but not the republic itself. This is yet another indication that the old reality was confronted by a new reality, not just by an individual.

Caesar took into his army the troops that had surrendered to him at Corfinium and quickly moved southwards. On 9 March he arrived outside Brindisi with six legions, three of them from Gaul and three newly enrolled. The consuls had already crossed over to Greece. He waited only for the return of the ships.

Intermediaries again went back and forth. But Pompey still refused to meet Caesar in the absence of the consuls. Caesar had started on the siege-work when Pompey finally set sail on 17 March.

Italy had fallen into Caesar's hands almost without a fight. However, he failed in his intention of ending the war quickly if it could not be avoided. For all his audacity he had been too weak to press forward fast enough. He was now obliged to fight the Pompeians across half the Roman empire. Apart from his own governorships in Gaul and Illyricum, all Rome's provinces were in the hands of his enemies. Moreover, they had control of the seas, while Caesar did not even possess a fleet.

Caesar was thus unable to pursue Pompey, and his opponents won time to train their legions, prepare them for war, and above all reinforce them. In Labienus they had a military man who was intimately acquainted with Caesar's strategy and tactics. And Pompey himself was no mean general. Everything pointed to a long and arduous struggle. In future Caesar would be dealing with quite different opponents, no longer with the brave, but essentially naïve Gauls, whom he had repeatedly outwitted.

He sent two of his junior commanders to occupy Sardinia and Sicily; the latter was especially important to him as a source of grain, since Pompey threatened to cut off his supplies. One of the two commanders was Curio. Caesar provided him with three legions and ordered him to cross over to Africa after winning Sicily.

He himself decided to go first to Spain, in order to defeat or win over the large Pompeian army that threatened Gaul and Italy. He is reported to have said that he was going to the army without a leader and would then turn to the leader without an army. On the way he intended to halt in Rome for a few days.

There his supporters were already in charge of events. In mid-February the praetor Lucius Roscius had brought in a bill conferring full citizenship on all freemen of Gallia Cisalpina. Caesar had spent years trying to achieve this. The population of the province had long been romanized, and many of its sons served in his armies. He set great store by the support he would gain through this law. It was passed on 11 March and was essentially the first measure that Caesar took in Rome, weeks before he actually arrived there.

On his arrival he sought the support of the Senate and at least a partial legitimation of his cause. Many senators had followed Pompey. He had summoned them all to Saloniki, where the magistrates and the Senate now resided, as Rome was now in enemy hands. Over half the Senate had stayed on in Italy, however, not because they were against Pompey or for Caesar, but because they had no wish to go abroad. Most of them remained on their estates, awaiting events.

Caesar posted notices everywhere, announcing that the Senate should assemble for an important session on 1 April; he attached particular importance to the presence of those *consulares* who were still in Italy. He was especially keen to secure Cicero's support. Early in March Caesar had written to him. 'Above all, because I intend to visit the city soon, I am anxious to see you there, so that I can benefit from your advice, your influence, your rank, and your help in all matters.'

He now met him on his way to Rome. Cicero reports that he had expected Caesar to be more conciliatory. But he was quite direct. When Cicero explained that he could not come to Rome, he thought this amounted to a condemnation. Now the others would be inclined to hesitate, he said. Finally he said, 'Come then, and speak for peace.' Cicero replied, 'In the way I think right?' Caesar: 'Am I to tell you what to say?' Cicero: 'I shall propose that it is displeasing to the Senate for troops to go to Spain and an army to be sent across to Greece; and I shall deplore the unfortunate situation of Pompey at length.' At this point Caesar said testily, 'But I do not want anything of the sort to be said.' 'That is as I thought,' replied Cicero; 'that is why I do not wish to be there. Either I must speak in this way and bring up many matters that I cannot forbear to mention, or I cannot come.' The conversation closed with Caesar asking Cicero to think again. On parting Caesar threatened that if Cicero would not give him advice and support, he would get them from those who would; he was prepared to consider anything.

What these words meant became clear to Cicero in front of the door of his house: Caesar's followers were a 'dreadful rabble'. 'They included all the good-for-nothings of Italy.' Nothing but desperadoes; it was shocking to see them all gathered together. Caesar naturally had a magic attraction for those who, by conventional standards, acted improperly in Rome; the worrying thing was that he did not disown such riff-raff.

★   ★   ★

Now, after nine years as governor and commander in Gaul, Caesar entered the city for the first time. He had meanwhile become a stranger; moreover, he was in a great hurry, very active and full of energy, and made his presence felt, even where he was not present in person. It was an unusual sight: the mighty commander, victorious, perhaps even domineering, in a city that had been deserted by all its powerful citizens. Coming from Brindisi, where the consuls, Pompey, and many senators before him had departed from Italy, he was surrounded by a host of eager admirers and followers, palpably in command of his legions. And he was far too hasty to show much consideration.

Caesar addressed the people, justified himself, and sought to reassure everyone that he would see to the city's food supplies; he also promised to distribute money.

Antony and Cassius Longinus, the two tribunes of the people who had fled from the city on 7 January, convened the Senate. Attendance was poor; only three *consulares* had managed to bestir themselves. It was not a proper meeting of the Senate, but an assembly of a number of senators. Caesar made a speech in which he dwelt at length on the insults and impositions he had suffered from his opponents; he maintained that he had sought nothing unusual and been ready to make many concessions. In particular he pointed to his many efforts for peace, which the others had all rejected. He is said to have spoken with great moderation, though he inveighed against those who wanted to make war on their fellow citizens. He then invited the senators to 'take over the commonwealth and join with him in governing it', though he immediately added that 'if they refused out of fear he would no longer trouble them, but govern the common-wealth himself'. According to his own account, he proposed that a peace delegation should to be sent to Pompey. Pompey had admit-tedly said that such embassies were a sign of weakness, but this appeared to him, Caesar, to be a petty judgement. 'He for his part, just as he sought to take the lead in deeds, wished to excel in justice and fairness too.'

Caesar does not speak of his other proposals. However, he wanted the public treasure to be handed over and the commands of Curio and others confirmed; he may also have sought a resolution to the effect that if his opponents were not prepared to make peace, he himself would be entitled to arm.

The senators, however, could not be convinced. Some may have been opponents of Caesar, though if so it is not clear why they were present. Many may have been indignant at the way in which this insubordinate governor addressed them – in a moderate tone perhaps, but with unconcealed threats, arrogantly offering them the chance to support him, yet with no intention of really listening to them. We may also wonder whether Caesar, after nine years of absence, was able to strike the right note in addressing the fathers.

It was agreed to send a deputation to Pompey, but no one wished to lead it. Pompey had after all stated that he regarded all who remained in Rome as enemies. Perhaps they doubted Caesar's seriousness. In any case he found the Senate just as he remembered it – a talking shop. The negotiation lasted for three days; nothing else was resolved. At one point the majority seems to have been prepared to hand over the public fund to Caesar. But Lucius Caecilius Metellus, a tribune of the people, entered his veto.

In the end Caesar angrily broke off the proceedings. He had neither the time nor the inclination for further delay. He may have started with the best intentions, but then everything turned out to be just as it had been in 59; in spite of all that he had achieved in the meantime, his relations with the Senate had not changed. Ten years had passed, yet he seemed to be as much of an outsider as ever, if not more so.

If he had really believed that the Senate, given the freedom to choose, would be on his side, he must have been bitterly disappointed. But this is unlikely to have troubled him for long. If the Senate would not co-operate, this was ultimately a matter of indifference to him. There were other ways of getting what he wanted. 'In future everything will be decided by me,' he declared. Martial law was in force, and he was in control.

With a troop of soldiers he crossed the hallowed bounds of the city – where, properly speaking, his military command ended, never to be renewed – and went to the temple of Saturn by the forum. When Metellus, the tribune of the people, blocked his path, Caesar, the champion of tribunician rights, threatened to kill him. He added that it was harder for him to say this than to do it. He had the door of the temple broken open and seized the treasure, which the consuls had left behind in the panic of their departure. The mood in Rome was so hostile that he did not even dare to address the people, as he had intended.

Caesar's bad conscience is evident from his own account: he not only omits to mention the incident, but even states, at another point, that the consuls had left in such haste that they had opened the temple of Saturn but failed to take the money with them.

Full of anger towards the Senate and embittered over the intercessions, he set out for Spain. 'Whatever he thinks and says is full of rage and cruelty,' Caelius reports. Yet at about the same time Caesar was once more writing friendly letters to Cicero, forgiving him for his absence and seeking to strengthen him in his neutrality. It thus did not take long for his superiority to overcome his vexation.

With the soldiers Caesar was again in his element. From now on the praetor Marcus Aemilius Lepidus ruled the city on his behalf. He gave Antony command over Italy and the legions that remained there. For Caesar everything was back to normal: he could begin a new campaign.

The First Spanish Campaign and the Second Visit to Rome (April–December 49)

THE SIEGE OF MASSILIA · THE SURRENDER

OF THE SPANISH LEGIONS · THE

DUTY TO SPARE THE CITIZENS ·

THE MUTINY AT PLACENTIA ·

ECONOMIC MEASURES AND

ELECTIONS IN ROME

I N THE MIDDLE OF APRIL Caesar reached Massilia (Marseille) by the land route. The city informed the former governor that it had a duty both to him and to Pompey and wished to remain neutral. It closed its gates to him; shortly afterwards, however, it allowed Domitius Ahenobarbus, whom the Senate had appointed as Caesar's successor at the beginning of January, to sail into the harbour. While Domitius took over the defence of the city, Caesar began to lay siege to it. Seeing that there would be no quick success, he placed Trebonius in command and moved on with a large part of the army.

In the middle of June he arrived in Spain. His subcommanders had already freed the Pyrenean passes. At Ilerda (now Lérida) he encountered the Pompeian army, consisting of five legions and many auxiliary troops and led by two of the most able and experienced commanders. Caesar had six legions, but fewer auxiliaries; he was superior in cavalry, however, having formed new squadrons from the foremost and bravest of the Gallic tribes. For the first time since the civil war began, he had to reckon with heavy fighting. He therefore deemed it appropriate to borrow money from his officers and distribute it among the men. In this way he made them all beholden to him.

For a while the armies faced each other. Finally battle was joined. The Pompeians almost defeated Caesar's troops, thanks to the loose order of battle that they had adopted from the Spaniards. The outcome was undecided. Then floodwater destroyed the bridges over the river Sicoris (the Segre, a tributary of the Ebro), which carried supplies to Caesar's army. Their opponents, who had ample supplies and were able to block Caesar's supply-routes from all sides, felt sure of victory. When their reports reached Rome, the Romans began to celebrate.

Caesar, however, devised a solution. He had special ships built, as he had learnt to do in Britain; these were covertly transported on linked waggons to a point thirty-one kilometres away and used to ferry the cavalry across the river. He then built a bridgehead and finally a bridge. At the same time he had news of an initial success at Massilia. Fortune changed, and many Spanish communities surrendered to him.

He then had trenches dug and part of the waters of the Sicoris diverted, in order to create a ford. The Pompeians, fearing that Caesar's cavalry would cut them off from their supplies, withdrew to the south. Caesar pursued them and barred their route by means of extraordinarily daring and surprising operations. Wishing to avoid bloodshed, he deliberately ignored various favourable opportunities to do battle. The soldiers urged him to fight and became angry and rebellious when he refused, saying that if he forfeited such a chance of victory they would not fight when he wanted them to. They saw the reward of all their endeavours – victory and plunder – within their grasp. Once, when the enemy leaders were absent, the soldiers on the opposing sides began to confer with one another. Old friends and acquaintances met. One of the Pompeian commanders finally intervened with his Spanish bodyguard and ordered the execution of all the Caesarians he could lay hands on. His soldiers were compelled to swear an oath that they would not desert or betray the army and its leaders. Caesar, however, sent the Pompeian soldiers back unharmed.

A little later he had his opponents surrounded at an unfavourable point; their situation was desperate. After four days without water and corn, and without fodder for those of the pack-animals they had not slaughtered, a meeting was requested by Lucius Afranius, one of the Pompeian commanders. Caesar agreed to meet him, but insisted that the exchange should be conducted openly, within earshot of both armies. He gives a detailed account of what was said. According to Caesar, Afranius stated that they had discharged their obligation to their commander, Pompey, and been defeated; he now asked for mercy. Caesar replied by reproaching both commanders severely. He spoke of another duty that had been performed by him and his soldiers, and indeed by those of his opponents, but not by the two generals – the duty to spare the citizens and to see that 'every opportunity for peace remained open as far as possible.'

As in his exercise of clemency, Caesar was playing the advocate of the citizenry as a whole, concerned for the lives of the citizens and for mutual reconciliation. Again he was able to present his own interest as that of the whole, contrasting with the supposed partiality of his opponents. Moreover, he complained loudly about the injustice they had done him and still intended to do him. He alone was to be denied what had previously been granted to all Roman commanders – 'to return home with honour, or at least without disgrace, after deeds successfully performed'. Yet bitterness now gave way to dignity. All this he would endure; he did not intend to augment his power by taking over the Pompeian legions; indeed, he guaranteed that no one would be taken into his ranks against his will. All he required was that his enemies should depart from the province and disband their legions. The Pompeians readily agreed. Caesar then made his soldiers return all the plunder they had taken from the Pompeians. He had every item valued and repaid in money. So punctilious was he about the property of Roman citizens.

After forty days he had annihilated the Pompeians' best army with scarcely any bloodshed. He could now take over the Spanish provinces without further fighting. In Hispania Ulterior he had excellent connections, dating from his days as quaestor and praetor. He now summoned two large councils in Corduba (Cordoba) and Tarraco (Tarragona). He thanked and rewarded his supporters and returned the moneys the Pompeians had confiscated. However, he imposed heavy taxes on others. He then appointed the tribune Quintus Cassius Longinus as governor of Hispania Ulterior and put four legions at his disposal. In no time Quintus managed to produce a mood of rebellion in the province, which had so recently been well disposed to Caesar. Cicero had once wondered, 'What companions or aides is Caesar to make use of? Are the provinces and the commonwealth to be administered by men who cannot keep their inheritance in order for two months?'

On the return march, Massilia, having held out bravely for a long time, finally surrendered. The city was obliged to hand over all its weapons and ships, together with the communal treasure. But it retained its formal independence.

In northern Italy, near Placentia (Piacenza), the first mutiny occurred. The soldiers said they were exhausted; but above all they were dissatisfied with the lack of booty. They reproached Caesar for deliberately prolonging the war and withholding the reward he had promised them. They were baffled by a war that brought them nothing but unending hardships and was being fought for Caesar alone, a war in which the greatest possible mercy was shown to the enemy and their own reward constantly deferred. They had already marched hundreds of kilometres and crossed the Alps and the Pyrenees, all at the greatest speed. Now they were to set off for Brindisi, in southern Italy, and cross over to the east. Understandably they found this excessive.

Caesar, we are told, did not let it be seen how much he relied on them. Addressing his men, he announced that in accordance with established custom he would decimate the ninth legion, where the mutiny had broken out (this meant executing every tenth man), and send the rest home as unfit for service. This took them by surprise; it was not what they had bargained for. They were in fact devoted to him, and discharge with ignominy was contrary to their honour. They begged to be allowed to go on serving under him. Caesar at last reluctantly agreed, though he demanded that the most guilty men should be named. There were about a hundred and twenty of them. Of these he had every tenth man executed, the victims being chosen by lot. Among those named was one who turned out not to have been in the camp at the time in question. The officer who had reported him was put to death instead.

In Rome, meanwhile, Caesar had been appointed dictator by the praetor Lepidus on the basis of a special law, his main task being to conduct the elections. In the middle of December Caesar paid a flying visit of eleven days to the city. His supporters had largely prepared his programme. The elections went off as planned; he was elected consul, together with Publius Servilius Isauricus, the son of the man under whom he had once served in the east.

The consulate of 48 was the first for which Caesar could legally be a candidate. Before this, a law had been passed permitting the sons of those proscribed by Sulla to seek office. For the first time the Transpadani – the inhabitants of the Po Valley, recently admitted to citizenship – were allowed to take part in the elections. They probably came in great numbers to honour their old governor and benefactor. Presumably Caesar had also seen to it that the politicians condemned since 52 under the strict aegis of Pompey were given an amnesty and allowed to return to Rome. As dictator he also celebrated the Latin games, which the consuls had failed to do in their haste to leave the city. It was a rebuke to the consuls; it was also correct, as befitted the *pontifex maximus*. This was the old feast of the Latin league. The consuls customarily announced it on assuming office, and they were not supposed to join the army until it had been held. All the magistrates had to participate. It took place outside Rome, near Alba Longa, the home of the Julii, lasted for several days, and culminated in the sacrifice of a bull to Juppiter. Originally it had to be a white bull; a later decision of the Senate allowed the sacrifice of red beasts, but we do not know whether this applied before 49.

Caesar was chiefly concerned about the parlous economic conditions. No debts were paid and no money was lent, because a remission of debts was feared. The dictator ordered that the property of the debtors should be valued by assessors. Upon the creditors he laid the obligation to adhere to pre-war values, lest debtors who owned property should be compelled to part with it at a lower rate. He also ordered a lowering of interest rates. These measures were a compromise, but above all they were a clear indication that there would be no remission of debts.

This time Caesar refused to sanction a peace embassy to Pompey, proposed by his father-in-law in the Senate. His position had markedly improved, and he no longer needed the Senate for this purpose. To finance his forthcoming campaign he requisitioned the remaining votive offerings in Rome's temples; he also had corn distributed to the populace.

At the end of December he laid down his dictatorship and left the city. He seems to have started his consulship on the way to Brindisi, where he set sail on 4 January.

The situation had changed substantially. Caesar was now consul. His opponents could no longer claim to be the lawful rulers. The republic was with him. A little later, after his initial successes in Greece, one of his young followers tried to persuade Cicero to change his allegiance, or at least to remain neutral. 'You have done your duty to your party and the republic you approve of,' he wrote; 'what matters now is that we should be where the republic now is, rather than follow the old one and be in none.' It was a bold use of language to localize the republic in this way; it was no longer just a question of the legitimacy of the magistrates. Yet Caesar's claim is clearly mirrored in these words.

He claimed to represent the whole, and this was made all the easier by his opponents, who had brought in numerous foreign potentates and intended to conquer Italy with their aid, as though it were an enemy country; Cato alone did all in his power to spare the citizens. When two hundred senators had gathered for a meeting in Saloniki – at a place set up for the taking of the auspices – he secured a resolution that no Roman should be killed except in battle and that no town subject to Rome should be plundered.

The Pompeians did not adhere strictly to it; Caesar, on the other hand, did everything to mitigate the effects of the war. He had admittedly the great advantage of being politically and militarily in control of everything.

The Greek Campaign
(until September 48)

B Y NOW the western half of the empire was in Caesar's hands – Spain, Sardinia, Corsica and Sicily. Curio, however, had been unable to win the province of Africa. He had fallen bravely in battle after losing his three legions. The Pompeian fleet had also inflicted a heavy defeat on Caesar's ships in the Adriatic. Worst of all, his troops in Illyria had been forced to surrender.

Pompey, whom the Senate at Saloniki had meanwhile appointed supreme commander, had had almost a year to gather large forces in the east. He had brought five legions with him from Italy; four came from the eastern provinces, and two others were on their way from Syria. The allies had supplied archers, slingers and horsemen, and large sums of money had been raised. Pompey now used the time available to him to train his troops; he is said to have personally taken part in the weapon-training. At fifty-eight he was able to wield his weapons as an infantryman or a cavalryman and could draw his sword and return it to the scabbard while riding at full gallop. And in casting the javelin he excelled most of the young soldiers not only in the sureness of his aim, but in the force and range of his throw. Finally he had built a great fleet and assembled stores at the ports: he intended to wait for Caesar in the coastal towns while his own ships controlled the seas.

When Caesar arrived at Brindisi, twelve legions and more than a thousand horse were waiting. He later reported that on the march from Spain many soldiers had left the army and many had become sick: the insalubrious autumn weather of Brindisi had affected them after the healthy climate in Gaul and Spain. The old Roman calendar was about two months ahead of ours: early January corresponded to mid-November.

Before they embarked, Caesar told the soldiers that they had now reached the end of all their exertions and dangers. Hence they should not be afraid to leave their slaves and their baggage in Italy and go on board with light packs; in this way more of them could be accommodated on the ships. They should put all their hopes in victory and his generosity. On hearing this the soldiers are said to have cheered.

Caesar had insufficient ships at his disposal. He could cross with only seven legions. The wind was favourable, and, trusting to his swiftness and his good fortune, he crossed over to Greece. It was a hazardous undertaking, as he had only twelve warships for cover, while the enemy fleet consisted of a hundred and ten. But the Pompeian commanders probably did not expect the Caesarians to arrive so soon, and their ships were still in port. Pompey had not yet arrived; he intended to take up winter quarters along the coast near Apollonia (in the south of what is now Albania) and Dyrrhachium. Thanks to the speed of his operations Caesar forestalled him and took possession of Oricum and Apollonia, though Pompey cut off the route to Dyrrhachium. The two armies encamped opposite each other near Apollonia; Pompey was well supplied, but Caesar was not. Pompey's fleet had now learnt its lesson and cut Caesar off from the Adriatic.

Immediately after landing, Caesar once again made peace proposals. Both commanders should commit themselves on oath to dismiss their armies within three days. Full authority should then be restored to the Senate. Pompey would not even listen to Caesar's terms: 'What need have I of life and citizenship,' he said, 'if I appear to owe them to Caesar's mercy?' And indeed, after his flight and Caesar's victories it could scarcely appear otherwise. Things had changed considerably since the spring of 49; there was no longer room for both of them. There was now no justification for breaking off the war against Caesar, on which the Pompeians had decided with good reason.

With the armies encamped close to each other, Caesar had Vatinius shout the demand for peace negotiations to the other side. Why should not citizens talk with citizens and exchange deputations? The Pompeians said they would give their answer the following day. From both sides soldiers flocked together, and Labienus began discussions with Vatinius. Then suddenly a hail of missiles descended on the Caesarians. Vatinius escaped them, but others were wounded. As the Caesarians withdrew, Labienus shouted after them, 'Let us have no more talk of reconciliation, for only when Caesar's head is brought to us will peace be possible.' In this way the door was shut on Caesar's attempt to win over the opponents' men.

For a long time nothing happened. Pompey's army was superior, but he hesitated to attack Caesar. He obviously hoped to be able to cut off his supplies. Caesar, on the other hand, waited with growing impatience for the remaining legions to arrive; with a favourable wind, he thought, they should have been able to cross long ago, despite the superior strength of the opposing fleet. At times he became distrustful of Antony and his colleagues and wondered whether they were still firmly on his side. He decided to cross over in a small boat, disguised as a slave, and fetch the soldiers himself.

The boat ran into a violent storm, and the captain wanted to turn back. Caesar insisted that he should go on and shouted in his ear, 'Have no fear! You are ferrying Caesar and his fortune.' The scene is reliably attested.

Caesar had long known that fortune had supreme power in all things, and especially in war. He repeatedly speaks of it in his writings. He felt he was favoured by fortune, and indeed he was. 'Let him have his fortune!' remarked Cicero at the beginning of 49, showing that Caesar's relations with fortune were also known in Rome. They derived from his descent from Venus and had been confirmed time and again in love and war. But naturally one could not simply rely on fortune. In one of his comedies, Terence had said, 'Fortune helps the brave' (*fortes Fortuna adiuvat*), and the Latin language seemed to bear witness to the connection. Fortune could be helped along; indeed it had to be. Caesar once told his men that 'if everything did not proceed fortunately they must come to the aid of fortune with their zeal'; a similar idea is also to be found in Terence. Caesar also knew that not only fortune, but understanding, helps the brave. Suetonius writes, 'It is difficult to say whether in his enterprises he acted more with caution or more with boldness,' and Caesar was certainly brilliant at reconnaissance and planning. Yet he could also stake all on one card. From the beginning there was something of the adventurer and the gambler in him; at first this was more a product of wilfulness, but it was increasingly nourished by the experience of how little resistance reality often offered if one took a firm grip on the facts of a situation. This had been amply demonstrated in Gaul. Fortune was of course fickle, but she could also be faithful. He enjoyed dicing. It was in his nature; he would never have wished to be anxious and hesitant.

Being dependent on the legions still in Italy, he took an extreme risk. Yet in the end the wind and the waves were so strong that the ship had to turn back nevertheless. Shortly afterwards, however, his army leaders received a letter with strict instructions to bring the legions across to him. On 10 April, with a great deal of luck, the crossing was made. Caesar now had an army of eleven legions and about 1,500 horse; though still numerically inferior to Pompey, he could begin the great game.

He first sent troops to Greece to secure his supplies. Then he dispatched others to prevent a link-up between Pompey's legions, which were on the march from Syria, and the army already in Greece. The outcome of the battles varied.

Pompey now moved camp. Caesar followed him and offered to do battle, and when his opponent avoided a decisive encounter Caesar attempted an extremely bold manoeuvre. He marched towards Dyrrhachium, at first setting off in the opposite direction and then approaching his destination along narrow paths and through difficult terrain. Pompey was actually deceived, and when he at last discovered what Caesar was about he tried to forestall him by taking the shorter route. However, Caesar managed to arrive outside Dyrrhachium by dint of an almost uninterrupted night march that called for the utmost fortitude. Pompey was accordingly cut off from the town where his supplies and war material were stored.

Pompey stationed himself on a hill, at the foot of which was a reasonably good anchorage. He was therefore able to obtain supplies from the immediate vicinity and farther afield, while Caesar was still cut off from seaborne supplies. Moreover, the ships that he had meanwhile had built in Sicily, Gaul and Italy had not yet arrived. He could hardly obtain anything from the surrounding countryside. There was in any case a dearth of corn, and Pompey had had the whole region plundered.

In this rather desperate situation Caesar once more devised a bold and quite unusual strategy. As the land was hilly, he occupied various high points in a wide area round Pompey's camp, fortified them, drew lines of communication between them, and began to encircle his opponent. His aim was to prevent the Pompeian cavalry from blocking the roads by which he hoped to obtain supplies; he also wished to put the cavalry out of action by cutting it off from its sources of fodder. But above all he wished to deal a severe blow to Pompey's reputation by showing that this famous general, with his worldwide clientèles and a greatly superior army, was encircled and besieged, and probably had no stomach for fight.

Unable to avoid encirclement – except through a large-scale battle – Pompey confined himself to occupying numerous high points himself, in order to control as wide an area as possible. His fortifications finally extended for fifteen miles, compared with Caesar's seventeen. Pompey thus had the advantage of the inner line; with the help of his archers and slingers he could seriously impede the Caesarians' fortification work. He also managed to force them back at several points.

Caesar himself shows how 'new and unusual' this kind of warfare was – and not just because of the extent of the fortifications. As a rule the army with superior numbers and the more powerful cavalry besieged the other; it was the besieged who starved, not the besiegers. Caesar had of course successfully used a similar technique at Avaricum and Alesia. Remembering this, his soldiers were infected with his defiance and determination, however hard they had to work (for they too were now older and worn down by ceaseless physical exertions). They often had to make do with barley and pulse instead of wheat; even the meat that Caesar was able to obtain in large quantities now seemed a delicacy. Instead of corn they used roots for baking a kind of bread, and when the Pompeians scoffed at their hunger they tossed a few loaves across to them. It was on this occasion that Pompey remarked that he was dealing not with human beings, but with wild beasts. He took care that this bread disappeared at once, lest anyone should see it. Yet the pride of an army can stiffen as conditions worsen; the soldiers said they would rather live off the bark of trees than let Pompey escape. Above all they wanted the war to end.

Rations were bound to improve as soon as the new corn was ripe, but according to the Roman calendar it was still May, June or July – corresponding to our spring. Yet even the Pompeians, who had much better supplies, were soon unable to feed their animals. They apparently did not slaughter them, and so the camp was filled with the stench of rotting carcases. They were short of water, as Caesar had dammed or diverted all the streams. The Pompeian soldiers had to dig wells, and as they were in any case not used to fortification work, disaffection grew, and more and more of them deserted.

A particular difficulty for Caesar was that he had to secure himself on the outside. For Pompey could attack him in the rear from the seaward side. On the coast there was a large gap in the ring encircling Pompey's army. To make matters worse, Caesar had not enough soldiers to man his long lines of communication. Significantly he does not mention this when describing his unusual strategy, but it is obvious.

Pompey declared at the time that, if he knew anything of warfare, Caesar's legions would not be able to extricate themselves without serious losses from the position that they had rashly taken up. He probably did not know Caesar well enough to realize that his audacity and shrewd calculations enabled him to do many things that seemed impossible. Yet Pompey was not entirely wrong: he was right in judging that Caesar had overstretched himself.

At one time Caesar had hoped to be able to take Dyrrhachium by treason, but the enterprise failed. Then Pompey succeeded in bringing out his cavalry, which had almost run out of fodder. It was intended that the cavalry should attack Caesar's rear and above all cut him off from his supply routes. Caesar, however, blocked the way from the town to the hinterland, so that Pompey eventually recalled the horsemen. Yet the horses had scarcely any fodder.

After weeks of siege and various minor encounters, Pompey decided to attempt a large-scale break-out. He was helped by the defection of two prominent Gauls, who advised him about the customs of Caesar's army and the weak points in his fortifications. At one point on the seaward side there was still a gap, in the part of the dyke lying farthest from the camp. In a well-planned operation Pompey sent sixty cohorts (about 36,000 men) to attack at this point from without and within. He finally stationed other troops at the gap in the dyke, covered by countless slingers and archers. Caesar's men, unable to withstand the onslaught, finally turned and fled. Reserves were quickly brought up, but they could not halt the fleeing soldiers. There was a wild surging mass with no clear direction. The enemy pressed home his advantage. When the Pompeians were already approaching Caesar's nearest camp, Mark Antony brought up new cohorts and finally stabilized the front.

Smoke signals sent from hill to hill had alerted Caesar. He approached hurriedly with twenty-five cohorts that he had had to withdraw from another position. But Pompey was already setting up a large camp by the sea for several legions; the encircling ring was breached. Caesar could do no more than build new fortifications close by.

While he was occupied with the fortification work, it was reported that at a spot not far away, hidden by trees, a Pompeian legion was occupying an old camp that had previously been abandoned. The site lay three hundred yards from the sea and five hundred from Pompey's new camp; he obviously intended to widen the breach he had already made. Caesar hoped to be able to overwhelm this legion. He took thirty-three cohorts and secretly led them up to the camp by a devious route. He then ordered them to attack in two divisions. The left-hand division, which he himself commanded, threw the Pompeians from the dyke, then turned to the middle gate, knocked out the *ericius* or 'hedgehog' (a beam studded with iron spikes) and entered the camp.

However, the right-hand division lost their way. They intended to break in through the side gate, but were diverted by a dyke linking the camp with the nearby river. At the river the soldiers finally made a narrow dam across the ditch, demolished the undefended dyke, and pressed on. The whole of the cavalry followed them.

Meanwhile Pompey had approached with five legions in close order of battle. The soldiers in the camp took fresh courage and mounted a counter-attack; his horsemen attacked Caesar's, who were just crossing the fortification over the narrow dam. Unable to spread out and fearing that their retreat would be cut off, they turned and took flight. The foot soldiers followed; everyone back-tracked in haste. Because the dams were narrow, many jumped from the ten-foot-high fortification into the ditch, and many soldiers were trampled to death. Now the other division withdrew.

Caesar placed himself in the path of the fleeing soldiers, seized the standard and ordered them to halt, but they streamed past him. The battle was lost. Pompey, surprised at his own success and fearing an ambush, did not pursue the enemy. Although he encircled Caesar's fortifications, he then withdrew. 'Today victory would have gone to our opponents,' Caesar commented, 'if they had had someone who knew how to win.'

He had lost nine hundred and sixty soldiers and two hundred horse. Most of them had not fallen in battle, but been crushed or trampled during the flight. Caesar's troops had lost thirty-two standards in the battle. Pompey was acclaimed *imperator* in accordance with time-honoured custom. Labienus persuaded Pompey to hand over the prisoners to him. He addressed them as comrades and asked them scornfully whether it was usual for old soldiers to flee; he then had them all massacred, as if they had to be punished for Caesar's defeat. This was one of several occasions when the terrible hatred of Caesar was visited on his soldiers, as it could not touch him. The hatred was all the more malignant when Caesar manifested his superiority. It was as though his opponents sensed their own weakness.

Victory was celebrated in the Pompeian camp. The soldiers ascribed it to their bravery. Many thought it was only a matter of time before they won the final victory. Various senators thought likewise, and the less they knew of warfare, the more convinced they were. Pompey himself was cautious; the others, Caesar remarked, did not know how unpredictable the fortunes of war were.

Caesar spent a sleepless night in which all his mistakes presented themselves to him. Why was he waging war at a place so unfavourable to himself and besieging a superior enemy whose fleet had command of the sea, while he himself lacked so much? The defeat had unnerved him: fortune had become fickle. He could not continue to plan as he had done before, for he could not be certain whether his soldiers could be relied on to endure difficult situations after such a setback. He therefore decided to raise the siege and shake off the enemy as quickly as possible.

He called the army together and warned the soldiers not to take the previous day's events too much to heart. He reminded them of all the battles they had won. They must be grateful to fortune and come to her aid with their zeal if things were not going well at present.

He saw the matter statistically, as it were, setting this one misfortune against many previous successes in an attempt to convince himself and his men that fortune was still with them. Fortune, after all, was with the brave.

He explained to them that the defeat could have had many causes, but their failure was least to blame; by their bravery they could make up for it – as they had at Gergovia. Then Caesar publicly reprimanded and demoted some of the standard-bearers.

According to Caesar's account, the soldiers wanted to attack immediately. But he had not regained sufficient faith in their fighting strength. He therefore marched his troops to Apollonia. Pompey at once set off in pursuit, but the Caesarians moved with such skill that his vanguard only briefly caught up with their rearguard. Pompey then ran into difficulty because some of his men, having left much of their personal baggage behind in the haste of the departure, returned to the camp without permission and were absent when it was time to move on.

At Apollonia Caesar saw that the wounded were tended. He then turned eastwards towards Thessaly with the bulk of his army. He wanted to link up with the troops he had previously sent there and force Pompey to do battle far away from the Adriatic. If Pompey were to cross over to Italy, however, he wanted to set off there through Illyria with his other troops. In fact Pompey's senatorial allies urged him to return to Rome, but he would not re-cross the Adriatic without having beaten Caesar.

He therefore continued to dog him, but tried to avoid battle, hoping to wear him down by large-scale manoeuvres. Caesar's defeat caused the Greeks once more to favour Pompey. Accordingly, Caesar decided to punish a town called Gomphi, the first in Thessaly to close its gates on him. After his soldiers had taken it by storm, he gave them licence to plunder. It may be imagined what took place there. All that is recorded is a surprising and rather amusing side-effect of the pillage: a plague had spread among Caesar's soldiers, but copious consumption of looted wine restored them to health.

At the beginning of August the two armies were again encamped opposite each other in the plain of Pharsalos. Caesar offered battle and stationed his army first in a particularly favourable position, then closer to Pompey's camp. Pompey, however, did not venture out of the safety of his camp. He wanted to continue his tactics of attrition, assuming, no doubt rightly, that time was on his side. Yet he was no longer fully in control of events. The august senators in his camp, contemptuous as ever of Caesar, did not understand Pompey's reluctance to defeat him and suspected that he was motivated solely by the wish to retain supreme command as long as possible. Finally, in a council a war, they demanded that he should do battle. Pompey, who had always praised doctors who refused to accede to the patients' wishes, finally gave way.

On 9 August 48 Caesar was about to move off again, in order to wear down the enemy army by the tactic of daily forced marches, when he saw that they were making ready for battle. He at once called his troops together and reminded them again how he had tried to win peace, in order spare the lives of the citizens and preserve the commonwealth's armies. But since his opponents gave him no choice, the battle must be fought. At last the day had come when they could fight against men, not against hunger and privation.

Pompey's army was far stronger than Caesar's. A good forty-seven thousand men were ranged against twenty-two thousand. On the left flank was the cavalry, outnumbering Caesar's by seven to one; what is more, all the archers and slingers were assigned to it. Caesar had drawn up his army in three lines of battle. When he surveyed the enemy order of battle he quickly detached six cohorts and formed them into a fourth line with which to confront Pompey's cavalry, rightly assuming that it was meant to circumvent his right flank and fall upon his rear.

The distance between the two armies was great enough to allow both to charge. Pompey, however, intended to await Caesar's attack, hoping that he would be better able to withstand it with closed ranks, when the Caesarians were out of breath after running the double distance. Caesar criticized this in his account, 'because in everyone there is something of an inborn urge and desire for deeds, which is fanned by eagerness for battle. Commanders should not restrain this urge, but encourage it. It is not for nothing that from ancient times the trumpets have sounded and all have raised battle cries; it was believed that this frightened the enemy and spurred on one's own men.'

Experienced as they were, Caesar's soldiers charged first, but halted before rushing upon the enemy lines. Out in front, a veteran who had volunteered to serve again under Caesar, is said to have encouraged his old comrades by shouting, 'Follow me and serve your leader as you have intended. Only this one battle remains; after it he will regain his honour and we our liberty.'

The enemy put up stiff resistance. Pompey's horsemen forced Caesar's back, but the six cohorts of Caesar's fourth line of battle charged furiously against them. Caesar had ordered them not to hurl their spears, but to strike at the faces and eyes of the horsemen; this was something that these fine, elegant dancers, anxious for their good looks, would be unable to endure. He was proved exactly right. The horsemen turned and fled. The cohorts then threw themselves against the archers and slingers, most of whom fell. Next they circumvented the enemy's order of battle and attacked from the rear. At the same time Caesar moved the fresh troops of the third line to the front. The Pompeians fell back along the whole front, and by about midday the battle was won. Pompey, shocked by the flight of his cavalry, had retired to the camp.

Yet despite the noonday heat Caesar did not let up. He had the camp stormed. Shortly before this Pompey had ridden out of the back gate. In the evening, surveying the battlefield strewn with the dead and wounded, Caesar observed: 'This is what they wanted; after such great deeds I, Gaius Caesar, should have been condemned if I had not sought help from my army.' It is noteworthy that he was still exercised by the question of who was to blame for the deaths of so many citizens.

On the following day the rest of the enemy army surrendered. Naturally Caesar once more showed clemency. No one was to be harmed or lose his property. Only those who had fallen into his hands a second time were put to death. However, each of his men was given the right to intercede on behalf of one prisoner. He invited the soldiers to join his army. Although he had suffered only minor losses in the battle, his legions were already much depleted.

It is said that Caesar instituted a special search for Marcus Brutus, the son of his mistress Servilia, as he was concerned for his safety. He was overjoyed to find him. To have Brutus as a supporter was particularly important to him, because he was Cato's nephew and a very persuasive advocate of the old republic.

Pompey's correspondence was found, but Caesar had it burnt without looking at it. The Caesarians also found many silver dishes, tables already laid, and lavishly decorated tents and arbours. The august senators had devoted more effort to preparing to celebrate the victory than to winning it. They had also shared out the booty before it had been taken. There had been debates about future consulates, and above all about Caesar's office of *pontifex maximus*. Caesar, however, whose soldiers had been able to sustain themselves like wild animals, had clearly lived more as an equal among equals and demanded, if anything, more of himself than of his soldiers.

The battle of Pharsalos may have decided the outcome of the civil war, but the war was far from over. Pompey's fleets still controlled much of the Mediterranean and had just retaken the island outside the entrance to the port of Brindisi. Another fleet had destroyed Caesar's ships in the strait of Messina. Africa too was still in the hands of his opponents. Pompey might try to raise another army, either in the east or in the province of Africa. There had been unrest in Rome and Italy too; it had been put down by Caesar's supporters, but it could easily erupt again.

In his account of the battle Caesar tries to indicate how decisive it was. When the outcome was known, it turned out that on the same day remarkable signs had been witnessed. At Elis, in the Peloponnese, the statue of Victoria, which faced that of Athene, had turned round to face the door and threshold of the temple. At Antioch in Syria the battle cries and trumpets of an army had twice been heard; they were so loud that the citizens had armed and rushed to the walls. The same had happened at Ptolemais (now Akko, in Israel). At Pergamon the kettledrums had resounded in the remotest rooms of the temple. In the temple of Victoria at Tralles in Asia Minor attention was drawn to a palm that had grown up at the time from the joins between the stones.

Before the battle Caesar had sacrificed to Mars and Venus, and the augur had predicted a great change in conditions. Just before dawn a bright light is said to have appeared over Caesar's camp; flames rose from it, as from a torch, and descended on Pompey's camp. Caesar himself is said to have seen it while inspecting the guards. This is how the ancient world recognized 'historic events'.

Caesar reports only the facts. Yet did he not sense that at Pharsalos he was favoured not only by fortune, but by Venus, his ancestress, and the gods in general? We have no evidence of his religious faith. Signs conveyed by lightning did not concern him unless they suited his purpose. Suetonius reports that once, when a sacrifice was being performed and the sacrificial animal escaped, he still continued his march. Later, when he tripped on landing in Africa, he discerned a propitious sense in this unpropitious omen and declared, 'Africa, I am holding you!' And when he heard that in Africa the name of Scipio was invincible (it happened to be the name of the enemy commander) he jestingly sent for one of his own soldiers, a man of particularly ill repute who bore the same name, and kept him by his side throughout the fighting. On the other hand, he took the trouble to celebrate the Latin games. Was this only for the sake of propriety? Why did he so often invoke the immortals? In 69, when he stated, in his funeral oration for his aunt Julia, that kings were subject to the power of the gods, was this just a form of words? And when he chose the battle-cry 'Venus Victrix' before the battle of Pharsalos, did he do so out of superstition or because of the beliefs of his men?

Was it part of his down-to-earth attitude that he could ignore superstitions and attend to the matter in hand? Did the serious attention he paid to ceremonial derive partly from a taste for parody and possibly also from a feeling that things should be done properly? Was it part of his fortune that he saw in it the hand of friendly gods, to whom he readily rendered what was due to them?

Once, when a soothsayer had examined the entrails of an animal and told him they boded ill because there was no heart, he declared that if he wished they would in future bode well; it was not to be regarded as a wonder if an animal had no heart. If he had drawn strength in his youth from the consciousness of being descended from Venus, was religion transformed in his later years into a consciousness of his power, a power that was in harmony with fortune, Venus, and the other gods? What in his youth had been a mental game may in later life have become a real experience. Was it not obvious that the gods existed, given the unmistakable evidence he had of their favour and active help? He had every reason to call upon them and revere them. There was of course nothing to indicate that he shared the common superstitions about them. Was this perhaps because he took them too seriously? Perhaps his religion was totally rational, based on what he had learnt from experience? Seel speaks of Caesar's 'direct affinity to the numinous, to the demonic, to fortune, daring and high risk'. Indeed, there is much to suggest that Caesar had a highly personal religion. May it not be that the more isolated he became, the closer he felt to the gods?

The Campaign in the East
and Sojourn in Egypt
(September 48 to September 47)

AFTER THE BATTLE OF PHARSALOS, Caesar's highest priority was to pursue Pompey and prevent him from renewing the war. He was less afraid of his other opponents, who did not command such wide and powerful clientèles. He also intended to secure his rule in the east. Not that he had already set his sights on a monarchy. We have no evidence of this. But now that the whole of the empire had been drawn into the struggle, the logic of the civil war required that, whereas Pompey had hitherto organized everything to his own advantage, Caesar should now try to push aside or win over those who held power and to enlarge and consolidate his own support. Here too he must establish his power and make it felt: moreover, he had to mete out punishments in order to fill his coffers.

As the remaining months of 48 would hardly suffice for all this, Caesar had sent his supporters in Rome not only news of his victory, but instructions that the Senate should appoint him dictator for one year.

This prolonged his legal authority without requiring him to go against convention by seeking a new consulship. The Senate had no choice but to accede to his wishes. It even authorized him to make war and peace as he saw fit. He was also allowed full discretion in dealing with citizens and confederates who had taken the field against him. Whether the Senate also conferred on him the right to hold the office of consul for five successive years is not clear. If so, he made no use of it. Finally, he is said to have been granted the honour of sitting on the bench of the tribunes of the people, in order to document his special relations with the plebs and their representatives.

Rome awaited events. No one spoke publicly against Caesar, for fear of informers, but he seems to have had few supporters, except perhaps among the common people. As a precaution, the statues of Sulla and Pompey were removed.

Caesar himself pushed ahead as far as possible every day with his cavalry. One legion followed him in shorter daily marches. He at once sent other troops across to Asia Minor. In September he arrived at the Hellespont, having passed through Macedonia.

As he crossed the strait, a Pompeian squadron is said to have appeared suddenly. Instead of trying to flee, he made straight for the flagship and called on the commander to surrender; the latter naturally begged for mercy, so that Caesar saved his life and at the same time won a whole flotilla. Such stories, in which presence of mind combines with daring in dangerous situations, are told about all great men. They are part of the image one forms of them, yet one can never wholly rule out the possibility that they are true.

On arriving at Ilion (Troy), the town from which his supposed ancestor Aeneas had set out for Rome, he exempted it from tribute; Alexander the Great had formerly done the same. He was fêted wherever he went, and towns that did not lie on his route sent deputations.

At Ephesos he was honoured as 'god and deliverer of humanity', as can be deduced from the monument that was set up shortly afterwards. The inscription reads: 'To Gaius Julius Caesar, son of Gaius, pontifex maximus, imperator, twice consul, manifest god (*theos epiphanes*) descended from Ares and Aphrodite and common deliverer of humanity'. He had made sacrifices to Mars and Venus before the battle of Pharsalos. The two are associated in Greek myth, and Caesar must have known this. As the god of war, Mars was clearly well disposed to him; so was Venus, who was increasingly regarded as the goddess who conferred success. Mars was the ancestor of Romulus and Remus, the founders of Rome, and Caesar now represented the city. There were thus various reasons why the Greeks, wishing to honour the descendant of Venus, should have believed him to be descended from Mars too.

As Hellenism spread in the east, divine honours were easy to come by and had already been bestowed on several Roman governors. They indicated simply, in religious terms, that the honorand had bestowed such benefits on the town as a god might bestow or that he appeared to possess more than merely human abilities. Such abilities could easily be ascribed to Caesar, a rebel who, with no great army, had defeated the most powerful Roman the Greeks knew, and the whole Roman republic. His victory seemed a miracle and could be properly understood and described in religious terms.

It is impossible to know how such honours affected Caesar. Did he treat them with scorn? Was he indifferent to them? Did they seem to him useful or appropriate as an expression of submission? Did he discern a kind of truth in them, in as much as the superhuman quality they attributed to him matched his own sense of superiority? All these factors might have coalesced in a complex involving an element of pious awe. In any case such honours somehow determined the way in which he was regarded in future and to some extent the way in which he presented himself to others.

It was at all events to the Greeks' advantage to pay such honours to the new master and thereby put themselves in good standing with him. After all, he proved to be a benefactor to the province. He reduced the burden of imposts after the large confiscations they had suffered under the Pompeians. Caesar proudly points out that he preserved the great shrine of Artemis at Ephesos from the confiscation of its costly votive offerings. Yet the likelihood is that he seized them himself; the war was expensive, and his soldiers had earned their rewards.

Caesar spent only a few days in the province of Asia; on hearing that Pompey had set off for Egypt, he crossed to Rhodes, in order to sail from there to Alexandria. He was accompanied by two whole legions and eight hundred horse. By now, however, the legions were reduced to barely a third of their original strength and amounted jointly to little more than 3,200 men. Caesar writes, however, that he thought the fame of his deeds allowed him to embark upon the long journey, even with such a small force. He believed that 'every place was equally safe' for him.

It was an ill-considered belief and soon belied. It is true that he had often staked all on one card: when he broke through to his troops during the rebellion of Vercingetorix, for instance, or when he set off across the Adriatic to fetch his soldiers, or more recently at the Hellespont, when he could well have fallen into the hands of his enemies. Sometimes he was very precise in his calculations, at other times reckless. In his own account he usually speaks only of his calculations; of the rest we learn only from other sources. On this occasion, however, he freely admits to having acted rashly. One wonders whether fame and success had made him more sure of himself; whether, after his endless efforts, he had begun to tire of caution, of reckoning with the mass of possibilities that fortune constantly had in store for him. Or had he become more fatalistic? Did he perhaps want to tempt fortune? Was he beginning to be bored by his good luck?

At all events he was in a hurry. The war had dragged on and on; it must be brought to an end. Caesar was now fifty-two and had waged war for nearly eleven years, fighting one battle after another and moving from camp to camp. Time and again he had crossed Gaul on foot, on horseback or on a litter, then repeatedly returned to Italy and Illyricum. He had gone on to conquer the whole of Italy, Spain and large parts of Greece. Now he had to set things to rights in the east. When would it end? Impatience, which was in any case a marked trait in his nature, must have driven him on, making him careless of his own safety.

★ ★ ★

When Caesar landed at Alexandria on 2 October, Pompey's head was brought to him. Pompey had arrived three days earlier, and the courtiers of Ptolemy XIII had agreed that it was best to eliminate him. Caesar turned away at the sight. Presented with Pompey's signet ring, he is said to have wept. He no doubt remembered their old friendship, the former glory of the man who had won so many victories and subjected so many nations, and was shattered by his tragic end. Yet it was probably also the abrupt ending of their contention that robbed him of his composure. It was this contention that had kept him in suspense for almost two years. Now he had lost his enemy; it all seemed to be finished. He had certainly hoped to be able to pardon Pompey. He showed a chivalrous respect for his mortal remains and concern for his surviving followers.

He now entered the city – probably giving little thought to the propriety of doing so – in the purple-edged consular toga, preceded by his lictors with fasces and axes. He wished to take up residence in the royal palace. The inhabitants received him with hostility. Such a procession, they shouted, was an affront to the dignity of the royal house. There were a number of altercations, in which several of his soldiers were killed.

At this time Egypt was partly independent and partly dependent, and the more dependent it was, the more it insisted – or the more the inhabitants of Alexandria insisted – on its independence and freedom. Apart from the Parthian realm in Mesopotamia and farther east, the kingdom on the Nile was the last of any size in the Mediterranean world, as Rome had destroyed and annexed all the others. The Ptolemaic dynasty came from Macedonia; the first of the Ptolemies had been governing the country when Alexander the Great died, and later made himself king. At times their rule extended far beyond Egypt, taking in parts of Asia Minor and many of the Greek islands. At the beginning of the first century BC, Cyprus and much of Cyrene (now Libya) had been ruled by Egypt. Yet even then the monarchy was plagued by grave structural weaknesses, foreign entanglements, and above all a series of disputes between pretenders. At first the throne had passed to the king's eldest son, but the Ptolemies later adopted the Egyptian custom of sibling marriage, so that the kingdom was inherited by the eldest brother and the eldest sister. Court intrigues gave rise to serious and at times bloody and costly quarrels. Moreover, the Greek colony in Alexandria, which was formally independent, was inclined to rebellion. It not only wished to play a special role under the otherwise absolutist rule of the Ptolemies, but seems to have evolved a special pride as a guardian of Greek traditions and a centre of Greek intellectual life; this pride seems to have spread to the city's non-Greek inhabitants, who had relatively little part in its cultural life. The first Ptolemaic kings had been anxious to appear particularly Greek in the alien world of Egypt and had founded a library that was to become the greatest in classical antiquity – a kind of academy in which Greek poets and scholars could live and work next to the royal palace.

In the first century BC the country came more and more under the influence of Rome. The Senate did not wish to assume responsibility in Egypt, but it interfered in disputes over the royal succession and was at times directly involved in them.

For this reason the Alexandrians in particular were ill-disposed to the Romans. The last king had been obliged to spend millions on gaining the Senate's recognition and then being restored after his subjects had expelled him. All this had cost the country dear. In the years before 49 there had been several bad harvests. Pompey had nevertheless demanded – and obtained – substantial deliveries of corn. And now Rome's consul appeared, as if at war with Egypt, flaunting the panoply of Roman might and requisitioning part of the palace, in order to demonstrate the country's dependence on the Roman super-power.

Now that he was here, however, Caesar could not brook such rebelliousness. And he could not leave, as the winds were unfavourable. Moreover, he had to collect large sums of money owed by the late king. Finally, he thought he owed it to himself and his position to restore political order in Egypt and sort out the power relationships – both here and in the east generally – to his own advantage. He may even have wished to conquer Egypt for Rome. At all events, he envisaged a fairly long stay and sent for reinforcements.

In accordance with the late king's will, his daughter Cleopatra VII and her brother Ptolemy XIII had succeeded to the throne. In 48 Cleopatra was twenty-one and Ptolemy thirteen; for all practical purposes the country was ruled by the courtiers. They had lately come into conflict with the queen and driven her into exile. Cleopatra now attempted, with armed support from Syria, to return to Egypt. Her brother's army faced her at Pelusium, a fortress on the eastern frontier.

It was this conflict that Caesar declared himself willing to settle. He instructed both parties to dismiss their armies and meet him. Alexandria, however, was in the hands of the king. The most Cleopatra could do – and probably only by stealth – was to send intermediaries. Feeling that this was not enough, she resolved to meet Caesar in person.

One evening at dusk a small ship entered the harbour of Alexandria and moored close to the palace. A Greek from Sicily disembarked with a long bag, tied about with straps, which he carried into the palace, in order to deliver it to the Roman consul. From this bag at Caesar's feet the young queen emerged, attractively attired, as may be imagined; one source describes her appearance as both majestic and pitiful. She apologized charmingly for choosing this unusual way of coming to visit Caesar. But how else could she have come? And had not Caesar summoned her? Plutarch reports that the ruse won Caesar's heart and that he was utterly captivated by her grace and charm.

Cleopatra has since captivated the imagination of humanity – the sensation-seekers during her lifetime, and later the poets. It is hard to form a precise picture of her. But of some features we can be reasonably sure. To judge by extant portraits she was not really beautiful; she had a fairly long, retroussé nose and very full lips. Yet she must have enthralled Caesar, as she later enthralled Mark Antony. Suetonius tells us that among foreign women she was to Caesar what Servilla was among Roman women: the one he loved most. Indeed, he loved her so much that for her sake he not only involved himself in a highly dangerous war – he may not have foreseen this – but stayed with her for weeks afterwards, when there was a pressing need for him to depart. During this time he sailed up the Nile with her in her large and superbly appointed barge, which was decorated with fine frescoes; they would have sailed as far as Ethiopia, had not Caesar's soldiers refused to go on. And it was certainly not merely for political reasons that Caesar summoned her to Rome in 46 and kept her there for nearly two years. Suetonius tells us that 'he also allowed a son born of their union to bear his name.'

Cleopatra was a highly educated woman and spoke many languages as well as her own. Politically she was uncommonly gifted, cunning and capable of any intrigue. As a child she had witnessed the difficulties and indignities suffered by her father, and his desperate, often unscrupulous efforts to assert his rule: she was up to every trick. At eighteen she assumed the reins of government. She knew what she wanted, but she was not popular with her people and had difficulty in outwitting her brother's experienced courtiers. She seems to have possessed the imperious pride of the Ptolemies and was at home – and not just superficially – in the sophisticated Egyptian culture they had adopted. She was versed in every fashionable refinement, and a work on cosmetics was even named after her. Yet this daughter of the New Dionysos, as her father had styled himself, was celebrated as the goddess Isis, and she must have had about her something of the mystique of an age-old culture. Even when – indeed precisely when – she was clearly acting the part.

Our sources would have us believe that she was not only beautiful, but had a charming voice. 'It was a delight to hear its tone; her tongue was like a many-stringed lyre.' She is said to have had a bewitching manner and radiated an irresistible fascination that had everyone spellbound.

She exploited her physical charms in the service of her political interests. And in her relations with Caesar politics and love seem to have been interfused.

This is probably what enthralled him – the politics of charm, which became the charm of politics. Here they could meet, recognize and understand each other. It may be that the twenty-two-year-old queen was the first and only person, since the death of his daughter Julia, who had understood Caesar, that she not only amused him and allowed him to conquer her, but knew how to pierce the shell of isolation that increasingly surrounded this man of fifty-two, to tempt him out of it and release him from it – with such insight and affection, such subtlety and grace, that he could perhaps even learn from her and allow himself, in some measure, to be conquered by her, as by no other. He was superior, and she was calculating. Both had to work hard to assert themselves; both were ruthless, proud and self-sufficient. They clearly discovered a mutual affinity and continued to take the same delight in it as they had in its discovery. After the first divine honours conferred on him, Caesar must have been intrigued to meet a member of a dynasty that had been accustomed for generations to such honours. The notion of divinity must have amused him, but also given him pause for thought. This was a further stage in his quest for self-understanding.

Caesar may have begun to wonder what was the point of all his endeavours. If so, he was probably the more inclined to open his heart to a woman who understood him.

Not least important are the festivities with which Cleopatra regaled Caesar. Suetonius tells us that their banquets went on till dawn. 'There had never been such merriment in his camp as there was during this Alexandrian episode,' says Mommsen. Yet we do not know what part the male company of the camp was allowed to play in this Alexandrian gaiety. Probably none. For in Alexandria Caesar soon found himself in extreme difficulties.

Ptolemy, learning that his sister had won Caesar's favour, stirred up the populace. Its fury soon boiled over and the palace was stormed. Caesar's soldiers were hard put to withstand the attack. At a popular assembly he explained that under the will of the late king his two elder children, Cleopatra and Ptolemy, were to rule Egypt, while the two younger children were to rule Cyprus. Cyprus, a former Egyptian possession, had been annexed to Rome by Clodius in 58. So desperate was Caesar's plight that he now ceded it to Egypt, but this did nothing to improve the situation.

For almost six months Caesar was besieged in the palace precincts. He had taken Ptolemy prisoner, but immediately afterwards Ptolemy's military commander ordered the king's army to march on the palace. Caesar succeeded in keeping the harbour entrance free by occupying the lighthouse on the island of Pharos. He also set fire to the Egyptian men of war in the harbour. The fire destroyed not only the granaries and naval arsenals by the harbour, but Alexandria's famous library; more than four hundred thousand papyrus rolls were burnt. More than once the Romans found themselves in desperate straits. At one stage they were almost cut off from the underground drinking water system, but Caesar remedied this by having wells dug. The Alexandrians hurriedly built a new fleet with whatever wood they could find in the columned halls and public buildings, but Caesar defeated this too. He then tried to take the offshore island and the kilometre-long mole that linked it to the shore. He had just built a bridgehead at the southern end of the mole when the oarsmen from his ships went ashore to join in the successful fight by shouting and hurling stones. Difficulties arose when the opponents attacked the oarsmen in the flank, causing them to panic, flee to their ships, and start casting off. At about the same time Caesar's soldiers took flight, fearing that the ships were about to leave without them. There was a mad scramble; the ships were overloaded and some sank. Caesar's own ship could no longer be manoeuvred from the shore. He saved himself by swimming to another that was already seaborne. In doing so he lost his commander's cloak, which the enemy kept as a trophy.

The Egyptians were at first led by Arsinoë, Cleopatra's young sister, as Ptolemy had been taken into custody by Caesar. They demanded Ptolemy's release and intimated that this was to be the first step towards an agreement. Caesar released the boy, who at first wept and begged to be allowed to stay; after his release he of course took his place at the head of his army.

At last relief arrived in the form of auxiliary troops from Cilicia and Syria, among whom were three thousand Jews. Ptolemy marched against them; Caesar followed by sea, taking a circuitous route. In the ensuing battle the king was beaten and shortly afterwards lost his life in the Nile. Caesar quickly seized the opportunity to lead his cavalry to Alexandria and ride as victor into the enemy-held part of the city.

Alexandria was one of the greatest cities in the world, with a population of more than half a million, comprising Greeks, Egyptians, Jews and others. It was a turbulent place, and to enter it with only a few squadrons was undoubtedly hazardous. It surrendered, however, and Caesar was able to ride unimpeded through the fortifications to the Roman-occupied quarter. It was a triumphal event. Cleopatra no doubt staged a splendid reception, for it was her victory too. This happened on 27 March 47.

So long had the Alexandrian war lasted – a war that Caesar had taken upon himself largely for the queen's sake. He had long been cut off from the outside world. On 14 June 47 Cicero wrote that Caesar had sent no letters to Rome since 13 December 48. Meanwhile Cato and others had raised a new army in Africa. There was unrest in Spain, where one of Pompey's sons soon fomented a rebellion. In Rome there was violent agitation, as in the previous year, for a remission of debts. The veterans in Campania were threatening to mutiny. Antony, Caesar's deputy, could not control the unrest – or hesitated to take decisive action to suppress it. The

Pompeian army from Africa was expected to land at any time. Caesar's army in Illyria had suffered a crushing defeat.

The news did not necessarily reach Caesar immediately in March or April. However, considering his long absence, he must have guessed that not everything was as it should be. After all, the Pompeians had still not been decisively defeated. Most importantly, Roman rule in Asia Minor had collapsed. Soon after Caesar's departure from Rhodes, Mithridates' son Pharnakes had set out to regain his father's kingdom. In December 48 Caesar's legate engaged him in battle at Nikopolis in Armenia and was defeated. This at least must have been known to the commander of the Syrian auxiliaries when he arrived in Alexandria. Moreover, Caesar could not fail to wonder why so far only one of the legions that he had urgently summoned from Asia Minor had arrived.

Nevertheless he stayed in Egypt for the whole of April and May in order to accompany Cleopatra on the long trip up the Nile. In this he evinced the same composure – or the same ability to concentrate, despite any emergency, on matters of immediate concern – as he had in 52, when he had allowed the Gallic rising to go unchecked until it suited him to leave Italy. Such composure – or such a capacity for concentration – seems almost superhuman, especially in one so restlessly active. Yet he was taken up with new impressions, with Cleopatra's politics and his own role in them. He probably valued the respite, far from the camp and the endless tension, with a companion part human and part divine. He must have been preoccupied in a way that was quite new to him and experienced that mixture of elation, relaxation, and freedom to reflect that comes to us when we are suddenly released from a tension we have lived under for years. Caesar probably needed Cleopatra at this time. Meanwhile, he had the opportunity to compose himself for the tasks that lay ahead.

It seems that about this time he wrote his book on the civil war, which was designed chiefly to convince the upper strata of Roman society of the justice of his cause and the seriousness of his efforts for peace.

In late May or early June he set off for Syria, leaving three legions behind to protect Cleopatra. They were commanded by Rufio, a brave and seasoned officer who, as the son of an emancipated slave, was unlikely to harbour political ambitions. In Syria Caesar quickly made his dispositions, dispensing rewards or punishments, appointing or dismissing rulers, granting privileges or demanding tribute, filling his coffers and settling disputes. The Jews were allowed to rebuild the walls of Jerusalem. Antipatros, the father of the later king Herod, was rewarded with Roman citizenship for sending auxiliary troops to Egypt.

Caesar crossed to Cilicia, held a council at Tarsos in order to settle the affairs of the province, then marched quickly through Cappadocia to Pontos. In Zela he encountered Pharnakes and his army. On 2 August they met in battle. Pharnakes led his troops so thoughtlessly to the hill on which the Caesarians were building fortifications, that Caesar at first did not take it seriously and failed to call his men to arms. Then the king's troops began charging up the hill. At last the Romans fell into line and were able, with great difficulty, to inflict a crushing defeat on the enemy. The campaign was over so soon that Caesar remarked to a friend, 'I came, I saw, I conquered' (*veni, vidi, vici*).

The very next day he departed for Bithynia, from where he set off shortly afterwards for the province of Asia. Everywhere he quickly set things to rights and finally returned to Italy, by way of Athens and Patras. On 24 September he landed at Taranto. From there he went to Brindisi, where he met Cicero. After Pompey's defeat Cicero had crossed over to Brindisi, but could not leave the town without Caesar's permission. Cicero was very anxious about approaching Caesar, not because of his clemency, but because of the circumstances in which he might exercise it. Yet on seeing Cicero, Caesar at once dismounted and greeted him, and the two spoke long and amicably.

Caesar had probably at first intended to go straight to Sicily and thence to the enemy-occupied province of Africa, which corresponded roughly to present-day Tunisia. Some legions had already assembled on the Sicilian coast. He now realized, however, that for various reasons his presence was urgently needed in Rome.

Two Months in Rome
(October to November 47)

CAESAR'S FOLLOWING · ANTONY ·
CAESAR DECIDES ON A CHANGE OF
PERSONNEL · THE DEBT PROBLEM ·
FINANCING THE WAR · MUTINY
OUTSIDE ROME

T HE PROBLEMS that made it so urgent for Caesar to return to Rome after almost two years' absence had arisen partly from the civil war, and at least some of them had to be solved in preparation for the African campaign. Behind them, however, lay a difficulty that was to dog him from now on: he had hardly any supporters who were both loyal to him and also capable of acting independently and effectively in the performance of tasks arising from politics and the civil war. In Gaul he had needed only military subordinates to whom he could issue clear and well-considered orders. In Rome until 49 it had been sufficient for him to have able tribunes of the people – and on one occasion a consul – to act on his behalf. They were apprised of Caesar's wishes, perhaps in a form that allowed them to react to various eventualities. It was for them to decide on the appropriate tactics, either after seeking further instructions or after consulting with other supporters. The rest was a matter for Caesar's diplomacy.

Now, however, there were mutinous legions to appease, economic problems to solve, and civic order to maintain; it was also necessary to bring skill and tact to bear on all those questions of personal politics without which Caesar's position in Rome would be insecure. He must not only win over and reward neutrals, but above all

414

take into account and reconcile the claims of his supporters.

He alone could do this. And it was a task that he reserved largely for himself. In some respects he could still rely on a number of independent senators who put themselves at his disposal, for example Servilius, the consul of 48. They were able to represent his interests with a fair degree of autonomy, acting in accordance with the conventional understanding of their function and with the support of the Senate, insofar as it was present; they did so out of a sense of duty. They were often moved not by opportunism, but by a desire to serve the commonwealth. Paradoxical though it may seem, Caesar's most reliable supporters in the administrative sphere were former Pompeians and neutrals, and they would have supported him in the rebuilding of the commonwealth.

Yet the closer he came to victory, the more generously he had to reward his own followers. They were implacable in demanding what he had promised them and what they thought they had earned. And he would have been the last to deny the justice of their demands. On one occasion he declared publicly that he would show gratitude even to bandits and murderers if he needed their help in defending his position. This is what made him so popular with his friends and at the same time alarmed many of his more serious supporters when they thought of the 'riff-raff' among his followers.

Among his supporters – apart from the soldiers and all the others with an eye to gain – were the lowlier spirits who wished to serve Caesar loyally, but were not exactly imaginative or able to act independently in performing tasks of major, let alone overriding importance. On the other hand there were those who saw themselves as little Caesars, spurred by similar ambitions and intent on securing leading positions; many of them were talented, even highly gifted, but self-sufficient in a way that was not necessarily to Caesar's taste. They were true sons of the aristocracy, but to some extent outsiders; they attached themselves to Caesar largely in the hope of becoming, with his help, what Roman nobles had always wanted to become – *principes*, the leaders of the Senate and the republic. Depending on the circumstances under which they happened to live, they had all the assurance of the ruling class.

Yet they had had an easier life than Caesar; they had not had to assert themselves in isolation against all the rest, but been able to join him. Their rise was hardly due to merit. They therefore found

415

it hard to conform. They were more interested in rights than in duties.

Foremost among them, after the death of his friend Curio, was Mark Antony. He was a tall, handsome man. According to Plutarch, 'his well-shaped beard, wide forehead and curved nose gave him a manly appearance that made him resemble Heracles, as represented in paintings and statues.' He was in fact an outstanding 'second man', although he clearly regarded himself as the 'first man'. He had great abilities, but was too soft, too good-natured, and also too lascivious; in any case he had not enough initiative or inward assurance, not enough concentration or stability, to play an independent, superior role. It may be that he owed both his charm and his weakness to the fact that he had never quite grown up.

In his wild youth Mark Antony seems to have shared the role of 'second man' with Curio. Curio having shown him the way, Antony plunged himself deeply into debt, determined to cut a dash and live life to the full. For a while he seems to have attached himself to Clodius, but is said to have felt uneasy in this role. It is more likely that his mother, reputedly one of the best and most virtuous women in Rome, wished to draw him away from bad company. At all events, he went to Greece to study rhetoric and train for military service.

He then became a cavalry commander in Syria, where he showed himself to be a bold and dashing officer, scornful of danger, skilled in tactics, and exceptionally popular with the men. This remained his real strength. He served under Caesar in Gaul and during the civil war, and was from 49 onwards his best officer. Twice he managed to prevent the soldiers from taking flight, and at Pharsalos he commanded the left flank. Earlier he had succeeded in bringing the second half of Caesar's army across the Adriatic.

Aristocrat though he was, Antony thought nothing of drinking away the night with the soldiers, boasting and bragging like an overgrown teenager. He impressed everyone with his extravagance and generosity. Yet at the same time he knew how to give orders and carry the men with him.

In politics he could assert himself forcefully, both as tribune in 49 and in other capacities. Yet he had no taste for 'slowly boring through hard boards with passion and precision': he lacked the necessary assiduity and was too easily distracted; as Caesar's deputy in Italy he had clearly been a failure. On the one hand he had been indolent and yielding, on the other ruthless and arbitrary, and there were many matters about which he had not troubled himself.

His chief concern seems to have been the pursuit of pleasure. He was the master, and he enjoyed living the life of a potentate, devoting his nights to theatrical performances, wine and love, then sleeping it off during the day or going around with a hangover; one morning he even vomited in the popular assembly. He surrounded himself with actors, musicians and ladies of easy virtue, and it caused widespread disapproval when they then had to be lodged at the houses of the leading citizens. Especially remarkable were the golden dishes he used when entertaining, and his occasional practice of harnessing lions to his chariot so that he could drive around like Bacchus. Cicero once remarked that he was taken up more with festivities than with stratagems. He could also be chivalrous, unless he forgot himself or was severely provoked.

Just as in youth he had been dependent on Curio, he was later dependent on Caesar and, from about 47, on Fulvia, the widow of Clodius and Curio, whom he married at that time. Fulvia was an ambitious woman, eager to lead a leader and command a commander. Plutarch tells us that Cleopatra found Antony already tamed.

It is not clear how loyal Antony was to Caesar. For all his dependence he probably had enough determination and detachment to be able to decide on a different policy. In 47 the tribune Publius Cornelius Dolabella, another of Caesar's bright young men, began to agitate for a remission of debt, and as a result disturbances and street fights erupted in Rome; but Antony hesitated for a long time before intervening. He may simply have been too indolent, or he may have wished to keep all his political options open. It is true that he finally cracked down with such brutality as to forfeit the sympathy of the populace. When Caesar's legions in Campania became restive and indiscipline spread, Antony at first did nothing. There are indications that in 46 he was involved in plans for a coup against Caesar. Later we find him engaged in a highly equivocal policy to establish Caesar's authority and at the same time to unmask him as a tyrant. When Caesar returned in 47, Antony fell into disfavour, but Dolabella was received amicably.

★ ★ ★

Caesar now decided on a change of personnel. He had himself and Marcus Aemilius Lepidus elected consuls for 46. Lepidus was to represent his interests in Rome and Italy; Caesar could be sure of having a loyal, honest and diligent deputy, but nothing unusual must be allowed to happen. For Lepidus, the son of the faint-hearted rebel of 78, seems to have been a weak and colourless man, possessed of no real energy. The historian Velleius Paterculus describes him as 'one of the hollowest men, devoid of any virtue that could have earned him the indulgence that fortune showed him for so long'. Caesar increased the number of praetorian and priestly offices, certainly not just for practical reasons, but in order to provide rewards for more of his supporters. For the same reason he had elections for the remainder of 47 held at the beginning of October. Publius Vatinius and another of his loyal followers became consuls; they had been his most important helpers during his consulship in 59.

In order to tackle the problem of debt among the populace, Caesar decreed a certain level of rent for a full year. The landlords had to carry the losses. But he again made it clear that he would not consider a remission of debt. It would have been impossible, not only because he wanted the political support of the upper classes, but because he needed some of them as creditors. For his main concern was to fill his war chest. At this time he is reported to have said more than once that in order to govern one needed only two things, soldiers and money; and armies could be maintained only with money. War was extremely costly, and even after two and a half years there could be no question of paying out the rewards that had been promised at the beginning of the civil war.

Caesar therefore required the towns of Italy to contribute gold wreaths and statues, such as were customarily presented to victors; the contributions were probably to be paid in cash. He also took out many loans. Having spent his whole fortune on the war, he said, the citizens must now come to his aid – as if it had been their war. They could scarcely demur, though there was scant prospect of repayment – and none was ever made.

Caesar then proceeded to auction the property of those of his opponents who had not been pardoned. Antony and others reportedly put in high bids, assuming that they would not have to pay. But they had miscalculated: the money was ruthlessly collected. Only Servilia is said to have obtained large estates at knock-down prices.

Cicero said that the price had been reduced by *tertia*: literally this meant by a third, but it was also an allusion to Tertia, Servilia's daughter, who reputedly brought Caesar and her mother together.

It was particularly difficult to assemble the army for the new campaign. Caesar's legions in Campania were unwilling to cross to Sicily. They wished to be released and demanded that Caesar should pay them the bounties they had been promised. They ran riot and harassed the local population. Deputies sent by Caesar were greeted with a hail of stones; two senators were even killed. The soldiers did not want to hear any more promises. Had Caesar not stated at the end of 49 that the war was nearly over? More than a year had passed since the battle of Pharsalos, which was to have decided the war.

They finally mobilized themselves and marched on Rome. Among them was the tenth legion, Caesar's dearest and bravest, which he relied on most and had always posted in the most difficult positions. When they arrived at the Campus Martius, Caesar went to meet them. There was an embarrassed silence. They were profoundly affected by the sight of their old commander, who appeared utterly composed, cold and silent – looking somewhat lonely, perhaps, and visibly older. Very diffidently their spokesmen explained that they had only come to ask for their release. They actually expected him to agree to all their requests; after all, he needed them. Caesar, however, addressed them not as comrades (*commilitones*) but as fellow citizens (*quirites*). He generously agreed to release them. Of course they would receive the promised reward, but only when he returned from Africa and celebrated his triumph with other soldiers. This provoked a reaction similar to what had happened at Placentia in 48. The soldiers are said to have been filled with shame and remorse and to have clamoured for Caesar to take them with him. Finally he agreed. This time he even refrained from punishing the ringleaders. It would be interesting to know whether his forbearance was due to weakness or to the realization that he had already made excessive demands on the men. Perhaps he had become more yielding since Placentia; perhaps his imperious will was no longer as strong as it had been then? He finally set out his plan for their settlement: unlike Sulla he did not propose to dispossess whole communities, but merely to allocate individual plots of land. The prosperous circles of Roman society took this scene as an occasion for criticizing Caesar's generosity. Dolabella and Antony had also provoked their displeasure.

At the beginning of December Caesar left Rome; according to our calendar it was mid-September. On 1 January 46 his third consulate began.

The African Campaign,
the Death of Cato

THE CROSSING TO AFRICA · THE

DIFFICULTIES OF THE WAR ·

VICTORY AT THAPSUS · EPILEPSY? ·

THE FURY OF THE SOLDIERS · CATO'S

FAREWELL TO HIS SON · ONE OF THE

MOST REMARKABLE POLITICIANS

IN WORLD HISTORY · RETURN TO

ROME VIA SARDINIA ·

'THE DIFFICULTIES INCREASE, THE

NEARER ONE COMES TO ONE'S GOAL'

WHEN CAESAR ARRIVED at Lilybaeum in Sicily (now Marsala) he would have liked to set off for Africa immediately with a newly raised legion. Only adverse winds prevented him, for he is unlikely to have been troubled by an augury that warned him to delay the crossing until after the winter solstice. He pitched his tent on the beach, as though he wished to be as close as possible to the future theatre of war. News came that his enemies were vastly superior in numbers, having assembled ten Roman legions and four belonging to the Numidian king Juba, countless horsemen and light infantry, a hundred and twenty elephants, and large naval forces. But he was extremely impatient: the enemy army was threatening Italy and encouraging his opponents, and there was unrest in Spain. His victory must at last be complete, so that everyone would accept it. Being militarily inferior, he once again banked on the element of surprise.

He finally embarked with six legions and two thousand horse. As he did not know where he would be able to land, the ships' captains were given no clear instructions; they therefore sailed off, hoping for the best. It was a reckless, irresponsible and costly undertaking, defying all the rules that Caesar observed strictly at other times. He acted with the courage of despair, out of contempt for his opponents or trusting blindly in Fortune – almost as if challenging her. Most of the ships were scattered by the wind. Caesar himself landed with all of three thousand men and a hundred and fifty horse in the region of Hadrumetum (now Sousse).

In view of the enemy's overwhelming superiority and the problem of obtaining supplies, Caesar at once sent to Sardinia and the neighbouring provinces for troops, armaments and corn; he also ordered the ships to sail out and bring together the parts of his fleet that had been dispersed during the crossing.

It would be pointless to give a detailed account of the campaign. It took place in a small area and followed much the same course as Caesar's earlier campaigns. But some details are interesting.

We have a description of the war, the *Bellum Africum*, from the pen of a senior officer who was able to observe Caesar closely, without necessarily being privy to his plans. We learn that the young soldiers were at first frightened and desperate, but found solace in the 'wonderful cheerfulness' of their commander, who seemed full of assurance and self-confidence. We read that when the opposing army paraded in front of his camp, Caesar directed his troops not from the wall, but from the commander's tent. He was content for messengers to put him in the picture and convey his orders. He also believed that his enemies were not sufficiently confident of victory to attack his camp: his name and reputation alone were bound to dampen their courage.

Logistics again posed a problem, but Caesar's soldiers were inventive enough, for instance, to feed the horses on seaweed, washed in fresh water, when fodder ran out.

Caesar was very anxious and sent repeated instructions for the rest of his troops to be shipped across; on one occasion he even sent orders that they should be dispatched forthwith, in spite of the season and the weather conditions. On the very same day that the courier had left, Caesar complained of the dilatoriness of the army and the fleet; all the time he looked angrily out to sea. Once, as in 48, he is said to have set off himself to fetch the missing legions, only to encounter the reinforcements on the high seas. Is it the reporter who makes Caesar seem so much more anxious than he does in his own account? Or did Caesar's anxiety arise from the difficulties of planning a campaign overseas? Or had he really become more restless and incautious, as he already seems to have been during the Egyptian campaign, unable to wait for the war to end? Cicero later wrote that Caesar was especially angry with those who had caused the African war, because they had prolonged his military labours.

At last the ships began to arrive. Caesar had given orders that only the soldiers and their weapons, 'neither baggage nor slaves, nor anything else that soldiers would normally bring with them, were to be stowed in the ships'. He then found that one of his senior officers had requisitioned a ship solely for his slaves and pack-animals. The author of the *Bellum Africum* calls this a 'trivial incident'; it casts a revealing light on the armies of the time. As the officer in question had had a hand in fomenting the Italian mutiny, Caesar dismissed him and others and sent them back to Italy with only one slave apiece.

With great difficulty, and by dint of skilful tactical manoeuvres, Caesar had prevented his opponents from attacking him with their greatly superior forces. On one occasion his army was actually encircled, but he was able to break out.

It stood him in good stead that the fame of his uncle Marius still lived on in Africa and that the opposing generals obviously made hardly any promises to their soldiers; hence many defected to his side. And a number of towns surrendered.

He was at great pains to prepare the army to meet the new enemy. 'He drilled with his troops,' we are told, 'not like a commander with an army of veterans that had been victorious in the hardest battles, but like a gladiatorial fencing master with new recruits, showing them how far to step back from their opponents, how to face the enemy, how to resist in a confined space, now advancing, now yielding, now feigning fresh attacks; he almost went so far as to prescribe from what distance and in what manner the javelins were to be used.'

With their skill and their highly versatile and deceptive tactics Caesar's opponents obliged him to operate slowly and deliberately, which was contrary to his custom. In order to practise fighting against elephants and accustom his horses to their smell, appearance and trumpeting, he had some sent over from Italy.

On 6 April 46 the decisive battle of Thapsus took place. It differed from all Caesar's other battles in that the soldiers started the fighting against his wishes, as victory seemed assured. The *Bellum Africum* does not make it clear why Caesar himself held back. We learn that Caesar, seeing that his men could not be restrained, uttered the battle-cry 'Felicitas' and was among the first to charge the enemy at a gallop. This too was a departure from the rule. Plutarch knew of other sources according to which Caesar took no part in the battle. 'While he was drawing up his army in battle order he is said to have been overcome by his usual sickness.' He reportedly felt an attack coming on and had himself carried away.

The 'usual sickness' was epilepsy, from which various sources say he suffered. We have evidence of other sudden attacks. After two thousand years no diagnosis is possible, and nothing can be gleaned about the exact nature of his illness or its physical and psychic consequences. An epileptic attack might explain why Caesar lost control of the army at Thapsus. The author of the *Bellum Africum* may have been mistaken about his riding into the fray. Perhaps the horseman leading the charge in the commander's cloak was not Caesar himself; this may have been one of the best-kept secrets, though it need not have been kept from everyone or for ever: behind Plutarch's sources there may have been a close confidant of Caesar's.

The soldiers charged so fiercely that victory was soon won. The enemy's elephants were subjected to a hail of missiles; the whistle of the catapults and the impact of the stones and lead bullets caused them to panic, turn tail and trample down the units ranged in serried ranks behind them.

Many enemy soldiers surrendered, but Caesar's veterans, in their fury, could not restrain themselves from falling upon them. In the *Bellum Africum* we are told that the victims appealed to Caesar. (At this point, then, he was present; perhaps he had just recovered from a fit and was therefore weakened.) Their appeal was unavailing. Caesar's men massacred them all; indeed, they even turned on their own senior officers and killed or wounded several. They are said to have reviled them as the *auctores* – 'instigators' – but we are not told of what. Of the war, perhaps; but more probably of its continuation and all the attendant tribulations and sufferings. Were not these – in the eyes of the simple soldier – connected with the generosity and mercy that their officers enjoined upon them? And so, not daring to turn against Caesar, they turned against the officers.

Caesar's soldiers clearly disapproved of his clemency: when the Egyptian king Ptolemy seemed to be making a fool of him, they did not conceal their gleeful contempt for the kindness he had shown him. The distinction between civil disputes and war was too subtle for them. They could not understand Caesar's policy; all they knew was that they were repeatedly called upon to pay for it. They took no pleasure in a war without plunder, so different from the one they had fought in Gaul – a war that dragged on interminably and in which the strains were probably worse than the bloodshed. First and foremost they were soldiers, not civilians. They had endured untold hardships, and in the end they had been held in the rear; Caesar had demanded too much of them. Now, while still in Africa, he lost no time in releasing his veterans. He seems to have settled a number of them in two colonies there.

After the battle of Thapsus he first moved northwards to Utica, where his opponents had a well-supplied fortress under Cato's command.

After Pompey had been forced to leave Italy in 49, Cato refused to have his hair or his beard cut. In January 49 he had been placed in command of Sicily. As the Caesarians were approaching, he voluntarily withdrew, believing that he could not hold the province and that prolonged fighting would only ruin it. After the defeat at Pharsalos he imposed a further privation on himself as a sign of mourning: from now on he would eat only in a sitting position and lie down only to sleep. He crossed over to Cyrene (now Libya), whence he reached the province of Africa after a twenty-seven-day march. The water was carried by donkeys, and members of the Psyllan tribe accompanied the army in order to cure snake-bites by sucking out the poison. Throughout the march Cato walked at the head of the army. The Pompeians wanted him to assume supreme command in Africa, but he yielded precedence to Metellus Scipio, Pompey's father-in-law, who was a *consularis*: Cato had not risen above the rank of praetor.

Hearing of the defeat at Thapsus, he conferred with the Romans at Utica, declaring that he was willing to fight, but would not compel others to do so. When it became clear that there was little will to resist, he saw to it that ships were available for all the Pompeians who wished to reach safety. He also made sure that there was no plundering: order was meticulously maintained in his town.

Cato was resolved not to surrender to Caesar. Only the defeated, he said, needed to beg, only the wrong-doer to apologize. He had remained undefeated all his life, and in as much as he had wished to defeat Caesar he had now done so and proved himself superior – through the purity and justice of his cause. Caesar had been defeated and found guilty, for he could no longer deny that he was at enmity with his native city. Cato judged everything as though a law-suit were to be conducted, in which the sole issue was rectitude. Was this because he had always been guided more by right than by might – which meant that the more he tried to force them together within the traditional framework the farther they drifted apart? If so, this was symptomatic of the split reality of the age. Or was he merely trying to vindicate the way he had lived?

He enjoined his son, who was with him at Utica, to beg for Caesar's mercy. Asked why he would not do so himself, he is said to have replied, 'I grew up in freedom, with the right of free speech. I cannot change my ways in my latter years and accustom myself to servitude. But it is right that you, having been born and brought up in such conditions, should serve the divinity that governs your destiny.' He was clearly aware of the relativity of the age. He seems to have seen that those who had grown up in an age of defeats, of senatorial and republican weakness, had less rigorous principles; on the other hand, they were blessed with more possibilities and therefore more adaptable. They could order their lives differently; it was up to them. He would not and could not interfere with them; he had to concede them their right, which to him was the right of the higher authority that governed all human life. In his case, however, to yield would have been to reduce life to mere survival. His conception of life was too lofty and too unambiguous for this. Caesar might no longer respect anything but human lives and perhaps human welfare, but to Cato republican liberty was essential to any worth-while existence. He was too upright, too strong-minded, too candid and courageous, to wish to forgo it. Yet even for his son there was still something to be respected; there was at least one thing that his father could expect of him. He should avoid politics: 'In present conditions it is impossible to engage in politics in a manner befitting a Cato, and to engage in them in any other way would be disgraceful.' His son fell at Philippi, the battle in which Caesar's murderers were defeated.

Cato took his life by falling on his sword. His son rushed in, and the physician bound his wounds, but no sooner was he left alone than he ripped off the bandage and bled to death.

Lucan wrote: *victrix causa diis placuit, sed victa Catoni* (The victorious cause pleased the gods but the defeated cause pleased Cato). The immense authority that Cato enjoyed during his lifetime, as the embodiment of the republic, became even greater through his death. No one sensed this as strongly as Caesar – as witness the unbridled and impotent hatred with which he pursued him, even in death, with at times obscene invective – a hatred that presumably derived from his inability to understand the source of Cato's authority. Pompey was great by virtue of his achievements, but in many ways weak. The leading senators all allowed themselves to be defeated somehow, and in the end Caesar despised them. But if he despised Cato, his contempt must to some extent have rebounded on himself. So hard, so Roman, was the stuff of which Cato was made, so unshakable his faith in the republic he served. Cato was one of the most remarkable politicians in world history, a man whose adherence to principle verged on the bizarre and made him in many respects a Quixotic figure. Yet for the Senate, which represented a tradition of immense political wisdom, he was the most respected and, in many ways, the most powerful politician of the age, enjoying the highest authority, though not the highest rank. Caesar may not have known, but must have sensed, that when he confronted Cato he was confronting Rome, that Cato gave the lie to his pretensions. The *res publica* alone could have produced such a man. Yet this also demonstrates the difficulties that Caesar faced. It may be that Cato did more than Caesar to determine the history of these years, in that he determined the way in which Caesar acted.

There was a logic in Caesar's reaction to Cato's death. We are told that he addressed him directly, as if standing in front of him, and his words presumably reflected his true feelings: 'I envy you this death, for you envied me the chance to save you.' He had wanted the final triumph, but now he saw that his opponent had robbed him of it. Only later did it become clear what power Cato exercised even after his death. When Caesar arrived at Utica, the inhabitants had already buried him with full honours.

After a short stay at Utica, Caesar hurriedly departed for Numidia and took over the kingdom of Juba, an opponent of long standing who had defeated Curio in 49. Juba had fallen in a duel with a leading Pompeian; it was a form of suicide. From Numidia, or part of it, Caesar carved out the province of Africa Nova and appointed Sallust as governor. Sallust proceeded to exploit the province so shamelessly that he was put on trial in Rome and acquitted only after paying out large sums in bribes. Yet in his historical work he affects to represent ancient Roman morality.

In Numidia Caesar was probably courted by neighbouring kings. We hear that he was well disposed to Eunoë, the wife of Bogud, king of western Mauretania (part of what is now Morocco), and gave many costly gifts to her and her husband.

He then set about stabilizing conditions in the old province. He imposed heavy fines on all who had sided with his opponents, both individuals and communities. He pardoned most of his opponents, but executed a few; others met their deaths in various ways, but it is not clear whether this was actually contrary to his wishes. Some succeeded in fleeing. Scipio's correspondence was found in his camp, but Caesar had it burnt, like Pompey's, without reading it.

In mid-June Caesar sailed to Sardinia, 'the only one of his estates that he had not yet inspected', as Cicero sarcastically remarked. Here too punishments were meted out. From there he sent most of the fleet to Spain, where unrest had spread and one of Pompey's sons had placed himself at the head of the rebellion. Moreover, a Pompeian adventurer in Syria had murdered Caesar's governor and taken over the province. Yet Caesar no doubt hoped to deal with these troubles by proxy. For him the war was over. On 25 July he was back in Rome.

'The difficulties increase, the nearer one comes to one's goal,' we read in Goethe's *Elective Affinities*. Caesar had meanwhile grown more and more impatient to see the end of the war. Could it be the kind of impatience that, when the goal is ambitious, makes the difficulties seem all the greater? But what if one has to ask oneself whether the goal of securing one's own honour really justified all the expense of such long and bloody wars? In Gaul the expense may have been commensurate with the goal; Caesar probably thought the effort worth while, knowing that it served his highest ambitions. With the same superiority that he had shown in Gaul he had again mastered every situation, in spite of a number of reckless decisions. Yet was not the endless warfare bound, sooner of later, to seem pointless? Was it only his veterans who were anxious for it all to end? And above all, was not the goal bound to become higher, the greater the cost of attaining it? After all, he must have felt a need, consciously or unconsciously, to find a reasonable equation between the value of the goal and the cost of its attainment.

Caesar may have been overcome with fatigue, as often happens when we come to the end of seemingly endless labours, when what has absorbed us for so long is finished, when we have reached our goal, but cannot yet enjoy it it. 'I have lived long enough for both nature and fame,' Caesar later said in Rome. We are told that he said it 'all too often'. He may have said it earlier, on the way back to Rome, or even before he left; and he may have had similar feelings as his impatience grew more urgent.

This would partly explain his long sojourn in Egypt, and his need for someone with whom he could speak intimately about the purpose of all his endeavours, about the purpose of his life. For this cannot have been clear to him. And Cleopatra would have known how to respond, not only theoretically. It was about this time that he invited her to Rome; she seems to have arrived there in the autumn.

As he travelled to Rome Caesar was probably tired and enervated. He may have felt liberated and had certain expectations. He may have composed himself in preparation for what was to come and been tempted to resignation, yet at the same time he was aware of a high obligation to himself, to his war, to his loyal followers, and probably to the commonwealth. But an obligation to do what?

14

Failure after Victory

'IF YOU DO WRONG YOU WILL BE
KING' · PESSIMISM IN ROME ·
SENATE DECISIONS ON POWERS AND
HONOURS · DEMIGOD · DICTATOR
FOR TEN YEARS · ALIENATION AND
ISOLATION: COULD HE GAIN
POWER OVER CONDITIONS?

THE CIVIL WAR, launched for the sake of Caesar's honour and safety, had the side-effect of making him the master of Rome and the empire. At his triumph his soldiers joked: 'If you do what is right you will be condemned; if you do wrong you will be king.' Was this what he had wanted? Had he bargained with it? Was he really prepared for such booty? Did he know that new tasks of an almost incalculable magnitude were beginning to pile up – that in fact he had not reached his goal, but only a new beginning? Might this knowledge have lain behind his remark about having lived long enough?

Roman society naturally went out to receive him at the gates of the city when he arrived, accompanied by a few troops and a large civilian escort and preceded by his lictors, their fasces adorned with the victor's laurels.

This time he was in Rome not to make hurried preparations for war, but to stay there indefinitely. Rome's good society flocked to meet him, partly out of a need to put themselves in good standing with the victor, partly out of curiosity.

No one knew what Caesar intended. He had obviously revealed nothing but the date of his arrival and perhaps a few wishes regarding specific powers and honours. There was much pessimism; the Romans feared a monarchy. They used the term *regnum*, which is commonly translated as 'kingship', but which in Rome was understood as tyranny and had connotations of usurpation, arbitrary rule and flouting of the law.

The pessimism may have contained an element of deprecation – a silent hope that nothing so bad would really happen. But it all depended on Caesar; it was an oppressive situation, in which Rome had to wait anxiously and see what he deemed right.

Unlike Sulla, he had no cause that united him with the Senate. He had of course repeatedly declared that the republic should be free and that everything should be decided by the Senate and the people; but would this still apply if their decisions displeased him? Many declarations are made in the quest for power, but how many remain valid once it is won? Precisely because he had no cause, and of course because Rome's ruling class regarded his war as unjust and viewed him with the utmost distrust, his victory amounted to subjection. Unlike Sulla, Caesar had not eliminated his opponents, and so the victor had to be on his guard against the vanquished. Politically he was in a minority. And he was very isolated. Whatever honours were heaped on him, he could not withdraw from the scene; he could not remove himself, as Sulla had done, from a reconstituted republic, but had to assert himself, defend himself, and consolidate his position.

For all his generosity and charm – indeed because of it – no one could be certain whether he intended to view the Senate and the citizens in political or merely in social terms – as constituent parts of the republic or merely as a society, which could go on enjoying life, welfare and honours, but would no longer represent the whole of the commonwealth, with a voice in its affairs and the power to make decisions.

Cicero later declared that 'no one should be able to do more than the laws and the Senate.' This was the essence of republican wisdom. For the time being, however, Caesar could do anything. Rome had to wait and see what he wanted. It was advisable to stay behind cover.

★   ★   ★

Before Caesar's return the senators had done only one thing, acting perhaps partly on direct or indirect intimations from him: they had resolved to grant him not only his triumphs, but a number of extraordinary powers and honours. The *supplicatio* for his victory was to last for forty days. His triumphal chariot was to be drawn by white horses and preceded by seventy-two lictors. It was then to be placed on the Capitol, before the statue of Juppiter; on it was to be a bronze statue of Caesar standing on a globe (symbolizing the world); the inscription was to describe Caesar as a demigod.

The Senate further resolved to appoint Caesar dictator for ten consecutive years and elect him *praefectus morum* ('overseer of morals'), a hitherto non-existent office that clearly carried powers of censorship. In future he was to be allowed to sit on the magistrates' bench, deliver his opinion first at all meetings of the Senate, and give the signal for the opening of all games. In addition he was to have the right to nominate all magistrates. Several of these resolutions had been given the force of law by the popular assembly.

Though not unimaginative, they were not necessarily an expression of state wisdom, and they need not be construed as part of a coherent plan. Sitting in the Senate with the older senators were many new members, supporters of Caesar. Yet even they did not see themselves merely as instruments; at least after Caesar's death many of them clearly tried their best to perform their role as old-style senators. But it is doubtful whether there was a leading group that possessed real authority and had a coherent view of the direction that Senate policy should take. Presumably the Senate's willingness to pass these resolutions was prompted partly by admiration for Caesar and a wish to do him homage and partly by fear, distrust and insecurity. A more interesting question is what directives it had received from Caesar; this cannot be answered. All we know is that he accepted most of the powers and honours he was offered. Whether he pressed for them or let himself be pressed to accept them, they must have been to his liking, though he is later said to have had the word 'demigod' deleted, perhaps after mature reflection.

To the extent that these honours represented him as the ruler of the world, they demonstrate that his *dignitas* had risen immeasurably. The senators were consistent: they took his claim to be as absolute and as superhuman as Caesar himself had done. The magnitude of his achievements seemed to transcend everything known to Roman experience – or indeed to human experience generally. They acknowledged it accordingly. Caesar was to have no cause for complaint. Even if the senators were not acting on his express wishes, they realized that he expected no less. They had to recognize the stature of a man whose wishes were known, who was powerful and feared, proud and pre-eminent. The fact that he was descended from Venus and obviously enjoyed her special favour may have made it easier for them to bestow on him the status of a demigod.

He probably had to be granted some powers, lest he should feel free to act outside the law. To make him dictator was unnecessary, since he was already consul, but it may have seemed a practical move: when his consulship ended he would not have to seek a further term. Yet it is unclear why the appointment had to be for ten years. There was no precedent for this; only Sulla had been dictator for more than six months, and he had been charged with the task of restoring order in the republic. No such charge was laid upon Caesar, and nothing in conventional Roman thinking suggests that he would have needed ten years to fulfil it. If the Senate was not simply complying with Caesar's known – or supposed – wishes, may it have hoped that such a generous conferment of powers would deter him from setting up a monarchy?

If so, it was giving him monarchical powers, but making it clear that they were exceptional. Essential to the office of dictator was its exceptional character. The powers it customarily entailed related chiefly, if not wholly, to military affairs. They were not precisely defined, but depended on a given situation, as did those of all Roman magistrates. What distinguished the dictator was that he had no colleague who could restrain him by entering a veto; how far he could be restrained by a tribune of the people was a moot point. It was expected that any arbitrary action he took would be restricted to specific areas. The laws were not suspended under a dictatorship, and the dictator knew that he would subequently have to resume his place in the system of oligarchic equality. Hitherto the Romans had thought it proper, in emergencies, to confer such comprehensive powers on an individual because they ultimately had faith in the republican constitution. In Caesar's case they presumably did so because the established institution of dictatorship would legitimate the power he already possessed and thus to some extent restrict it. Conversely, the Senate may have hoped that after being granted such powers Caesar would be better disposed to it and vouchsafe it some role in decision-making.

Whereas the dictatorship related chiefly to the conduct of current business, the supervision of morals involved the very foundations of public life. This had traditionally been the province of the censors, embracing the composition of the social orders, the census classes and the electoral districts, as well as roads, aqueducts, public buildings, and the collection of communal revenues. An important duty of the censors was to ensure that the members of the various orders and classes were worthy of their political and social rank. Given the rapid loosening of all ties during the latter years of the republic, Caesar's function as a supervisor of morals had a reforming character. Politically – in the narrow sense – it enabled him to purge and extend the senatorial and equestrian orders.

As far as we know, he made no use of the right to nominate magistrates, except perhaps on one occasion. On the other hand he conducted most elections, and it was hardly possible to be elected against his wishes. All the same, there might be a choice among various candidates whom he found unexceptionable. This no doubt suited him, as it spared him the unpleasantness that would have been bound to arise had he shown any bias.

The Senate's decisions gave Caesar nothing that he could not have taken if he had wished, but merely documented what he had achieved and what he had become. Through their legal form they imposed certain limits, however generously drawn. Essentially they reflected the situation that resulted from Caesar's victory in the civil war and testified to the difficulty of dealing with him. They point to a mutual establishment of positions, beside which the question of who prompted them, while still interesting, becomes less important. Caesar's isolation switched, as it were, from the horizontal to the vertical plane. Previously he had stood outside society, but now he stood above it; the alienation between him and the senators was greater than ever. This was the source of all the apprehensions; these now became expectations on the part of the senators *vis-à-vis* Caesar and corresponded to the expectations that Caesar had of them. These expectations and the further expectations they generated probably intensified to such an extent that Caesar – the dictator and demigod, exalted above all the rest – was able to enter into a special relationship with his society, one of mutual repulsion. When someone has been declared a demigod and is treated as such, many factors may contribute to his actually becoming one.

Could Caesar now find a place in the republic? Or could he devise and create a whole new order, higher and more embracing than the old, which would induce society, for his sake, to accept monarchic rule? Or could he make it clear to this society that he needed its help in solving the present crisis, so that the republic could then survive without him? Victory in the civil war had given him power within existing conditions. The question was now whether he could gain power over these conditions.

Caesar in Rome

IT IS LIKELY that Caesar's victories in the civil war led to an awareness of the complex task of reorganizing the commonwealth. Many things had to be set to rights.

The future position of the victor had to be determined; his primacy, indeed his leadership, had to be securely established. His soldiers and supporters had to be rewarded – materially, socially, and perhaps politically. An accommodation had to be reached with neutrals and opponents; at least it had to be made clear to them how things would be ordered from now on.

The war had destroyed or disrupted essential conditions of life. The old ruling class had suffered severely. The work of political institutions had been seriously impaired. The economic situation posed intractable problems. And, not least, there was much confusion in the empire.

One wonders whether Caesar too did not ask himself how the deplorable conditions of the pre-war period might be remedied or mitigated within the framework of a new order.

They affected almost all parts of society and the empire – high and low, rich and poor, soldiers and civilians, citizens and provincial subjects. There was a wide discrepancy between law and practice, needs and possibilities, expectations and demands. Many conflicts arose. Above all, people no longer knew where they stood; this had been true even before 49. Too much was uncertain, too much was possible. Everything hung in the balance.

The central task was political. How was Rome to be governed in future? How was policy to be made and the opposing forces reconciled? How was power to be shared out? Should the old institutions be restored, perhaps in a modified form? Or was the republic to be superseded by a monarchy? If so, how was this to be structured? What part were the leading personalities, the Senate, and the various social strata to play in it?

Everything else depended on how the political problem was solved. For the peace and stability of the commonwealth were at stake. Without political consolidation there was a threat of renewed civil strife, and whatever had been established or restored would necessarily be endangered.

Caesar was responsible for creating the new order. He was the victor and had been granted extensive powers by Senate and the people. Nothing could be undertaken unless he willed it. Although he had not been charged with the task of restoring the commonwealth, the responsibility inevitably lay with him.

In formulating the problems in this way, however, we must not forget that we view them from the perspective of our own age. It is unlikely that Caesar and his contemporaries saw them as we do. While some things must have been obvious to them, others are probably obvious only to us, since we know – or think we know – that the republic was superannuated, given the prevailing power relationships and the demands that were made on it, and that only a monarchy could solve the crisis. However, the monarchy that brought the solution was of such a kind as to do minimal violence to republican ideals and the self-image of Roman society; it appeared in conservative guise, as a restored republic, and was able, within limits, to constitute an alternative to what already existed.

We are inclined to assume that the Romans must have pondered the question of the political forms under which they wished to live and to overcome the difficulties facing the city. We are inclined to look for coherent conceptions and all too easily assume that men are naturally willing to develop institutions that will be adequate to all their problems.

In doing so we perhaps presuppose a detachment from prevailing conditions that contemporaries could not have had and we only seem to possess. It is therefore wrong to begin by outlining the problem as we see it – *ex eventu* and in the light of modern ways of thinking, with their potential for self-delusion – and then to look for answers to the supposed problems in the words and actions of those who lived at the time. This is perhaps to place them in an alien context. We should rather be prepared to envisage the possibility that Caesar and other members of his class – admittedly from differing perspectives – did not see the central political problems at all. Indeed, it is not easy to determine to what extent they recognized the crisis of their age. At all events we must not confuse their knowledge with ours.

However, we have one source that informs us at least about Cicero's expectations, which were probably shared by most of the senators, if not by the vast majority of the citizens. It is a speech that Cicero made in the Senate during these months, when Caesar had allowed one of his keenest opponents, Marcus Claudius Marcellus, the consul of 51, to return to Rome. Cicero thanked and praised Caesar; he said at the time that Caesar had triumphed over his victory. He then addressed himself to the tasks confronting Caesar. Caesar had informed the Senate of an attempt on his life; a hired assassin was said to have been seized in his house. It was later said that Antony had been implicated in the plans; they may well have been prompted by dissatisfaction with Caesar's policy, which was in many respects quite moderate.

Cicero now recalled Caesar's remark that he had lived long enough for himself and his fame. Cicero disputed it: he had certainly not lived long enough for the city. Whatever the war had cast down must be raised up again: 'Tribunals must be set up, loyalty and faith revived; the licence that was rife on all sides must be checked; a new generation must be brought on; everything that had dissolved and drifted apart must be bound together again by strict laws.' The dignity and justice of the republic, and everything that guaranteed its stability, had been sorely shaken. Not even the foundations had been laid, though Cicero credited Caesar with the intention of laying them. Caesar alone could heal the wounds dealt by the war.

Cicero reminded Caesar that his fame exceeded that of any other man; for that very reason he had not lived long enough if so much remained to be done. 'If, after your victory, you leave the commonwealth in its present state, you must be anxious lest your divine capacity for achievement should engender more amazement than praise.' In that case his fame would be carried far and wide, but find no lasting abode. It would find one only if Caesar discharged his obligations to the citizens, the city of his fathers and the whole human race.

Cicero's expectation was thus directed to the restoration of orderly life under the law after the convulsions of war. He did not say a word about any possible change. He probably did not exclude certain modifications. He left open the question of Caesar's future position. The Latin technical term for the task he outlined – *rem publicam constituere* – had earlier been used to describe that entrusted to Sulla. And Cicero had already indicated, in his work of 54, *De re publica*, that in times of emergency only a dictator could help. He had of course said that the dictator would subsequently resume his place in the ranks of the *principes*.

Yet Cicero clearly voiced his doubts: Caesar was again thinking only of himself. His reference to 'the whole human race' suggests that he may have known of Caesar's thoughts at the Rubicon. It is not impossible. But it need not be assumed, for everyone knew that Caesar's war had affected the whole world. So why should he not be responsible for everyone's welfare? All the more remarkable, then, is Cicero's assumption that the only way to persuade Caesar to take on the task that faced him was to remind him of the fame he would earn if he performed it – and forfeit if he ignored it.

He clearly had no great hopes of Caesar's responding to his plea. In the letters he wrote during these months he praised Caesar for becoming daily more reasonable and approachable. He did not fear him: admittedly, Caesar had all Cicero's utterances – and those of others – reported to him by his spies, but he was subtle enough to be able to distinguish between Cicero's real utterances and any that were falsely attributed to him. On one occasion Cicero expressed the opinion that Caesar was still the best hope; everything else was hard to endure.

Yet in Cicero's opinion the Romans were not living in a *res publica*, and it was unlikely that Caesar would do anything to change this. 'Even if he wishes the commonwealth to be as perhaps he too would like it to be, and as we must all want it to be, he has still not done what he could do – so deeply has he involved himself with many people.' Even Caesar, he said, did not know what would happen, for he was dependent on current conditions. He was not the master of his decisions. For there were many things that the victor must do, willy-nilly, to please those who had made his victory possible. On another occasion Cicero wrote that Caesar did not even ask his own people for advice. But he had to accede to their wishes. At this time Cicero was on the best of terms with Caesar's closest confidants, but unable to report on his plans. He found everything obscure and uncertain. He was utterly in the dark.

Caesar thus seems to have left everything open. We have every reason to suppose that he never contradicted the expectation that the old republic would be restored. Yet at the same time he probably kept his own intentions hidden, except for minor details, even from his closest associates.

Admittedly, he once said, 'The *res publica* is nothing – a mere name without body or shape.' But this was probably said in a state of emotion. We have similar observations by Cicero, who wrote in 54 that the commonwealth had 'lost not only all its sap, all its blood, but even its colour and its earlier shape'. True, Cicero wrote this in a mood of regret. However, Caesar's words – reported, moreover, by an opponent – must be taken with a grain of salt, if only because at that time the term *res publica* meant not only the 'republic', but the 'legal order' or the 'commonwealth' in the most general sense. Even if Caesar denied the republic, he could not deny the legal order and the commonwealth. On the contrary, Cicero was careful to avoid giving Caesar the impression that he did not regard the present state of affairs as a *res publica*.

Caesar admittedly said that Sulla was an illiterate because he laid down his dictatorship. And indeed there is little to suggest that Caesar intended to follow his example. Yet this was also connected with the fact that everything depended on him. More than once he said that if anything happened to him the *res publica* would have no peace, but be subject to even worse civil wars. This was because he could not at once restore Rome to a state of order that would enable it to do without him. This was a difficulty that may have greatly exercised him.

Nevertheless, Roman society was anxious to see the emergence of a stable order. It was no doubt willing to render unto Caesar the things that were Caesar's – indeed it had no choice. Yet it expected that even if the old order were not reconstituted, at least some kind of legitimacy would be restored. There had to be guarantees of security, areas of responsibility and competence.

Cicero, for instance, later expressed a wish for peaceful conditions and 'some state of the commonwealth that, if not good, is at least stable'. This seems like a move away from the old republic, but it still includes 'honourable concerns and endeavours'. Otherwise life would have little to offer. Yet 'honourable endeavours' must include the chance to win success in political disputes without dependence on another's whim or favour. He writes: 'If *dignitas* means holding sound views on the commonwealth and winning approval for them from the good, then I have preserved my *dignitas*.' One indication of this was that Cicero's morning reception – a custom observed by all members of the nobility – was particularly well attended. 'If, however, it means being able to put whatever you think into action or at least to express it freely, then we have no trace of *dignitas* left; it is an achievement if we can rule ourselves in such a way as to endure with decency what partly exists already and partly threatens to befall us.'

This was an unworthy state of affairs. No one person could be allowed to control everything. It was customary for all members of Roman society to attend to their interests independently, though within a whole network of connections. This was the foundation of rank and honour. Social life was largely identical with political life. Caesar was therefore expected to restore the framework of an order that was both natural and stable, with its own specific weight, even if it accorded Caesar great influence – which was already implied by the conferment of the dictatorship. He was to have supreme power, but not absolute power. It was unacceptable that everything should be left to the uncertainty of his decision-making. Rome also needed institutions that would allow everything to be settled among the citizens, not through favour from on high. Caesar could not put an end to the whole of social life as Rome had known it hitherto. These are the kinds of expectations that must have been addressed to Caesar, and not just by Cicero.

It should be added that even if Caesar was intent upon setting up a monarchy, he still had to create some form of institutional order that would be largely self-supporting.

In any case he must somehow declare his intentions. However, if he was aware of this problem at all, he put off dealing with it. He had more pressing things to attend to.

After his return Caesar made conciliatory speeches to the Senate and people. They should have no anxieties: he did not wish to set up a tyranny, but to consult with the senators. He did in fact consult with them, but not very often.

Not least among Caesar's preoccupations was the celebration of his triumphs and the other festivities he planned. Rome was to be held spellbound for many days by the grandest celebrations. What he and his soldiers had achieved, who he was, and what constituted the basis of his claim, must be made manifest and palpable. Much of this – not least his policy for public buildings – must have been planned well in advance, for the dedication of his forum took place during these celebrations in September 46.

He had four triumphs to celebrate: over Gaul, Egypt, Pontos and Africa. His victory over Pompey and his armies was omitted. Yet in celebrating his African triumph Caesar could not forbear to represent the death of Cato and others in large pictures; this gave much offence. He must have represented them implicitly or explicity as slaves of the Numidian king Juba.

The triumphal processions began on the Campus Martius, passed the Circus Flaminius, the Velabrum, the Via Sacra and the forum, and ended at the temple of Juppiter Optimus Maximus. At first long columns passed through the densely lined streets with booty, trophies and costly treasures, as well as large paintings of battles, lists and maps. It was later stated that Caesar had fought fifty battles and killed 1,192,000 opponents; the figures were presumably arrived at by adding up the numbers that he proudly displayed during his triumphs.

Then came the prisoners, most notable among whom were Vercingetorix, Cleopatra's sister Arsinoë, and Juba's son. To lead a woman in triumph was scandalous. All the same, Caesar spared her life, as well as Juba's, but Vercingetorix was executed as a perfidious rebel after languishing in prison for nearly six years.

After the prisoners came the long line of lictors, their fasces entwined with laurels, followed by the commander himself, riding on a chariot drawn by three white horses and conventionally attired in a purple toga, a laurel wreath on his head, an eagle sceptre in his hand, his face coloured with red lead: for the victor was supposed to represent Juppiter, whose power had made Rome's armies victorious. Over him a slave held the golden wreath from the temple of the supreme god and constantly repeated in his ear, 'Remember you are human.' Finally came the soldiers, who by tradition sang satirical songs, one of which ran 'Citizens, guard your womenfolk: we bring the bald adulterer.' There were also allusions to Caesar's youthful liaison with King Nicomedes of Bithynia. These offended him – so much so that he protested against the imputation and swore that it was untrue. The effect is said to have been ridiculous, but he no doubt took it seriously. The other jibe – that by doing wrong he would become king – he let pass.

The act of triumph had probably once had a mainly ritual character, but the aspect of honour had long since become predominant. When a human being represented Rome's supreme god, he acquired an aura of supernatural glory. He therefore had to be shielded from the danger of hubris and divine envy. This was provided by the *bulla*, an amulet he wore, by the slave's reminders of mortality, and by the soldiers' songs. None of these was missing at Caesar's triumphs. Yet never before had a commander so obviously represented the god. After all, the placing of his triumphal chariot on the Capitol was meant to document his close relations with Juppiter.

Each procession took up a whole day. During the Gallic triumph the axle of the triumphal chariot broke in the Velabrum, in front of the temple of Fortuna built by Lucullus. In token of propitiation Caesar ascended the steps of the Capitol on his knees. The incident may have made him truly uneasy. Having apparently come so close

443

to the god, was he not bound to regard the mishap as a kind of warning?

Finally, the commander sacrificed white bulls to Juppiter in front of the great temple. He also replaced the laurel wreath in the lap of the god's statue. It belonged to the god, not to man.

The triumph was followed by great spectacles. The city was splendidly adorned. There were gladiatorial games, and feasting at twenty-two thousand tables. Caesar had now at last to redeem the promise he had made at the time of Julia's death. Suetonius reports: 'So that the excitement should be as great as possible, he even had the food for the feast prepared in private houses, although cooks were employed to prepare it. On his instructions well-known gladiators were forcibly removed from the arena and reserved for him as soon as there was a risk of the public's requiring them for a dangerous fight. He saw to it that future gladiators were not trained in a fencing school or by fencing masters, but in private houses, by knights skilled in arms and even by senators, whom he urgently requested – as is attested by letters – to look after the individual trainees and give them personal instruction during practice.'

Caesar had theatrical performances staged in various districts and in all the languages spoken in the city. He had had the circus enlarged and the arena surrounded by a moat; here there were displays of trick riding, turns involving teams of horses, and a demonstration of the tactics used by the British with their war chariots. For five days there were bloody fights involving five hundred men on foot, twenty elephants and thirty horsemen on each side. For this purpose two camps had been built in the arena: the populace should see how Caesar's soldiers had had to fight, and the soldiers should recall their tribulations with pride. As Caesar had waged war not only on land, an artificial lake had been made on the Campus Martius, where two fleets met in battle. Nothing like it had ever been seen in Rome, though there may have been Hellenistic models. The battles were fought by prisoners and men under sentence of death; it was after all a matter of life and death.

It had for some time been an ambition of prominent nobles to introduce the Romans to fresh wonders of nature. Caesar was the first to import a giraffe, probably from Egypt, and the public duly marvelled at it.

Particularly ingenious was the system of awnings that covered the whole of the Forum Romanum, the Via Sacra as far as Caesar's house, and the slope leading up to the Capitol. They are said to have been of silk. Their purpose was to provide protection from the sun for spectators at the games, and no doubt also for those attending the triumph, which thus ended in a vast artificial interior.

The citizens could thus marvel at contrivances of the utmost ingenuity, brilliant feats of organization and staging. Much was entirely novel – above all the fact that so much was organized simultaneously. In this too Caesar was superior to everyone else: it was *his* mind that lay behind it all and seemed to be at work in many places at once, transforming and controlling the city like a gigantic machine. There must have been an almost universal feeling of wonderment, at least among the mass of the citizenry – Cicero and his like could not be impressed by such means.

According to Suetonius, such a multitude gathered from all parts that most had to sleep in tents on the streets and squares; 'and some were crushed or fainted in the throng, including two senators'.

As with earlier games, one sees what extraordinary technical resources could be mobilized in Rome at this time; what Caesar achieved here bears comparison with the techniques he used in war. One should in any case not underrate the technical achievements of antiquity. One detail will suffice: the Twelve Tables already mention gold teeth or teeth fastened with gold. At least among the affluent citizens, and in public displays and warfare, a high degree of inventiveness was evident. It was here – and not, for instance, in the production of goods – that technical imagination was concentrated.

On 26 September Caesar consecrated the temple of Venus Genetrix and its sacred precinct, which was intended as a forum to provide additional offices and courtrooms. From as early as 56 or 55 Caesar's agents had spent huge sums on acquiring numerous plots adjacent to the old forum. It may be that after Caesar won control over Rome the Forum Julium was realized on a grander scale than was provided for in the original design.

By consecrating the temple of Venus Genetrix, the mother of Aeneas – from whom the Romans were descended and who was supposed to be his ancestor too – Caesar was ascribing his victory to the goddess. This demonstrated that the victory was closely linked to the person of the victor, not just to the magistrate of the Roman people. There had been a precedent at the time of Sulla, when the Victoria Sullana was celebrated instead of the Victoria of the Roman people; not long afterwards came the Victoria Caesaris.

In this temple the dictator later set up a statue of Cleopatra opposite that of the goddess. Was this undoubtedly unusual consecration perhaps meant to honour the mother of his only son? Interestingly, the statue still stood there in the second century A D. Augustus thus did not have it removed after his victory over the queen. Pliny tells us that Caesar also allowed an armed statue to be set up in his honour in the forum. It was probably the first time that this had happened publicly in Rome.

It is reported that after dinner on the last day of the triumphs Caesar paid a visit to his forum, wearing slippers and garlanded with flowers of all kinds. The celebrations were probably still in progress everywhere. The crowd escorted him home. The route was lined by elephants carrying burning torches in candelabra. Might this superbly staged conclusion to the festivities have come as a surprise to him?

The vast expenditure that Caesar incurred at the time and the shedding of so much blood – the fact that he 'was not yet sated with blood' – are said to have been much criticized. There was displeasure too at the great man's capricious insistence that the sixty-year-old Laberius, a member of the equestrian order and a writer of pantomimes, should appear in one of his own pieces; this was incompatible with his rank. Laberius avenged himself by declaiming, 'Hither, O citizens! We have lost our liberty!' or 'He whom many fear must go in fear of many!' Caesar compounded his arbitrary behaviour by rewarding him generously and renewing his knightly status with the gift of a gold ring.

After this the booty was shared out. Very large sums were allocated to Caesar's veterans, but small amounts also went to the urban populace. When the veterans complained indignantly that they should have received it all, Caesar personally seized one of them and led him away to be executed. He had two others ritually sacrificed by the priests on the Campus Martius and their heads displayed in front of the Regia, the official residence of the *pontifex maximus*. The religious background to this, if there was one, remains obscure.

After his return to Rome, Caesar was preoccupied with the settlement of his veterans. His agents purchased the land, but he reserved the right to decide on all controversial matters. He began to plan and carry out a comprehensive social programme. He reduced the number of those entitled to subsidized grain from three hundred and twenty thousand to a hundred and fifty thousand and seems to have revised the qualifications for inclusion in the list. He instituted special payments for prolific families in order to make up for the losses caused by war. Many members of the Roman plebs were to be settled in provincial colonies; at the time of his death eighty thousand had been settled. To prevent unrest, he suspended the civic associations that Clodius had reintroduced in 58.

He reduced the term of office of provincial governors, reformed the lawcourts, increased certain penalties, and sought to attract physicians and teachers to Rome with the promise of citizenship. He decreed that one third of the farm-workers on large country estates should be freemen. He brought in a sumptuary law restricting the use of litters and the wearing of purple garments and pearls, and laying down precise regulations regarding food and funerary monuments.

He also introduced the Julian calendar. This reform had important consequences. The Romans had hitherto had a lunar year, involving the intercalation of an extra month every two years. In the confusion of the late republic this had often been neglected, so that the harvest festival no longer took place in summer and the wine-gathering was not celebrated in autumn. Caesar introduced a solar year of three hundred and sixty-five days, with a leap-year every four years. In February 46 he had already inserted the normal leap-month, and between November and December two others followed, amounting in all to sixty-seven days.

Apart from these and other reforms, work on many major and minor projects, and the conduct of day-to-day business, he was occupied with reviewing the cases of former opponents in the civil war, with auctioning confiscated property and distributing rewards to supporters.

He had taken on an immense load. He could rely on his well-organized office, but made most of the decisions himself. He worked unremittingly. Even at the theatre he read and replied to letters and supplications; some took this amiss. It was only with difficulty – under conditions that Cicero found unworthy – that one could gain access to him. Shortly before his death Caesar is said to have remarked, as Cicero sat in his anteroom, 'Can I still be in any doubt that I am the most hated man, when M. Cicero sits and waits and cannot speak to me when it suits him? And yet he is the easiest to win over. All the same, I am sure that he thoroughly detests me.'

What Plutarch writes about Gaius Gracchus could apply equally well to Caesar: 'He did not let himself be wearied by so many important affairs, but dispatched each with unusual speed and energy, as though it was the only one, so that even those who hated and feared him most could not help admiring his industry and quickness.'

Some of the matters to which he addressed himself were certainly necessary, especially the provision for his veterans and some of his other supporters. Some answered to the demands made by Cicero in the Marcellus speech. Yet in other cases one wonders whether Caesar actually did what was important, and not just everything possible; much of it was no doubt useful, but not really urgent, compared with the task of political consolidation.

Rome was governed literally like a conquered city. True, the commandant was personally charming, well-mannered and gracious, and saw that law and order were maintained. Yet he acted as he saw fit. Others had no right even to be informed of his intentions. They were in his hands. Rome still had a Senate and magistrates, but they were not free in their decision-making. The *comitia* still existed, but they could vote only with Caesar's permission and sometimes – when he was away – they could not vote at all. There were also lawcourts. But in all matters the decisive authority lay with Caesar alone.

He was wholly taken up, at least at first, with administrative, organizational and social problems. It was these that engaged his attention and interest. They were often best solved by one who enjoyed monarchic authority; and Caesar was adept at solving them.

Here his old passion for achievement came into its own; in this he could demonstrate his superiority over everyone else. Moreover, his mode of operation was one that he had practised for more than twelve years: he planned, gave orders, and made dispositions on a grand scale. He now ruled the empire as he had once ruled his army, the theatre of war, and his provinces. It called for tremendous effort, but the effort was commensurate with the greatness he had assumed. A petition addressed to Caesar at this time – or perhaps somewhat later – urged him: 'Therefore, by the gods, look after the *res publica* and walk straight through all the difficulties, as is your wont.' He did just that.

He boasted about his laws, claiming that some of them held the commonwealth together. If he felt an obligation to the citizens, he doubtless thought that he must act in his accustomed manner, in accordance with his nature.

Yet this did nothing to promote relations between the citizens as citizens, to solve the political problem, or to meet the expectations generated by the republican independence and liberty that was enjoyed by the upper classes and to some extent by the lower orders. The result was that the old one-sidedness reappeared in a new guise. Caesar, like Pompey before him, was concerned with the problems facing the commonwealth, while the others were preoccupied with the commonwealth itself: it was after all they who constituted it, and they wished to go on doing so, each in his own right, in accordance with the old rules. But Pompey would have done things differently, had he been able to.

Perhaps it was Caesar's insensitivity to institutions and his detachment from society that prevented him from appreciating these problems. He was aware of the tasks that had to be performed, but otherwise he was aware only of persons. In the Senate he saw only the senators. Some of them supported him; to many he had shown mercy; and a few he actually esteemed. He cultivated Cicero as a brilliant man of letters; several others, such as Servius Sulpicius, the consul of 51, and Marcus Brutus, served him as governors, to his entire satisfaction. As an institution, however, the Senate meant nothing to him. He could not see it as the embodiment of supreme authority, the assembly of the worthiest citizens, with responsibility for the whole of the commonwealth, but only as a committee that deliberated endlessly, often to no effect, and had opposed him as late as 49, when he had offered his collaboration. The difference between Caesar and Sulla is plain: Sulla may have had scant regard for the senators and restored the oligarchy in spite of the oligarchs – but he respected institutions.

Caesar's world, in which only tasks and persons counted, was dominated by two principles – care (*cura*) and competition. This accorded with his thinking in terms of *dignitas* and fame. No one understood care better than he, no matter what problem it was to be applied to, so why should he not concentrate on what he did best? Why should he not follow his nature and outdo the rest? He may have had no interest whatever in the institution of monarchy, but only in the power and opportunities available to the autocrat – in other words, in his own position. If institutions meant little to him, the question that presents itself to us as a simple alternative – republic or monarchy –cannot have been important to him. Or did he realize that although the old order was no longer practicable, nothing new could be established in its place? If so, he had no choice, for the time being, but to do everything himself.

This might well explain the mutual misunderstanding that separated Caesar not only from senators like Cicero, but from large parts of the body politic. He felt that he was doing his best to master all the problems. What more did they want? Hearing their pleas for the restoration of the *res publica*, he probably felt like someone who was continually baling out the water to keep the ship afloat, while they carped about the flaking paint or the new steps to the captain's bridge. By nature impatient and inclined to be reproachful, he could easily give offence, as when he declared that the *res publica* was nothing. The others probably thought (to stay with the nautical image) that he was preoccupied with a thousand miscellaneous problems, but not with stopping the leak that imperilled the ship. Because Caesar's expectations did not coincide with society's, the outcome was mutual disappointment. And it was likely to increase.

One may of course wonder how far Caesar's ambition was reliant on a public. Yet it is quite improbable that the consciousness of his unprecedented achievements and the praise they deserved was sufficient for him, and that he was indifferent to the judgement of Roman society.

In his earlier years in Gaul he may have believed that he must first lay the foundation for his fame; he may have had an imaginary public in mind. Failing to win the recognition that he felt to be his due, he could resort to a kind of conspiracy theory and blame his opponents. He may have convinced himself that influence and fame depended solely on achievement and not on his attitude to others. He was therefore not reliant on instant recognition. In 49, however, he seems to have been very disappointed by the Senate and the people.

It is hardly conceivable that, at least after his victory, he did not think of some body of authoritative opinion, apart from himself, for whose applause and approval he secretly yearned. As long as he had opponents and could assert himself against them all, he was sustained by their resistance. But now he had reached his goal. His former faith in the future must have been spent. The recognition he had probably gambled on – the reward that would compensate him for all his efforts and justify his career as an outsider – must now be due. It had surely not all been for nothing. There was no cause, no ideology, no mission, to which he could cling or behind which he could take refuge. Had he had such a cause, vanity might have craved recognition, but lack of it would have simply confirmed his superior judgement. He did not want to rule by force, as an oppressor. Applause from his own court was scarcely sufficient. Among the others, he was bound to attach most importance to the senators, the leaders of Roman society; he might despise them as a political force, but he could accept their judgement, at least if it was favourable to him. Supreme *dignitas* meant supreme recognition. And this should be accorded not only to his achievement, but to his nature, to himself. Was he not bound to be convinced by himself, having used up all his energies for Rome? Could he fail to be profoundly disappointed when all this found no echo in Rome's good society? Of the reform of the calendar, which Caesar saw as a significant achievement, Cicero could only joke that even the stars in their courses now obeyed Caesar's commands.

If Caesar was so disappointed by the Romans, must his pride not have inclined him to ignore their judgement and defy them by being, in spite of them, what he had wanted to be for their benefit – to remain true to himself and to be from now on the only authority? If he really did not care whether he 'ruled over reluctant subjects', as one of our sources puts it, such indifference presumably belonged to the last period of his life.

After his victory in the civil war, Caesar stayed on in Rome for over five months. Then, during the second leap-month, he had to leave hurriedly for Spain. He arrived there at the beginning of December. He had not even had time to conduct the elections. His colleague had him elected as sole consul for 45. The dictator can hardly have needed this office. But his election deferred the decision as to who should become consul and – more importantly – who should not. It also obviated the uncertainties of the electoral result.

For Caesar probably had good reason to wish to conduct all elections himself. The popular assembly was not to be relied upon when it exercised its right of election. Caesar was safe only when he was present in person. Instead of praetors, special delegates known as praefects were appointed.

The Second Spanish Campaign, the Dispute about the Dead Cato, the Decision to make War on the Parthians

VICTORY AT MUNDA – THE LONG
JOURNEY HOME – THE ANTICATO –
CAESAR'S TRANSFORMATION IN
SPAIN – CICERO'S ATTEMPT AT A
PETITION – THE PARTHIAN CAMPAIGN
AS THERAPY? – FRESH HONOURS –
EXPECTATION OF AN ASSASSINATION
ATTEMPT

IN SPAIN the Pompeians were in open revolt. The governor of Hispania Ulterior, appointed by Caesar, had been deposed and Pompey's elder son Gnaeus proclaimed *imperator*. The province had been devoted to his father and grossly exploited by the Caesarians, and in a very short time he was able to raise thirteen legions, though not of the highest quality. He was joined by Labienus and other leaders of the army defeated in Africa. These forces were all the more formidable as Caesar could no longer call upon his battle-hardened veterans, and some of his troops were tied down elsewhere.

As usual Caesar completed his journey with great speed. On the way he whiled away the time by composing a long poem entitled *Iter* ('The Journey').

In Spain he immediately had a number of successes, as his opponents overestimated their position and, unlike him, did not wage war with the utmost energy. On 17 March a battle was fought at Munda. The enemy army offered strong resistance; the soldiers, especially those in the best legions, knew that they could expect no quarter, for they had formerly left the Pompeian army and been taken into Caesar's service.

Seeing his troops pushed back and beginning to flee, Caesar barred their way and asked them whether they were not ashamed to let him fall into the hands of young cubs like Pompey's sons. He then jumped off his horse, seized a shield and pressed forward into the front line. He is said to have shouted that this would be his last day, and the last day of the campaign for his army. He cursed his fortune for preserving him for such an end. Then, under a hail of missiles, he faced the enemy, ignoring the danger, and called after his soldiers that he would not follow them. They should mark what kind of a commander they were deserting, and they should remember the place. More out of shame than bravery they finally halted. Bogud, the Mauretanian king whose wife Eunoë had been Caesar's mistress, then forced a decision by attacking the flank of the enemy camp; this caused the enemy to flee, first on one side and then on much of the front. Fortune was forced back on to Caesar's side. He later told his followers that he had often fought for victory, but never before for his life.

A few towns still remained to be taken. At the beginning of April Gnaeus Pompey was killed. The rebellion was over.

Caesar, however, spent about two more months in the country, rewarding friends, punishing opponents, and collecting money. Among other things he confiscated the votive offerings in the temple of Heracles at Gades (Cadiz). He also made some more far-reaching arrangements. Various colonies were founded for the settlement of veterans and members of the urban plebs. Several towns were granted Roman citizenship. Some were given new constitutions, which included the provision that freedmen – who were not normally entitled to political honours – could become members of the town council. All this made an important contribution to the permanent Romanization of the Spanish provinces.

In June and July Caesar made similar arrangements in his old province of Gallia Transalpina. He decided to set up a colony for his veterans at Arelate (Arles) and build the naval port of Forum Julii (Fréjus). Various Gallic communities were granted the Latin right, preliminary to the conferment of full citizenship; this meant that the holders of high offices obtained Roman citizenship for themselves and their descendants.

He was in no hurry to return to Rome. In August he was in northern Italy. Afterwards he spent some time in the country, perhaps with Cleopatra. At the beginning of October 45 he returned to Rome for his triumph.

Roman society had not hoped for much from the war. A bleak alternative emerges from Cicero's correspondence: slavery if victory went to Caesar, murder and manslaughter if Pompey won. Gaius Cassius, one of Caesar's murderers, wrote: 'I will die if I am not full of anxiety and would rather keep the old mild master than take my chances with a new one, who is cruel.'

During the ten months or so of Caesar's absence Rome had been outwardly peaceful. Yet under the surface certain changes had taken place that affected Caesar's relations with the Romans.

A particularly important event seems to have been the literary war over the dead Cato. At Brutus' request, Cicero had dedicated a long memorial necrologue to this champion of the republic. We have one single quotation from it, to the effect that Cato was one of the few men who were greater than their reputation. Cicero must have praised him as the embodiment of true Roman manhood. For all his political circumspection he could not forbear to describe Cato's suicide as the magnificent consequence of his republican convictions. At about the same time Brutus published his own 'Cato'.

When Caesar received these writings, he instructed his confidant Aulus Hirtius to compose a reply, but after the victory at Munda, he too set about writing an extensive work. His 'Anticato' was a positively obscene composition, a virulent attack on the man who until 49 had enjoyed supreme moral authority in the Roman republic. Caesar caricatured him as eccentric and self-seeking, a drunkard and a miser, who had even sold off his wife. Nature, he said, had made Cato different from everyone else. Yet even the officer who wrote the *Bellum Africum* for Caesar had described Cato as a highly serious man of unparalleled integrity. Caesar's unworthy invective can be explained only as deriving from boundless fury and profound hatred. He viewed his dead opponent with utter incomprehension. Hence, in his account of how the war broke out, he had credited Cato with only selfish motives. Applying his own criteria, he could find no good in him; the fact that others could – and apparently rated him higher than himself – was extremely irritating, especially as they included not only the pusillanimous Cicero, but also Brutus, who had so readily taken up his offer of reconciliation and been received with such kindness. Their views presumably found a strong echo in Rome.

Caesar construed their praise of Cato as an attack not only on himself, but on his judgement as to what counted as Roman manliness. If extraordinary virtues were attributed to Cato, this seemed to put him in the right and Caesar in the wrong. If this were so, he did not deserve to be the first man in Rome. So seriously was he affected by the opinions of Rome's republican society, whose ideal Cato had embodied. So keenly was he struck by his inability to defeat his opponent, even posthumously.

Of course, if society judged a man of Caesar's stature to be inferior to one whom he found pedantic and obdurate, wholly negative and worthless, this was in his view a massive indictment of society. The revival of Cato's fame was thus a political reverse for Caesar, a bitter experience for one who thought he had at last reached his goal. It was compounded by the knowledge of how easily this opponent, who had so often been beaten, could still challenge his cause, even militarily, under the leadership of inexperienced youths. There was no doubt that he lost the literary battle too – and he was unaccustomed to losing. The real result of this experience seems to have been an even greater estrangement between him and the society he belonged to. For in Spain Caesar clearly became a different man. At least there seems to have been an inward transformation that made him indifferent to the judgement of Roman society. It appears to have become clear to him that he could gain no foothold in it. Various ties that had bound him hitherto were broken. He withdrew more into himself, however charming and open he remained on the surface. He was determined to pursue the course he had set himself, which was the right one, even if it was not understood; to have done otherwise would inevitably have exposed him to self-doubt. Relations between Caesar and Rome became increasingly tense and acrimonious.

Towards Cicero he maintained the outward forms of courtesy. At the beginning of his book he praised Cicero's artistic style. To this extent he regained his superiority. For it pained him not to excel in the literary sphere too. He consoled himself with the thought that his military duties allowed him too little time to train himself fully in the art of writing. Cicero, however, was apprehensive. He felt obliged to do something to regain Caesar's favour. He wanted to write a political tract that would combine respect, praise and expectant exhortation. Naturally he would not recommend the restoration of the old republic – which he regarded as right – but propose a solution that implied a position of supremacy for Caesar. But he would plead for the creation of stable political conditions, a legitimate order in which not everything depended on a nod from the dictator. He would write 'as a right-thinking citizen who nevertheless adapts himself to present conditions'. Yet this presupposed that Caesar should devote himself in Rome to the consolidation of the common-wealth. He was advised by Caesar's confidants that for this very reason he should refrain from such a suggestion. It would not commend itself to the dictator. Caesar thus obviously knew what was expected of him; he repressed the knowledge as best he could and did not wish to be reminded of it. He had had enough of such pleas from Roman society.

Caesar had long had other designs. In Spain he had decided to lead a campaign against the Parthians, whose mighty empire began beyond the Euphrates, on the borders of the Roman province of Syria, and extended far into what is now Persia. In 53, when Crassus had launched a war against them, they had inflicted an annihilating defeat on the Roman army. They had now penetrated into Rome's eastern provinces. To make matters worse, Syria was in the hands of a rebellious Pompeian. It is not easy to judge whether Caesar's intervention was really so urgently necessary or whether it was bound to seem so on the basis of reports he received from the theatre of war. And was there no one, apart from Caesar, who could conduct such a war?

However, Caesar let it be known that he did not wish to leave until he had 'settled things'. He now seemed to be planning exactly what Cicero had wished to recommend. Those in charge of his office must have been surprised, but they could do no more than guess at his intentions. In another of Caesar's letters he says that after his return he wishes initially to stay in Rome to make sure that his laws will not be flouted – the sumptuary law, for instance. This suggests that he too saw the crisis chiefly in moral terms.

All the same he proceeded to arm for the Parthian war. Yet he decided to spend the winter in Rome. Whether he thought this would give him enough time to 'settle things' – or indeed what he meant by this phrase – is not clear. He may have used it merely as an initial response to the demands for consolidation. One of our sources suggests that he planned the campaign as a kind of 'therapy'. According to Strasburger, this is 'at first sight absurd, yet perhaps not wholly wrong'. We may wonder whether it was not intended as a flight from Rome, whether the tasks that faced him there had not by now become so onerous that he was reluctant to tackle them.

His planning was masterly. It was to be a campaign of conquest in the style of Alexander the Great. Caesar's intention was not only to force the Parthians into submission, but first to secure the northern border of Macedonia against the Dacians. The Dacian king Burebista had extended his dominion to the south of the Danube, from what is now Romania. Possibly Caesar intended to make conquests here too, but in any case he wanted to drive Burebista back within his borders. The last thing Rome needed was a strong neighbour in this area. He then planned to cross into Asia Minor and move against the Parthians, and finally to march round the Black Sea from the Caucasus. He estimated that this would take three to four years; when he set out he would be fifty-five years old.

Perhaps he thought that after such victories he would at last carry conviction in Rome. Perhaps he thought that during his absence unrest would break out in Italy and he would be able to return as a saviour and put it down. It is more probable, however, that he paid less and less heed to the Roman citizenry of his day and more and more to posterity – that his will for achievement was increasingly driven on by an urge to escape from the transient. Yet he may also have been keen to leave Rome and set out on another campaign. Even if he was already weary of campaigning, was it not bound to appear in a different light after more than five months in Rome? Dealing with the soldiers, at which he was so adept, was certainly preferable to dealing with Rome's pretentious grandees. It is true that in the field he received letters from Rome, but Roman society could not force itself on him there; he could shield himself against it. The theatre of war demanded much of him, but had he not discovered in Spain that he grew in stature when put to the extreme test? Was not the sum of what he could accomplish in Rome trifling beside this – and wearing too, since it stretched him only mentally and psychically, not physically? After all, he had long regarded physical effort as therapeutic. The struggle with the enemy, however strenuous, might well have seemed easier than the daily shadow-boxing with the aristocracy, who ducked his blows, yet repeatedly came up again dissatisfied – opponents whom he had beaten but could never beat conclusively, who were easy to subject and hold down, but impossible to convince?

However, if he thought the Parthian campaign would help him politically, it must have been clear to him that he could not solve the internal political problems directly. And in any case he wanted to escape from them. His plan was thus an expression of helplessness in the face of Rome and her society.

Yet there was helplessness on the other side too. For the resolutions passed by the Senate during several sessions after news arrived of the victory at Munda reflected its inability to comprehend the phenomenon of this man in normal human terms.

It resolved that 21 April would in future be celebrated with circus games. This day counted as the anniversary of the city's foundation. The occasion for this resolution was the news from Munda, which had arrived on the previous day. The victory was to be seen as marking the refoundation of the city. This time the *supplicatio* was to last for fifty days. Caesar was accorded the titles *imperator* and *liberator*, and it was decided to build a temple of liberty and a palace for him. On all public occasions he was to appear in his triumphal robe, and from now on he was to be permitted to wear the laurel wreath, which properly belonged to Juppiter. This honour is said to have been particularly welcome to Caesar as it enabled him to conceal his increasing baldness. For he was still as vain as ever. He thought that Venus had granted him a kind of second youth, which was incompatible with receding hair.

Foremost among the honours accorded to Caesar was the decision to set up a statue of him in the temple of Quirinus, bearing the inscription 'To the undefeated god'. Quirinus was the old warrior god of the inhabitants of the Quirinal district of Rome; he had long been identified with Romulus. This too seems to point to Caesar's refounding of Rome. Another statue of Caesar was to be set up on the Capitol next to the seven kings, who stood there with Lucius Brutus, the republic's founder. The Senate further decreed that an ivory statue of Caesar should be carried in the solemn procession of the gods that belonged to the ritual of the circus games. Finally, the Senate gave him charge over all the city's military and financial affairs and made him consul for ten years. While the extravagant honours conferred on him after Thapsus may have been intended to confine him within the bounds of republican legality, the resolutions passed after Munda far exceeded all republican norms.

Not only Cicero mocked 'this processional figure', the housemate of Quirinus. And in July, when Caesar's image was first borne in procession, there was no applause whatever. Not even for Victoria – 'on account of the wicked neighbour who walked with her'. It was too much even for the populace lining the streets. And the fact that Caesar's statue stood next to those of the kings encouraged talk of his being a tyrant.

Yet this seems to have been the most important, perhaps the only important link between the senators and the dictator. All they could do was to shower him with fresh honours. Instead of finding a genuine power-base in Rome's leading body, he was raised to dizzy heights. What is more, he was content to let it happen. He saw no reason to refuse such treatment: he may even have engineered it.

We hear hardly anything of the rest of Roman society at this time. Caesar's ten months' absence may have produced a certain detachment. He may earlier have dazzled the Romans and kept them in suspense, but now it must have become increasingly clear, especially after the 'Anticato', how little they had to expect of him in the way of political consolidation. After his last victory the desire for normalization may have revived, but the uncertainty must have become more unsettling. What had he and Roman society to do with each other?

Nevertheless, some prominent senators still pinned their hopes on him. Marcus Brutus went to meet him in northern Italy, where he had recently served as governor. Caesar praised him, showed himself charming and accommodating, though he probably did not commit himself politically, and promised him the praetorship for 44. Brutus reported back to Rome that Caesar would join the 'good'. 'What glad tidings!' remarked Cicero, 'but where will he find them if he does not hang himself?' (To find them Caesar would have had to go down to Hades.) Cicero was disappointed, for he had reminded Brutus more than once of his two great ancestors, Gaius Servilius Ahala and Lucius Junius Brutus, who had freed Rome from tyranny. A little earlier, when Caesar was in Gaul, one of the leading Caesarians had tried to persuade Antony to join in an assassination plot. If Cicero now reckoned that the time would pass 'when one must truckle to the will of one man', he must have been thinking of this man's violent removal.

Cicero had long since withdrawn largely into philosophy and confined his political activity to begging Caesar to show mercy to his friends. His bad conscience at having reached a hasty compromise with the victor made him doubly assiduous in this regard. Did he wish to put as many others as possible in the same position?

Above all he was deeply despondent about the republic. In the past he may often have been confused by the complexity of the political situation, but he now had no doubts about Caesar: his régime was intolerable. When working on his tract for Caesar, he made the interesting observation that he found it an unworthy task, 'but in these matters we have already become callous and cast off all *humanitas*'. It is the only passage in Cicero's writings where *humanitas* means something like 'human dignity': elsewhere it means 'culture', 'refinement', 'serenity', even 'philanthropy'. And the dignity of the *consularis* was *dignitas*. Cicero was no longer concerned with social judgements, but with the human values underlying them. So profoundly was he affected by recent events.

In Rome for the Last Time:
from the Spanish Triumph to the
Ides of March 44

A T THE BEGINNING OF OCTOBER 45 Caesar entered Rome in triumph. For the first time he was publicly celebrating a victory over his fellow citizens. Again there were great festivities, ending with a popular banquet. This caused much displeasure. As the triumphal chariot passed the tribunes' bench, Lucius Pontius Aquila, a tribune of the people, pointedly remained seated, whereupon Caesar shouted to him angrily, 'Demand the republic back from me as a tribune, Aquila!' The reverberations of his anger were felt for days; whenever he made a promise he added sarcastically 'if Pontius Aquila will allow me to'. Gelzer surmises that it was the unfavourable public reaction that caused him to give another popular feast four days later. Caesar, however, declared that the first had been too frugal and saw to it that the second was exceptionally lavish.

He no longer behaved in his accustomed manner. The least challenge to his authority provoked him beyond measure. The Spanish revolt had after all challenged his authority; this was why he had meted out such severe punishments to all who had taken part in it. All had to yield to him. He was no longer prepared to show any consideration – except by continuing to spare the lives of his opponents.

He therefore broke with tradition and allowed two junior commanders a triumph, even though they were not legally entitled to it. He wished to reward them, and presumably took all the more pleasure in showing such generosity as it was in breach of the law.

He then played fast and loose with the consulate. He suddenly resigned his office and appointed two of his supporters as successors for the rest of the year. This unprecedented act demonstrated that Caesar regarded the consulate principally as an honour, a reward, a kind of booty. For in the few weeks that were left the new consuls could scarcely take up their official duties properly.

When one of them entered the theatre and was announced by the lictor, the audience protested that he was not a consul. Perhaps Caesar had merely appointed them, without having them elected. After all, the Senate had empowered him to do so.

On 31 December something quite scandalous occurred. Cicero reported it to a friend. At about seven o'clock the official chair of one of the consuls was set up at the beginning of the quaestorial election. It was then announced that he had died suddenly. Thereupon 'the chair was removed. Yet *he*, having taken the auspices only for the tribal assembly, conducted the centuriate assembly and then, at about one o'clock, had a consul elected to hold office until 1 January – that is to say, until the following morning. Note therefore that under the consulship of Caninius no one breakfasted. Nothing untoward happened during his consulate, however, for he showed a degree of vigilance verging on the miraculous and never closed an eye during the whole of his consulate.' Another source reports that Cicero urged his friends to hurry to congratulate him, 'otherwise he will have laid down his office before we reach him'. The fine wit of Cicero's *humanitas* helped him through this terrible period.

If Caesar's scorn had previously been directed at many features of the old institutions that may well have seemed meaningless, it was now directed against the meaning of Rome's highest office.

The citizens looked on in sullen silence, and according to Cicero 'a thousand similar things' happened. Gelzer writes, 'The republican institutions could not have been treated with grosser disrespect, and one has the impression that Caesar wished to use this occasion to demonstrate what the new age would be like.' This is extremely doubtful, however; it is far more likely that he wished to show his contempt, openly and unambiguously, for the old institutions and the society that cherished them. If society did not respect him, he would not respect society.

There was nothing new to be seen – only his contempt for the old. No new age had dawned: Caesar had simply entered upon a new phase in his life, not in his policy. He was no longer willing to impose restraints on himself. The last threads that had bound him to Roman society were severed. Was he now freeing himself from the question of a new order – if he had ever considered it?

Caesar's natural dynamism, without which he would not have been Caesar, was still strong enough to be effective. Yet it seemed to have become increasingly breathless. It is therefore not unlikely that he could no longer tolerate obstacles. As he no longer had the same *élan*, the same resilience, that would have enabled him to take a long run at the obstacles, he reacted nervously. He pushed them aside, since he could scarcely admit to himself that he had lost much of his power to assert himself.

Many things may have contributed to his ruthlessness. And who would deny that the old wilfulness, the desire to let himself be simply carried along, may have been one of them? Just as he seems to have tempted Fortune on occasion, in order to see how true she was, so now he experimented with republican sentiments, in order to see how far he could go in offending them.

The Senate exalted him into the sphere of the immortals, but there was murmuring everywhere. What was to be done? Did he want to take soundings? Or did he want to show that he would tolerate some things and not others?

On one occasion Caesar is reported to have said publicly that people should speak to him more circumspectly and take what he said as law. Whether he was serious or merely parodying the role that had been ascribed to him, whether he was indulging in mockery or whether power had gone to his head, is a relatively unimportant question, given that he now lived only partly in the solid world of human reality.

His actions were certainly consistent, irrespective of the accompanying circumstances. He wanted to use all the honours that the republic had to offer, which had previously entailed duty and service, as rewards for his supporters. Just as the consulate had for him been primarily a rank rather than an office, so he regarded senatorial posts as a kind of booty, so to speak, to be shared out among his loyal followers. He increased the number of senators from six hundred to nine hundred. Such a large membership would certainly make the house incapable of functioning properly, but it enabled Caesar to show favour to many. Most of the new members were the sons of respected knightly families from all over Italy. Many belonged to the circle of the allies who had enjoyed citizenship since 90; some may have had long-standing ties with Marius, Cinna and Caesar, and been members of the 'military class'. Yet among them was a number of junior officers, freedmen and Gauls, who owed their citizenship to Caesar; this must have incensed the older senators. Some of them were strangers in Rome, as was obvious to all. 'A good deed! No one should show a new senator the way to the curia' – this and similar slogans were plastered on the walls of the city. Ten praetorians whom Caesar did not wish to make consuls were arbitrarily given the rank of *consulares*.

Caesar gained access to another reserve of honours that cost nothing by having conferred on him the ancient royal right to nominate patricians. The patriciate, the oldest stratum of the nobility, had been frozen since 450 at the latest. No new members were admitted. Its ranks had become thin. Caesar now supplemented them with his supporters. He also doubled the number of praetors and quaestors, but this may have been due mainly to practical reasons.

In 44 he had himself and Antony elected consuls. If he had nominated the two three-monthly consuls without an election, the consequent indignation would have obliged him to hold another regular election. We have evidence that he would have forfeited his right of nomination. He would have offended not only the people, but the consuls themselves, as their dignity would have been spurious. It is interesting that in 44, when a tribune of the people gave Caesar the right to nominate half the magistrates, an exception was made of the consulate. Even Caesar's supporters were sufficiently conventional in outlook to want to have their offices conferred by the *comitia*. This applied also to the offices below the consulate, insofar as Caesar had doubled their number. His right of nomination was thus restricted to the newly established posts. It goes without saying that the electors themselves attached great importance to having a voice in these matters and exercising their influence. Elections, by which the highest honours in the commonwealth were allocated, were still considered vitally important. For this reason they were attended by the greatest controversy.

There seems to have been no lack of controversy in 45. For there was resistance to Dolabella, for whom the second consulship was intended. Caesar himself had encouraged him to stand. The details are not known. At all events, Caesar announced on 1 January 44 that he would stand down in Dolabella's favour before departing for the Parthian war.

Under existing law Antony and Dolabella were too young for consular office. Antony was about thirty-eight, hardly old enough to be a praetor; Dolabella is said to have been all of twenty-four years old and would have had to wait six more years before becoming quaestor. Caesar had hitherto respected the rules regarding age. Hence both candidates had not yet moved far up the official ladder. Now, however, he ceased to be bound by the rules.

In the summer Antony had gone to Gaul to meet him. Caesar had received him amicably and been reconciled with him. He needed Antony, as he was one of his ablest political associates. A little earlier Antony had been informed by Gaius Trebonius, who had until then been Caesar's governor in Spain and became consul in 45, of a plan to assassinate Caesar (as has already been mentioned), but he did not betray his knowledge.

Publius Cornelius Dolabella was one of the restless figures among Rome's gilded youth. Deeply in debt, bored, and no doubt sickened by the Senate régime, he had joined Caesar in 49. After the unrest he had stirred up in 47 during the agitation for a remission of debts, he had had to accompany Caesar on the African and Spanish campaigns and seems to have acquitted himself well. Caesar was clearly much taken with the young man. Antony, however, continued to oppose Dolabella, probably out of political jealousy, but no doubt also because Dolabella had been his wife's lover.

At the beginning of 44, with the Parthian war in mind, Caesar appointed the magistrates for 43 and 42, probably in the popular assembly, but again with no regard to tradition as far as the timing was concerned. The candidates for election included not only supporters of Caesar, but neutrals and former opponents. For 43 and 42 his choice fell on older men who seem to have had a regular career behind them.

During his last five and a half months in Rome, Caesar devoted himself for the most part to his organizational projects.

Much time was taken up by the settlement of his veterans. Countless decisions had to be made, and he took an active interest in everything. A particular difficulty was that a surprisingly large number of veterans insisted on being given land in Italy. The public domains, Caesar's personal property, and the additional lands he was able to purchase, were far from sufficient. He was therefore obliged, contrary to his original intention, to appropriate some land in order to appease those to whom he owed his victory and keep them off the streets. He was thus not prepared to risk various changes of plan, which were actually called for, lest they should lead to unrest. Caesar founded colonies not only in Gaul and Spain, but also in Africa, Greece and Asia Minor. He was especially proud of having restored the cities of Carthage and Corinth – as Roman colonies.

He thus had his veterans dispersed throughout the Roman world; this was, if nothing else, a way of safeguarding his rule. The same may be said of the granting of the Latin right – the stage preceding full citizenship – to the Sicilian communities and the conferment of Roman citizenship on towns and individuals. These were substantial moves towards the enlargement of the citizen body and the coalescence of the different parts of the empire. This may have been Caesar's objective, though there is no record of his intentions. All we know for certain is that he continued, on a grand scale, the old Roman practice of increasing the number of his supporters and strengthening their position.

He was concerned to ensure observance of the sumptuary law. He not only kept a check on the market, but had private houses visited by lictors and soldiers, who might even confiscate the food from the tables. This is bound to seem inconsistent in view of his own lavish lifestyle, but it was not: private and public luxury were two quite different matters. Benefactions, games and buildings were expected of the nobility. We do not know whether Caesar transgressed his own law in private. What he had done in his early years was irrelevant in the present context.

Countless other matters required his attention. Work was put in hand to improve roads and harbours. A great library was planned to house the whole of literature. Roman law was to be codified, so that what was 'best and necessary' could be collected in a few books. Laws were passed under which Caesar was made responsible for draining the Pontine marshes and driving a canal through the isthmus of Corinth. Having enlarged the empire, he followed the example of Sulla by extending the hallowed bounds of the city.

Many developments were inaugurated in Rome. The grandiose Basilica Julia was built in the forum. This building, which had several naves, presumably served as a business centre, like the one it replaced; it may also have accommodated public offices and lawcourts. Caesar's victories were probably commemorated in its internal décor. Caesar planned to extend his own forum by building a new curia. The old one, destroyed by fire in 52, seems to have been rebuilt and, in accordance with the Senate's wishes, bore the name of Sulla. The Senate now gave Caesar the right to build his own. Where the old curia had stood, a temple of Felicitas was to be erected in his honour. Thus, not only did a new Curia Julia arise on the site of the Curia Cornelia, but the Senate's most important building, the Senate House itself, was to be Caesar's work. The curia still stands by the forum. Though rebuilt in the early fourth century, it presumably corresponds to the original design, except that at first it was surrounded by a columned hall, as is shown by illustrations on coins. The curia, which was planned to stand next to the ensemble of Caesar's forum, thus took up part of the old *comitium*, the meeting place of the popular assembly. This was a bold and imperious invasion of the old centre of the city and the world – as though Caesar, not content to make his presence felt architecturally by his own forum, as he had at first intended, now wished to push aside the old structures that had grown up over the years and replace them by a self–contained complex. The old meeting place of the popular assembly had to make way for Caesar's new buildings. It was a powerful demonstration of his pretensions.

During the building of the curia the speaker's platform (*rostra*), which had stood on the *comitium*, was moved to the narrow west side of the forum and probably rebuilt. Caesar thus took a decisive step towards a unified design for the whole area, where there had formerly been an assortment of unmatching buildings. Having first taken his place beside it, as it were, by building his new forum, he now transformed the old one. When the *rostra* was moved, he had the statues of Sulla and Pompey re-erected; this met with universal approval. Was it his way of making amends before departing from Rome?

It is not known what progress was made during his lifetime on the marble voting hall on the Campus Martius, which he had planned since 54. During his dictatorship it was bound to acquire added significance – as an attempt to replace liberty by marble.

The artificial lake created for his triumph was to be filled in and a temple of Mars, larger than all the other temples, erected on the site. The Campus Martius was to be developed and its function as a place for assemblies, sporting activities and all kinds of games transferred to the Campus Vaticanus. To provide a better link with the city, Caesar planned to divert the Tiber to the foot of the Vatican hills. Enabling legislation had already been drafted in the summer of 45.

Finally, Caesar planned a huge theatre to the west of the Capitol. He had already started buying up the land, demolishing existing houses and temples and – to everyone's horror – having the wooden images of the gods burnt. Augustus eventually built the theatre, though perhaps on a smaller scale. This was the Theatre of Marcellus. Its ruins still stand.

Caesar was thus occupied with many ambitious and grandiose schemes that would perpetuate his memory. The old urge for achievement was now perhaps compounded by the fear that he might not have much longer to live.

He kept a whole team of contractors and workers constantly busy: he himself was of course far from idle. As Heuss observes. 'Wherever power could be converted directly into action his rule came into its own.' Everything had to be done in the shortest possible time. Yet such an immense programme of work, carried out under the pressure of extraordinary demands, had to run smoothly and according to plan. Sometimes it was too cumbersome to involve the Senate, however compliant that body had become. Caesar therefore often had his orders formulated as resolutions of the Senate and left it to his secretaries to invent the names of the petitioner and witnesses. From time to time Cicero would receive letters from remote parts, in which exotic personages, quite unknown to him, thanked him for petitioning on their behalf for the title of king. 'I am more likely to learn that a resolution, allegedly passed at my request, has reached Armenia or Syria than to have heard any mention of the matter.' Presumably Caesar resorted to this method chiefly in dealings with distant lands. He is said to have employed his slaves to settle many questions – relating, say, to the coinage or public finances – that properly fell within the competence of particular magistrates.

Caesar restored full rights to those of his opponents who had not yet been pardoned; this probably happened in 44. The widows of those who had fallen in the war and whose estates had been confiscated were given their dowries, and their children were allowed to receive part of their patrimony. In a sense this was the cornerstone of his policy of reconciliation. The dictator apparently wished to do whatever he could to right the wrongs of the civil war before setting out on the Parthian campaign.

This remarkable show of consideration seems at odds with his arbitrary and arrogant behaviour during these weeks. The salving of his conscience, the demonstration of his superiority, and the lordly gesture all combined in a new way, now that there was no longer any tactical reason to show clemency. He no longer had any enemies; he stood above everything, even above the past. Was he now taking leave of time?

During the Saturnalia in December 45 Caesar was in Puteoli (Pozzuoli), where he paid a visit to Cicero. 'What a difficult guest! And yet I do not regret it; it was actually very agreeable.' He came with a large retinue. An endless crowd of people. Cicero arranged for the soldiers to camp out in a field. Caesar walked across from a neighbour's house. 'After two o'clock he took a bath . . . Then he had himself oiled and came to dinner. Intending to take an emetic, he ate and drank freely and heartily, and it was indeed a grand and splendid dinner.' His immediate entourage dined at three long tables.

'Of course, he was not the kind of guest to whom one would have wanted to say, "Do me a favour and drop in again on your way back." Once is enough. Not one serious word during the conversation, but a lot of literary talk. Enough! He enjoyed himself and felt at ease.'

However charming, if difficult, Caesar appeared to many, and however restlessly active he was, we still hear of nothing that could be construed as a move towards the consolidation of the commonwealth. After his death, his close associate Gaius Matius declared, 'If he, with his genius, could find no solution, who is to find one now?' It was not just old republicans, then, but Caesarians too, who saw that there was a problem.

Did Caesar fail to see it? Or was he at a loss to know what to do about it? Or had he a very clear idea of the new order he wanted to establish and simply judged it inopportune to announce it at present? If so, he must have had the establishment of a monarchy in mind.

All the measures he introduced, however practical and far-reaching, are wholly understandable on the assumption that he thought only of consolidating his position as victor, that he thought only of himself and his associates – and of extending and proving his capacity for effective action. We have no evidence that he intended to set up a monarchy. Everything can be explained on the assumption that his sole aim was personal autocracy. To make such a distinction is not to quibble: it is quite clear that Caesar was interested chiefly in his own power and the opportunities available to him, and that he had little interest in institutions. Had not Cicero clearly hinted that only Caesar's desire for fame could induce him to take an interest in the institutional problem?

It is striking that in his last will and testament, drawn up on 13 September 45, he made his eighteen-year-old grand-nephew Gaius Octavius his chief heir, bequeathing to him his name and most of his fortune. Caesar had already had Octavius appointed *pontifex* and took him with him on the Spanish campaign. Though apparently sickly and anything but a hero, Octavius showed himself tough and courageous. Caesar was fond of him, enjoyed his company, and probably had long conversations with him. He then sent him to study in Greece, after which he intended to take him with him on the Parthian campaign.

Caesar may have recognized the many-sided talents that Octavius, who later founded the monarchy, was to display soon after he himself had died. Yet he could scarcely assume that the boy would be able to succeed him at once as sole ruler of Rome. True, Octavius might have been able to count on his adoptive father's supporters and inherit, with Caesar's name, his claim to authority and the opportunity to win exceptional power. But he could not inherit Caesar's special position. If monarchy was to prevail in Rome, someone else would presumably have to assume Caesar's role – and have the utmost difficulty in contending with Caesar's heir. Caesar's will was quite conventional, that of a Roman nobleman who sought an heir among his relatives. The Caesar who drew up the will was quite different from the Caesar who had fought the civil war and won control of Rome and her empire. Whatever view one takes of the possibilities of finding other heirs or enrolling Octavius as his real successor – and of much else besides – it is hard to avoid the conclusion that if Caesar had wished to set up a monarchy he went about it very ineptly.

In any case, the necessary preconditions for such a scheme hardly existed at the time. Caesar would have found little support among his followers. Those who wished to serve him without having much of a will of their own counted for little. And his more ambitious supporters were not prepared to submit to the inevitable restraints and obligations. During the civil war they had thrown in their lot with Caesar because they hoped to share in his victory. Now they wanted a say in how things were ordered. Caesar might occupy a pre-eminent position, but there was no significant role for them under a man who 'did not take even his followers' advice, but only his own'. What influence did that leave them? It was not a game they wished to have a hand in. Caesar might award the prize, so to speak, to some also-ran. They could play along and enjoy various advantages – after all, the life they led was in many ways enviable – but this was not a world in which they could feel at home. To them – and to everyone else – the present state of affairs could be no more than provisional.

In other words, there was no potent cause that would have required all existing criteria to be changed. If there had been pressure for change in Roman society – if there had been powerful expectations directed towards new political forms – Caesar's watchwords would have been available, and his monarchy (supposing he wanted it) could have locked into a powerful trend – but only if he had had a cause. He could have pointed to a suprapersonal goal and stated what needed to be done. Everyone would have known where he stood. And Caesar could have mobilized support in such a way that service to his cause would have become the decisive criterion for the position, rank and prestige of his supporters; they would have been the leading politicians in Rome.

However, because no such cause existed in this crisis without alternative, there was no basis for thoroughgoing institutional reform. No plans for it would have found any resonance. Hence, the crisis could not be recognized or made subject to political action. The dictator could not bring his own interests into line with the general interest or link his quest for power, respect and effectiveness to a universal desire for new institutions. After all, it is probably naïve to derive what we regard as 'statesmanship' merely from insight, 'vision' and detachment from party interests. To say that statesmanly action presupposes a certain distribution of power is not to belittle the greatness and personal merit of the statesman. What the statesman, with his 'vision', recognizes as the general good must have a substantial basis in the present, in the form of interests and opinions. There must be a measure of give and take between him and individual factions, so that, while bound to a particular cause, he has the freedom, the opportunity and the means to put into effect what he recognizes from this vantage point. What matters is not that he should simply be independent of the contending factions, but that he should not be tied to those that would force him into a corner and, in times of crisis, lock him into the crisis itself. Indeed, the existence of such factions, or at least the existence of a distribution of power that enables him to direct sufficient force to mastering central problems, is probably a *sine qua non* of statesmanly action.

Given the circumstances prevailing in Rome, whatever institutional plans Caesar may have had were bound to be fairly vague. This depended less on him than on the conditions of the age. Yet one cannot make a precise distinction here, because he had grown up in these conditions and developed in his own way. It was only by placing himself in opposition to the reality of the late republic that he could allow his brilliance and greatness to develop so freely and so amazingly.

The high standards of personal action and effectivencs that he created for himself in Gaul prevented him from setting out on the path that would in principle have been open to him – a principate agreed with the Senate. It was this path that Pompey had wished to follow. Caesar of course, as victor in the civil war, would not have had to endure the same humiliations or perform all the complicated manoeuvres by which Pompey had finally reached his goal. He would no doubt have had to face great difficulties, but there was an underlying readiness to collaborate that he could have turned to his advantage. The attempt might have failed. But Caesar clearly did not make it. It would have meant showing consideration to others and doing a good deal of listening and persuading; he might have had difficulty in bringing his supporters into line, and it would not have done to proceed by issuing orders and decrees. He would have had to show a degree of patience, of which he was even less capable now than he had been in the past. He might have felt that he was being untrue to himself.

Yet presumably no community can be reorganized in opposition to all the main forces if one reveals nothing of one's intentions. Caesar neglected the all-important issues and attended only to those that yielded to his masterly talent for planning. Yet because he left crucial questions unanswered, others were induced to seek the answers. There was a great vacuum, and gradually the situation became so unbearable as to set a curious process in train.

In late 45 or early 44 the Senate was still devising fresh honours for Caesar. They no longer related to his victories. He was to be allowed to consecrate the so-called *spolia opima*, the spoils taken from the enemy general slain by the Roman commander; Caesar did not qualify for this honour. The fasces of his lictors were to be permanently entwined with laurels. He was given the title of *pater patriae*; this was intended for deserving politicians and had previously been conferred only once. His birthday was declared a public holiday, to be marked by sacrifices. The month of his birth was later renamed 'Julius' (July). In all Roman temples and in all Italian cities statues of Caesar were to be set up. Two were placed on the speaker's platform, one commemorating him as the saviour of the citizens, the other as the liberator of the city from siege. A temple of the New Concordia was to be built, to seal the peace inaugurated by Caesar, and a temple of Felicitas – replacing the old curia – to celebrate his fortune. Work on the Curia Julia was already in hand. Caesar was granted tribunician inviolability. Later he was given the right to use a golden chair, instead of the ordinary official chair, at meetings of the Senate or at court hearings. He was to be permitted to wear the gold wreath of the Etruscan kings. He had already appeared on occasion in the high red boots of the Alban kings, which he liked and felt entitled to wear as a descendant of Aeneas. The senators were to swear an oath that they would protect his life, and he was given a bodyguard of senators and knights. His acts of government were declared valid in advance.

Other resolutions provided for four-yearly games to be held in his honour, as for a national hero. His image, borne in procession with those of the gods, was to be deposited in a holy place like the others. A gable was to be built on his house, as though it were a temple. A shrine was dedicated to him and Clementia. He was to have his own priest; Antony was appointed to this office. However, the cult of the new god was to begin only after his death. Unlike other mortals, Caesar was to be buried in the city. The resolutions regarding his deification were set in gold letters on silver tablets and placed at the feet of Juppiter Capitolinus.

Finally, Caesar's office as overseer of morals and – more importantly – his dictatorship were extended for life. He at first hesitated to assume the new dictatorship, but ultimately agreed shortly before 15 February 44. This put an end to the ostensibly provisional character of his *de facto* monarchy.

Much research has gone into examining the origin of the honours and powers conferred on Caesar and the significance of their combination. Yet in all probability the senators merely strung together whatever occurred to them, trying to outdo one another in framing ever new resolutions in session after session. The Senate was seized with a mania for honours. Caesar accepted nearly all of them and occasionally expressed his pleasure.

Various motives coincided. Many wished to flatter him and curry favour. Some may indeed have deemed him superhuman. As Goethe observed, 'Against another's great merits the only salvation is love.' In the ancient world the boundary between man and god was fluid. Whatever was extraordinary could be felt to be divine. Moreover, Caesar may have coveted at least some of these honours. Yet there was unquestionably another motive at work: Caesar was to be branded as a tyrant. One source states that he was adorned with honours like a sacrificial animal. It was thus also a case of unmasking by panegyric.

The continued exaltation of Caesar's position seems to have been the only function that the Senate was still free to perform, the only means by which the senators could vie with one another for his favour. Some must have had misgivings, but once the first step had been taken, they were inevitably impelled by ambition, apprehension and distrust. Each new honour generated fresh resolutions. The grander, the more potent, the more remote Caesar became, the greater their insecurity, their fear, their helplessness, their distrust. And the more the honours proliferated, the more urgent it became to win his favour by showering him with fresh honours. All resistance, all seriousness had long since vanished – and all sense of reality.

Caesar himself was caught up in the process. On one occasion the senators came to him in solemn procession to present a series of resolutions conferring fresh honours. He did not even rise to greet them. He was in his forum, allocating commissions to artists. This show of disrespect aroused universal indignation. He sent out apologies, saying that he had suddenly felt unwell. The scene was in keeping with his earlier remark that whatever he said should be taken as law.

Yet whatever honours Caesar had received, one was still missing – the title of 'king'. The question on everyone's mind was whether he wanted this too. After all that had happened, this might seem to have been just one more honour – the last step. Yet the last step might be the longest. For in the Roman republic kingship was punishable by death. Seen from below, the distance that still separated him from it may have seemed minimal. This question thus gave rise to suspicions that he could no longer shake off.

We cannot know whether he wanted the title. The likelihood is that he did not. He forbade his supporters to petition for it in the Senate. Nevertheless many believed – or professed to believe – that he sought it. The title was often shouted at him, for whatever motive, and more than once he was offered the diadem, the symbol of kingship. Each time he refused it, but each time the suspicion revived.

On one occasion his statue on the speaker's platform was adorned with a diadem. Two tribunes of the people had it removed. When Caesar rode into the city on 26 January, some bystanders greeted him as 'Rex'. With great presence of mind he riposted that he was Caesar, not Rex, which was a cognomen of his grandmother's family. But when the man who had first uttered the hated word was led away to be prosecuted, to loud cheers from the crowd, the two tribunes who ordered his arrest incurred the wrath of the dictator. They replied in an edict that their official freedom was threatened. Caesar convened the Senate and complained that he must either act against his nature or suffer a loss of *dignitas*. The Senate dismissed the tribunes and had them struck off the Senate roll. Caesar even demanded that the father of one of them should disinherit his son. The father refused, and Caesar let the matter rest. Once more it became clear how lightly he took the rights of the tribunes, in defence of which he had supposedly begun the civil war.

If Caesar felt that his honour was deeply offended because the tribunes of the people arrested a man who greeted him as 'Rex', this can only mean that he was insulted by the suspicion that he sought the title. Kingship was associated with arbitrary action against the lives of the subjects; and Caesar could claim that he was innocent of any such charge. But the distinction he drew between kingship and the autocracy he already enjoyed – associated with all kinds of divine, indeed royal attributes – was too fine to be really appreciated.

On 15 February he seems to have made the point again. A few days earlier he had accepted the title of *dictator perpetuo*. The feast of Lupercalia was now celebrated. Caesar sat in his golden chair on the speaker's platform in the forum, decked out in his triumphal robes and wearing the golden wreath. Cicero tells us that Antony, Caesar's fellow consul and his foremost follower, having taken part in the traditional run of the Luperci and wearing only a loincloth, 'placed a diadem on his head'. Caesar rejected it and had it sent to Juppiter, Rome's only king. Antony had it recorded in the official calendar that at the behest of the people Caesar had been offered the kingship (*regnum*) by the consul Antony and had refused it.

The incident can be variously interpreted. Caesar may have wanted the diadem, but rejected it because he did not get the necessary applause. Or he may have had it offered to him in full public view, so that he could be seen to reject it and thereby counter all the suspicions that were abroad. It has also been surmised that Antony acted without Caesar's prior knowledge – in order to gain his favour, or alternatively to discredit him. It is impossible to say which interpretation is correct. If Caesar had really wanted the diadem, however, it is likely that the applause could have been engineered. If so, he must have wanted to reject it. This seems most probable. Yet perhaps he did want it – but only if it was freely given. He is said to have bared his neck and chest, clearly indicating his respect for the Roman view that whoever sought the crown deserved to die.

But even that did not help. Rumour was rife. It was said that only the absence of applause had made him reject the diadem. It was also said that he planned to move the capital to Alexandria. This was a fantasy inspired by the presence of Cleopatra, who still resided in Caesar's gardens beyond the Tiber. It was rumoured also that legislation was being drafted to allow Caesar to marry as many women as he wished in order to beget male heirs. This was clearly directed against the son of the Egyptian queen. Caesar could no longer do anything to escape the suspicion that he sought the crown. His opponents certainly helped to promote it. It was bruited abroad that at the Senate meeting on 15 March a bill was to be brought in, on the basis of a sybilline prophecy, nominating him king of the provinces. It no longer seemed incredible. The mood was far too tense, reality too obscure, and anything was bound to seem possible.

One wonders why Caesar resisted only at this last line, which separated him from the crown. Why did he not call a halt to the honours before it was too late? Did he really take pleasure in them? Or did he merely go along with them willy-nilly? The former possibility seems the more likely. They contributed to his fame by documenting his outstanding qualities, his fortune, his achievements, his near-divinity. This was not of course the kind of fame that is both coveted and despised in the west, the *gloria mundi* that may seem vain within the wider horizons of modern times, which are open to transcendency and span the millennia. It was a fame that ensured that the memory of the great Roman would live on in the minds of the citizens; it was the ancient form of immortality. Though Caesar did not fear death, he certainly feared transience. This was one of the motives behind his deeds, the literary record he left of them, and the architectural monuments he set up to perpetuate his memory. Was this not also the reason for the unremitting exertions to which he subjected himself during his last months? This fear was bound to become the more oppressive, the less able he was to find a *raison d'être* in the present. Should not the senators at least confer marks of fame that could provide him with a guarantee for the future? According to the Greek historian Dio, 'Pompey wanted to be honoured voluntarily, . . . but Caesar did not care if he awarded honours to himself'. Perhaps in the end he was indifferent only to the cause of the honours, not to the honours themselves. In that case, the goal of immortality must have become all-important to him.

At the same time he may have found it quite in order that he should be placed close to the gods, who had shown him such incredible favour throughout his life and raised him so far above other mortals. The Christian virtue of humility had no place in relation to the gods of antiquity, and he had abandoned the republican virtue of equality. Was it so wrong for the senators to understand Felicitas and Clementia as divine forces and to represent him as enjoying the favourite of Victoria? And if they distinguished him with the purple toga, could it be displeasing to him if this made him increasingly unapproachable and inaccessible?

Finally, he might well have been gratified by the way the senators – both the successors of the old leaders of the republic and his own supporters, who now revelled in their new dignity – bowed ever lower, the higher they raised him. He probably despised them heartily. And he would probably not have been himself had he not given free rein to his contempt.

In any case Caesar was no longer willing to show any consideration. Sometimes he found it difficult to control himself. His powers were failing. He was not well. Cicero wrote that he would 'never have returned from the Parthian war'. May he not have been carried away by a kind of diabolical pleasure as he contemplated this undignified spectacle? Was he not bound to be tempted to join in and play the role in which he had been cast, yielding perhaps to his old wilfulness? Finally, he was in many respects no longer as particular as he had once been. In a few weeks he would be setting off on the Parthian campaign. And during these weeks, having concluded, as early as 46, that he had lived too long for himself and his fame, he seems to have been possessed by a kind of fatalism. One indication of this is that he dismissed his bodyguard, which consisted of Spaniards. It is hard to give much credence to the opinion of a later commentator who wrote that Caesar wrongly interpreted the honours bestowed on him as proof of his universal popularity. Rather, he chose to let it be known that he did not take his opponents seriously.

It is not blindness, wrote Ranke, that destroys men and states, 'but there is in them an instinct, favoured by their nature and strengthened by custom, which they do not resist, and which drives them on while they have any strength left.' Caesar was increasingly driven to be what he knew how to be – the first man of Rome, unmatched by virtue of his deeds and victories. This is what he wanted to remain. He had climbed ever higher and defeated everyone. Where was it to stop?

Whatever motives he may have had in letting himself be involved in the series of honours, the fact is that the more he stood above other men, the less he stood above events. He was the master of everything that was subject to his power and authority, but not of the republic, which he had subjected but not conquered. He no more understood the republic than it understood him; he could find no place in it, no foothold, no meaning. This fuelled the dynamism of what seems like – and perhaps was – a catastrophe.

While Caesar yielded to the plethora of powers and honours that were bestowed on him, and was perhaps intoxicated by them, a number of men agreed among themselves that he must be murdered.

At last it seemed clear that Caesar intended to go on extending his power until it amounted to a tyranny. The conspirators were guided by the time-honoured conviction that anyone who strives for tyranny must be killed. It was believed that Romulus, the founder of Rome, was killed because his rule lapsed into tyranny. It is true that he was at the same time a god, identified with Quirinus. Hence in 46, after the Senate's resolutions honouring Caesar, Cicero liked to picture Caesar as sharing the temple of Quirinus. There was also a well-known tradition that after the fall of the last king the Roman people had sworn that Rome should never again be ruled by a king. Republican consciousness derived essentially from the rejection of monarchy.

The plan to assassinate Caesar was older. Cicero had promoted it, and in the summer of 45 an attempt was made to win Antony over to it. The conspirators, who numbered about sixty, decided not to bind one another by oaths; among their number were former Pompeians and Caesarians. The leading personalities all stood high in the dictator's favour. Trebonius had been consul in 45; Decimus Brutus had been elected consul for 42; Gaius Cassius Longinus and his brother-in-law Marcus Brutus were praetors, and it was probably intended that they should become consuls in 41. Marcus Brutus became the real leader of the conspiracy. He was predisposed to the role by his claim to be descended from two men who had freed Rome from tyranny, by the fact that he was Cato's nephew, and by the weight of his own personality. It is one of the most curious links in the story that these three factors came together in him – and that his mother was Caesar's favourite mistress. Thus all hopes for the removal of the tyrant were concentrated on him; and at an earlier date he had represented himself on his coins as a champion of republican liberty. Yet he remained true to Caesar for a very long time, hoping that he would join the 'good'. His judgement had obviously been determined by the alternative: restoration of the res publica or illegitimacy and despotism. Since Caesar behaved so little like a despot he concluded that he was benevolent.

Brutus had a simple, upright nature, fortified by philosophy; like his uncle he had devoted himself to learning. We are told that in the camp at Pharsalos he made excerpts from Polybios' history and was always poring over his books. He was extremely conscientious; he was also straightforward in his thinking and not easily carried away. He examined everything carefully and was always prepared to make decisions that conflicted with his inclinations. It was only when he felt he had been deceived, when he experienced injustice or realized that he had made the wrong decision, that he became a forceful champion of what he now saw to be right.

Thus in 48 he placed himself quite deliberately and unequivocally on the side of Caesar. The image of the detested consul of 59 and the Gallic proconsul seemed to him false. Caesar had been treated unjustly. This alone was bound to endear him to Brutus. If he was affected by Caesar's charm, this was because his sense of justice had already made him susceptible to it.

It is not clear how far his mother had a hand in all this. With regard to Caesar she presumably had no real authority over Brutus. In dealings with his mother's lover he could only try to set aside his personal feelings – until he realized that justice required him to join him. The fact that Cato was his uncle does not seem to have troubled him, especially as he believed that Cato had been mistaken. During these years he had probably drifted away from Cato. His defence of the dead Cato was not so much a move against Caesar as a form of personal amends, an apologia for the man who had been defeated and whose death put him to shame. He at last turned decisively against Caesar when even he could no longer doubt that he was intent upon autocracy. This must have been, at the latest, when Caesar became dictator for life. Caesar once said of him, 'It depends very much on what he wants; but what he wants he wants utterly.' So it was now. And precisely because he had adhered to Caesar for so long and was exceptionally conscientious, his decision carried the greatest authority. Perhaps it was he who decided that the plans for a coup should at last go ahead in earnest.

At any rate he embraced the cause with a resolution that fully matched his inner stature. Yet the seriousness of his resolve imposed its limits: he insisted that only Caesar should be murdered, not Antony too.

The conspirators in fact believed that once the tyrant was removed they would have achieved their aim. They made no preparations for a seizure of power, without which they could hardly guarantee that the Senate and the magistrates would be able to take over the leadership of the republic. After all, Antony was consul and Lepidus was Caesar's deputy as dictator; they had several legions at their disposal, and apart from anything else Caesar had many veterans who were bound to try to avenge him.

In Cicero's judgement the conspiracy was planned with 'the courage of men but the understanding of boys' (*animo virili consilio puerili*). He was right. It was another example of the one-sidedness of the late republic: this group of upright, honourable republicans were firmly convinced that the republic would come into its own again when once the tyrant had been murdered. They had understood nothing of the conditions that made his ascendancy possible. They were thus right and wrong in equal measure – like Caesar himself. But according to their criteria they could not act otherwise. Their view of the republic was as unequivocal as their view of Caesar's tyranny. There were no conditions or reservations; all that mattered was the deed. Yet it did not bring back the republic; its only outcome was, as Caesar had predicted, the return of civil war, which was to continue

continue for nearly fifteen more years, during which Roman society had to undergo a grievous process of attrition. Only then was it possible for the dictator's heir – with much skill, patience and self-abnegation – to establish a monarchy.

It has been said that Caesar's murder was not just a crime, but a mistake. Yet if Caesar had died a natural death the result could hardly have been very different. And is it a mistake to act in the only way one can?

Brutus and his fellow conspirators were imbued with the same faith in the *res publica* as the citizens at large. However much they pondered and discussed the matter, however much they searched their consciences and were assailed by doubts – what made them act was the potency and clarity of the inherited political form: so much had it moulded them and given meaning to their lives, as they now perceived.

The fascination that Caesar had exercised over many of the conspirators now paled; they saw in him only the tyrant. Awareness of their duty shielded them from their knowledge of Caesar the man, of his generosity and clemency. They could not spare him, though many of them owed their lives to him.

And they were right. They may have been wedded to the old, but if Caesar had really known anything new, he had not disclosed it. And however beneficial some of his actions were, they had on the whole been destructive. He had not pointed the way to a new order, but only burdened the old with the civil war and the cost of his victory. In this way he speeded the decline of the inherited institutions.

The one curious aspect of all this is that the destruction was the work of a man who possessed not only extraordinary gifts, but immense superiority and personal charm, a man who embodied, to a higher degree than any other, all the potential of ancient humanity, Greek and Roman alike. The culture of the age had endowed him with all that a man needed in order to develop whatever lay within him, and he was sufficiently gifted to make it his own.

Yet seemingly he did not have to pay the price for such an endowment: he needed to adapt himself to this culture only to a limited degree. The beneficent tension that had once existed between individual interests and communal demands, and had already begun

to slacken even before Caesar's day, could not withstand his dynamism. He never understood the workings of the republican institutions. At first he despised them and, from his position of detachment,
saw through them well enough to be able to assert himself brilliantly
within them. Yet he failed to appreciate what purpose they served,
how they sustained themselves, how the forces within them were
held in balance, what they demanded of everyone. He then came into
conflict with the institutions and was rejected by them; only then,
perhaps, was he forced to do many things that he had not originally
intended.

And since this process took place within an aristocracy that ruled
the world, it was possible for an outsider like Caesar, given the
disintegration of the aristocratic order, to win supreme power and
combine the fascination, the clarity and the freedom of his position
with immense scope for action. This was compounded by his incredible successes, his wilfulness, his playful audacity, his air of superiority, and not least his elegant nonchalance, which makes the most
strenuous endeavours appear effortless and contributes so much to
the aesthetic appeal of his greatness. For ultimately our susceptibility
to Caesar's greatness derives from an ideal that we all secretly cherish
– the ideal of self-sufficiency, of being able to act as we wish and to be
what we choose to be.

This undoubtedly involved a fair measure of immorality. Cicero
later went so far as to say that Caesar harboured 'such a desire for
wrong-doing that he delighted in doing wrong even when he had no
cause to'. This is certainly unjust. If it is true, Caesar must have
missed countless opportunities for doing wrong. Cicero could of
course justly claim that Caesar 'trampled on all divine and human
ordinances'. And Cato could accuse him of offending against universal law. Yet Caesar's immorality is not as unequivocal as it appeared
to his opponents. He certainly seems to have been conscious of the
wrong he did. When he blamed his opponents at Pharsalos, he
revealed not only his conviction that he was in the right, but also his
awareness of the enormity of the battle and the agonizing question of
responsibility. In 46, when Cicero pleaded for Quintus Ligarius to be
pardoned, it is reported that beforehand Caesar arrogantly declared to
his followers that he was looking forward to hearing another speech
by Cicero, but that it would not alter his judgement. Yet when
Cicero pulled out all the stops 'Caesar's colour changed more than

once, clearly reflecting the turmoil of his soul. And when the speaker referred to the battle of Pharsalos, Caesar completely lost his composure; his whole body trembled and he dropped some of the documents he was holding. He was so overcome that he finally acquitted the man' (Plutarch). Nor was Caesar's clemency merely a manifestation of his superiority and his will to win; it probably rested on a sense of right and wrong.

Yet this sense of right and wrong did not prevent his trampling on Rome's institutions for personal motives, conquering Gaul and launching the civil war. It merely mitigated the human consequences. No moral objection was strong enough to inhibit him from resorting to such extremes in pursuit of his own interests. To this extent he must be convicted of immorality, or rather of blindness to the moral standards that were conventionally – and rightly – expected of a Roman aristocrat.

Yet if his freedom to flout tradition was justified only by his personal honour, this merely demonstrates the absence of the powerful modern forces for legitimation, which can both restrict and encourage the active outsider: there could be no appeal to a transcendent religion or a transcendent state – which in certain circumstances can legitimate any immorality. We, obsessed with legitimacy, may find Caesar's self-absorption monstrous, however mild, humane, and generous it really was. And even in ancient times such legitimacy was claimed by Sulla, who justified the terrible murders resulting from his proscriptions by reference to the good of the republic. Caesar was quite incapable of such action.

Caesar may have acted immorally, but what was much more important was that he was different from the Romans of his age – alien, inscrutable, and then at once repellent and fascinating. This was what made him guilty *vis-à-vis* the republic. Yet far more important than this was the fact that he built up his own reality beside that of the republic.

The combination of brilliance – personal, not institutional brilliance – and power that we find in Caesar is probably almost unique in the whole of history. Yet this made him strong in relation to the republic only while he had to win victories within it. Afterwards it became evident that his strength was also his weakness, and in the end a certain melancholy of fulfilment – and a sense of futility – may have descended on him.

Caesar's greatness, loneliness and failure belonged together: the elation born of his extraordinary achievements in war and politics, his contempt for others, and his helplessness when faced with the problem of political consolidation. He possessed unprecedented freedom, superiority and assurance, yet his existence was baseless, and as he grew older he probably found himself staring more and more into the abyss. His nature and his destiny were thus bound up with each other. For 'one need not trivialize life by forcing nature and destiny apart and placing one's misfortune apart from one's fortune. One must not separate everything' (Hofmannsthal).

If the Roman republic had had its day without the Romans' knowing it – if the citizens still believed in it – then it was responsible for its own demise, whether this resulted from a long process of attrition or from the actions of one outstanding man, who developed all the possibilities it had to offer and turned them against it. As history took the latter course, the fall of the republic was played out in the form of a drama. But drama is concerned less with guilt than with fate. And it was Caesar's fatal greatness that led to the catastrophe.

Although the conspirators had not let their intentions leak out, the danger of an assassination attempt apparently did not remain entirely hidden. Caesar was warned by loyal supporters that he should provide himself with another bodyguard, but he refused. Nothing, he said, was more miserable than having oneself guarded. Only someone who was always afraid needed that. It was better to look death in the eye than live in constant fear of it. He was superior even to death. On the evening of 14 March he was a guest of Lepidus, and while Caesar was signing letters the conversation turned to the question of what was the most pleasant death. Caesar said it was one that came suddenly and unexpectedly.

The conspirators had decided that the murder should take place during the Senate meeting of 15 March. Caesar intended to leave for the Parthian war on the 18th. The Senate met in Pompey's curia. Caesar arrived late, as he had been feeling unwell. He had at first wanted to cancel his attendance, but one of the conspirators persuaded him to change his mind. He was taken by litter to Pompey's theatre. On alighting, he is said to have been approached by a Greek scholar, Artemidoros of Knidos, who obviously knew something about the conspiracy. Artemidoros was carrying a scroll that he wished to show to Caesar. Seeing him about to hand it to his servant – like all other documents presented to him – he is said to have cried, 'Caesar, you must read it, alone and quickly! It contains important matters of special concern to you.' Caesar held on to the scroll, but never read it.

The augur Spurinna had prophesied that misfortune would befall him on the ides of March. He is said to have been standing by the door to the Senate. Seeing him, Caesar smiled with an air of mocking superiority and said that the ides of March had come and nothing had happened to him. Spurinna replied, 'They have come, but they are not over.'

The senators rose from their seats. Caesar made his way forward through their ranks. Some of the conspirators had taken up positions behind his chair, which stood at the foot of a statue of Pompey. Others approached him, as if wishing to support a petition that was presented to him. When he refused, the petitioner tore the toga from Caesar's neck with both hands. This was the agreed signal. Then they attacked him. The first blow did not go deep. Caesar was able to seize the dagger. The rest of the senators looked on in horror. Then the other conspirators drew their daggers. They had agreed that each must strike once. Twenty-three dagger-blows rained down on the dictator. For a while Caesar defended himself, trying to evade the blows. At last, severely wounded, he drew his toga over his head. No one should see him as he lay bleeding, powerless, dying.

Afterword

T HIS BOOK is intended as a scholarly biography. It is based on long and intensive research into the history of Caesar and the late republic that has appeared in various publications, but tries to draw as full and vivid a picture as possible.

Today Caesar and his contemporaries seem to us somewhat strange and remote. To try to make them familiar to the reader would be to falsify the picture; but for all their strangeness it is possible to make them less remote, and this is what I have attempted to do. What confronts us, in Roman garb, is a kind of action and thought that is thoroughly comprehensible within the context of the age, a characteristic manifestation of the general in the particular, such as one finds in all history, and a number of quite modern problems.

When trying to acquaint the reader with a remote age, one should not take its peculiarities for granted, but incorporate them into the presentation. If modern historical research is incomprehensible to a wider readership, this is largely because of the way in which a kind of professional blindness induces the writers to take the alien, if not positively exotic, features of past societies for granted without elucidating them further. Paradoxical though it may sound, it is only

487

when one is fully aware of the strangeness and remoteness of a past age that one can make it comprehensible to one's own. What is known as structural history is thus an indispensable requirement for historical narration, though it often employs difficult concepts.

Such concepts, which facilitate mutual understanding among academic historians – though they sometimes bear little relation to experience – must of course never appear without elucidation in a book like the present one, but be explained in the clearest and most concrete terms. I have tried to do this. The reader must be spared nothing that is relevant to the biography, but have access to everything, whether or not he has had a classical education. I hope that I have succeeded in providing such access, and that the reader who is more familiar with the period will not find my treatment tedious. For the picture of Caesar and his age that is drawn here is in many respects new.

A special problem for the biographer is the silence of the sources about many things that he ideally ought to know, such as Caesar's upbringing, his thinking, and the last months of his life. I have tried to make a virtue of necessity: where one might otherwise have lost oneself in an account of the hero's individual career, I have attempted to reconstruct the special features of the age – its educational practice, for instance – and so to describe the general conditions prevailing in Rome at the time. As history is always concerned with relating the present age to others, this seemed a useful approach. For the rest I have endeavoured to fill in any gaps by deduction and surmise – or at least to indicate what is missing by asking questions. For I was anxious to indicate the limits within which our knowledge of Caesar is confined.

Any historical account that seeks to present a full picture must after all start from an awareness of the framework that it has to fill. A writer whose choice of what to narrate is guided solely by sources that happen to have survived is likely to remain farther from the truth – perhaps not in detail, but in general – than one who is aware of the whole. And if the degree of probability attaching to the individual statements is indicated, as in the present work, the writer can take responsibility for everything, including the details.

It would have been incompatible with the purpose of the book to support every statement with chapter and verse. This would have entailed not only extensive quotation of source material, but a full critique of both the primary and the secondary literature.

★   ★   ★

We are fortunate in possessing two books by Caesar himself, the *Bellum Gallicum* and the *Bellum Civile*, which, together with a few letters, fragments of speeches and other writings, are of the utmost value as personal evidence. Cicero's correspondence and speeches bring us close to the events and everyday life of the period and contain some brilliant, if at times rather partial, contemporary judgements. Other literary sources from the period are the works of the poet Catullus and the historian Sallust. The present work also draws upon the evidence of archeological remains, especially contemporary coinage.

We have two biographies of Caesar from a later period, one by Suetonius, a dry, but reliable collector of details, and one by Plutarch, who was concerned mainly to present a portrait of his subject. Three later historians must be mentioned: Velleius Paterculus, Appian and Cassius Dio. In addition there are of course interesting observations elsewhere, for instance in Pliny's *Natural History* and Vitruvius' treatise on architecture.

Helga Gesche, *Caesar* (Darmstadt 1976), provides a synopsis of the sources and secondary literature, to which should be added Hinnerk Bruhns, *Caesar und die römische Oberschicht in den Jahren 49–44 v. Chr.* (Göttingen 1978). An admirable introduction to the sources is given by Matthias Gelzer in his *Caesar: Politician and Statesman* (tr. Peter Needham, Oxford 1968, from the 6th German edition, Wiesbaden 1960). The same author's *Pompeius* (2nd ed. Munich 1959) and *Cicero. Ein biographischer Versuch* (Wiesbaden 1969) have not been translated into English, but adequate substitutes are Robin Seager, *Pompey: a Political Biography* (Oxford 1979) and Elizabeth Rawson, *Cicero* (London 1975). On the image of Caesar in history one should above all consult Friedrich Gundolf, *The Mantle of Caesar* (London 1929, tr. of *Caesar. Geschichte seines Ruhms*, Berlin 1924) and Zvi Yavetz, *Julius Caesar and his Public Image* (London 1983, tr. of *Caesar in der öffentlichen Meinung* (Düsseldorf 1979). Yavetz's account is also important for the period from 49 onwards.

From the vast literature on Caesar and the late republic I should like to single out a number of works that I found especially important, in addition to Gelzer's, and from which I quote in the text. These are: Hermann Strasburger, *Caesar im Urteil seiner Zeitgenossen* (2nd edn. Darmstadt 1968); Theodor Mommsen, *Römische Geschichte*; Eduard Meyer, *Caesars Monarchie und das Prinzipat des Pompejus* (3rd edn.

Stuttgart and Berlin 1922); Alfred Heuss, *Römische Geschichte* (Braunschweig 1960; Propyläen Weltgeschichte 4, Berlin 1963); Otto Seel, *Caesar-Studien* (Stuttgart 1967); Hermann Fränkel, 'Über philologische Interpretation am Beispiel von Caesars Gallischem Krieg', *Wege und Formen frühgriechischen Denkens* (Munich 1968); Friedrich Klingner, *Römische Geisteswelt. Essays zur lateinischen Literatur* (5th ed. Stuttgart 1979). Of special interest is an attempt by Walter Jens to interpret the conspiracy against Caesar in a television play. The quotation from Ronald Syme is taken from the German edition of his *Sallust* (Darmstadt 1975); the quotation from Peter Sattler comes from his book *Studien auf dem Gebiet der Alten Geschichte* (Wiesbaden 1962)

Of my own previous works on the subject, above all in the context of structural history, I should like to mention *Res Publica Amissa. Eine Studie zu Geschichte und Verfassung der späten römischen Republik* (2nd ed. Frankfurt 1980, with a new introduction). Also of interest perhaps are three biographical sketches of Caesar, Cicero and Augustus published under the title *Die Ohmnacht des allmächtigen Dictators Caesar* (edition suhrkamp, vol. 1038 [1980]), and 'C. Caesar Divi filius and the Formation of the Alternative in Rome', in Kurt A. Raaflaub and Mark Toher (eds), *Between Republic and Empire. Interpretations of Augustus and his Principate* (University of California Press 1990), 54ff. The problems facing the biographer are treated in my essay 'Von der Schwierigkeit, ein Leben zu erzählen', in J. Kocka and T. Nipperdey (eds), *Theorie und Erzählung in der Geschichte* (Munich 1979), 229ff.

For exemplars for the illustrations I am indebted to the British Museum, London, the Ny Carlsberg Glyptotek, Copenhagen, the Münzen und Medaillen AG, Basle, the Gerd Rodenwaldt Gedächtnis Stiftung, the Free University Berlin, the photo collections of the archaeological institutes of the universities of Heidelberg and Munich.

The illustrations were kindly prepared by Luca Giuliani, with advice from Paul Zanker of Munich. Maps and reconstructions were drawn by Jean-Claude Lézin of Berlin. I am very grateful to them all, and also to others who have helped me and will understand if I cannot name them all here. I am especially grateful for the stimulation and encouragement I have received from many quarters, for my long conversations with the publisher, Wolf Jobst Siedler, and for a general interest in the subject – of particular importance to me – to which, perhaps not quite fortuitously, the present book was able to respond.

I wish to dedicate this book to my mother and the memory of my father.

In conclusion it may be of interest to describe how the crisis of the Roman republic ended – how the alternative that was so long wanting finally emerged. But first, perhaps, it would be as well to return briefly to the not altogether easy concept of the 'crisis without alternative'. This does not imply that in the crisis there was any dearth of 'alternatives' – that is to say, different courses that might be adopted as the situation changed – or that it was impossible to introduce this or that reform. Naturally there was much that could be done, much that could be changed – and indeed was changed. Yet one thing could not be done: no new force could be created that was capable of placing the obsolete and largely ineffectual inherited order on a new footing. As long as this remained impossible, even reforms tended only to prolong the crisis.

An order usually fails when the community is no longer able, with its help, to perform its tasks more or less satisfactorily, or at least without causing major damage.

The late republic could no longer do this. It could no longer contend with the social problems at home or the military and administrative problems abroad. For any attempt to solve them only increased the power of individuals, or at least the fear that they inspired among most of the senators. This led to fierce conflicts, to restrictions on what could be expected of the individual senator, to growing inefficiency and thus to the disintegration of the inherited order. The senatorial régime was not designed to cope with the problems that now faced it. Time and again it showed itself to be superannuated. Yet it could not easily be replaced.

Hence a great crisis arose, the main feature of which was that for a long time no alternative could emerge. An alternative did emerge in the end, but only after a long phase in which fundamental features of Roman society were worn down. It is to this long phase that the term 'crisis without alternative' applies.

The term is used here deliberately to counter the widely held view, first advanced by Mommsen, that a revolution was taking place at the time. The term 'revolution', if it means anything more than a period of unrest, implies the emergence of a new force that sets itself up in opposition to the old and overthrows it, so that the community is placed on a new basis. It is wrong to transfer the notion of revolution, informed as it is by nineteenth-century experiences and expectations, to Roman conditions. Such a transference is only apparently supported by certain statements in the sources that seem to point to a two-party system. Quite apart from the fact that political reality in Rome was much more differentiated (and only occasionally produced a substantial opposition between Senate and people), the people cannot be seen as a new force that set itself up against the old one represented by the Senate.

At one point, for instance, Sallust writes that in the late republic the nobility used its *dignitas* and the people its *libertas* as mere arbitrary factors – that is to say, as pretexts for pursuing their own arbitrary aims. Both *dignitas* and *libertas* entailed rights that could be exercised within a definite framework. Now that the framework was fractured, assertions of these rights began to proliferate, and grave conflicts ensued. Hence, according to Sallust, the whole was torn into two parts; the republic, lying between them, was rent apart: *res publica, quae media fuerat, dilacerata* (Jugurtha 51,5). This at any rate accurately describes one thing: the conflicts produced nothing new, but only the unleashing of old hatreds or, to express it in modern terms, the disintegration of the republic. The republic thus destroyed itself. It had to, for otherwise nothing new, no alternative, would have emerged. And it destroyed itself even though no one desired its destruction. Yet if an alternative was to take shape, it would have been necessary for discontent to accumulate and disaffection to be directed, sooner or later, against the order itself.

It can hardly be assumed that the Romans were intellectually too limited, too ignorant of political science, to be able to come up with new solutions. Rather, there were no positions from which it would have been possible to arrive at a proper appreciation of the whole and draw the proper conclusions. For where a new force is present, or at least in the making, there will also be people who try to find ways of helping it to achieve its ends, viz. through reform of the existing order.

This is what happened in modern European history when the liberal bourgeoisie, and subsequently the proletariat, won a place in the existing order and imposed substantial changes on it. Something not unlike this happened when the Greek peasants opposed the old aristocratic régime: new institutions were created, with the help of which they could regularly assert their will; thus the pre-forms of democracy were able to develop. In a different way the Roman plebs was able to win strong influence in the class conflicts. As a result, the capacity of the republic was enlarged; it was able to react successfully to many demands and satisfy many needs without any upheaval or even major disturbances.

In the late republic, however, no such movement can be discerned. What we observe is rather the contentment of all who were powerful or potentially powerful within the existing order and the powerlessness of all who were discontented – at least with their conditions, though hardly with the order itself.

This was because the whole existence of Roman society, indeed its identity, was so bound up with the inherited order as to block off any thought of change. The republican order was not a means to an end – such as the guaranteeing of work and prosperity – but the very element in which the citizens lived. Admittedly this applied above all to the ruling class of senators and knights. Yet the wider circles in Roman society and throughout the empire, those who suffered deprivation and hardship, could wish for nothing else: they had no possibility of striving for a change in their favour, as no institutions would have been conceivable with whose help they could have controlled the republic and its vast empire.

The oligarchic republic could thus be replaced only by a monarchy. Yet this became possible only when society had been so weakened by protracted civil wars as to recognize that things could no longer go on without a *princeps*.

The crisis without alternative is a special case of the transformation of politics into change. When considering any period, historians have to ask to what extent those living at the time were in a position to influence the processes of change taking place in their midst. At some periods this is not very important, because 'progress' takes place in the form of a process. In other periods change takes place very slowly. But when it becomes rapid and moves in an undesirable direction, undermining the foundations of an existing order, the urgent question arises as to how it is possible, in a given situation, to gain control of the processual changes. If this is to be possible (and is not effected from a monarchic centre), it must come about through the formation of factions: what is susceptible of change must also be

493

subject to controversy – which means that one side must be able to oppose the other with regard to the order as a whole. Thus in certain circumstances the factions may produce a switch from synchrony to diachrony.

A society's ability to place its order on a new footing should not be equated simply with its goodwill, its agreement that order is desirable. Rather, as I have said, it must be possible for factions to form within it, not only to represent their own interests, but to question the order itself and produce something new.

Theoretically, it is conceivable that, given the many abuses in Rome, some force might have taken shape on which great individuals like Pompey and Caesar could have relied. Yet this would have presupposed not only that all kinds of hardship and abuse were keenly felt, but that remedies could have been devised or, more precisely, that discontent could have been transformed into demands that were ultimately realizable, since the forces that raised the demands would have been ready to play their part in satisfying them.

Since this was not the case, only Augustus could provide an alternative.

The civil war broke out again shortly after Caesar's death and went on for many years. When Caesar's adopted son and heir finally won control of the empire in 31 BC, almost two decades of civil strife had passed and a new generation had grown up; above all, Roman society had been worn down.

The situation had changed fundamentally. Numerous interests and opinions could now be marshalled in favour of a new political structure. Resistance had weakened. The republic could no longer be defended, but at best be restored. And the victor was so well placed, so gifted and so flexible, that he was able to exploit all the opportunities available to him.

He had learnt much, not least from the fate of Caesar. At an early stage, being relatively weak in the struggle with Antony and therefore dependent on alliances, he had promised to restore the republic. Above all, he had realized that if he was to be strong he had to commit himself to a cause. And by now there were causes to which a man of extraordinary power and ambition could lay claim. His rival Antony had allied himself with Cleopatra, and Caesar's adopted son could set up the standard of Rome and Italy against him, as though control of the Mediterranean were threatened from the East. He could plausibly claim that victory for him meant peace, and peace was by now all that mattered: the civil war had wrought havoc, and there was much that needed to be restored – political institutions, morality, legal security, the old religion, the infrastructure, the economy. This was best tackled by an individual, especially as the Senate itself was in need of reform. There were thus plenty of tasks through which Octavius could legitimate his extraordinary power.

This was of course possible only if his power had a form that could also be legitimated – that is to say, embodied in the republic. Caesar's adopted son realized this. He pretended to restore the republic. It was only with apparent reluctance that he took on particular tasks, the first of which was to secure the border provinces. This meant assuming supreme command of almost all the armed forces. In Rome itself, however, he was merely the first citizen, and all his major powers and responsibilities were of limited duration. He was careful to avoid the appearance of wanting to establish a permanent power-base. Few honours, and seemingly minor powers, were conferred on him for life, largely in gratitude for the apparent restoration of the republic. Among them was the conferment of the name Augustus.

In this way Caesar Augustus created a limited alternative, insinuating himself, as it were, into the republic. He outdid the Senate in conservatism by committing himself to the restitution of tradition and its further evolution. This meant that in the future conception of the commonwealth the performance of various tasks should take precedence over the mere continuance of life within an inherited framework: the commonwealth was in other words to become an entity that could be shaped, rather than a shared ambience in which life could go on as before. Its very foundations, all that had formerly made up the special character of Rome, must be renewed. It was this that accounted for the strength of the first citizen, the *princeps*, a strength that conferred legitimacy – to say nothing of the immense power he already possessed. Accordingly, his régime carried such conviction as to win the approbation of Rome's ruling class.

The crisis could thus finally be compassed by political action and reform. Yet even now the power of the old order was still manifest, inasmuch as the alternative could be shaped only by one who respected tradition. The republic could be overcome only if it was restored. Peace and order, the guarantee of property, legal security, the assurance of sound government and necessary reform – and much else that Augustus achieved – were very important, but they were the piers, as it were, on which the bridge to the principate rested. The bridge itself had to be built in the name of the republic.

Augustus gained control over Roman conditions only through deliberate restraint in the use of his power within them, through self-abnegation. In order to legitimate his power, he had to conceal its extent. Only by pretending to be what he was not could he graft the new reality of the monarchy on to the old republic and impose the imprint of the new on the traditional forms.

It was one of the most difficult roles ever to be played on the world stage, and quite un-Caesarian. He had to be an actor, and he knew this. Just before his death he summoned his friends to his bedside and invited them, in words used by the mimes in certain pieces on leaving the stage, to applaud if they had enjoyed the farce. Beforehand, we are told, he had had his hair carefully combed and his sagging cheeks massaged.

His person and his role finally merged: as a player he had become an institution. He found himself by denying himself. Whatever Caesar and his nephew Octavius had in common as regards power and political skill, their personalities and identities developed in opposite directions, in accordance with the situation in which each grew up and the different ways in which each confronted it and was confronted by it. Caesar grew up in the unruly and apparently provisional republic. Though he had high ambitions, he was for a long time not really stretched, and when he at last achieved personal greatness it was in a world of his own making. Augustus, the dictator's heir, was thrust into politics at the age of nineteen and forced to make all kinds of compromises; he too was ambitious but, recognizing the difficulties he faced, he was prepared to sacrifice himself to his ambition. He was thus able, in a quite different situation, to found the Roman monarchy.

When Cicero declared that the republic was lost it was still in existence. When Augustus said it was restored it had come to an end. Yet it is typical of periods of decline that nothing is so predictable as the paradoxical: at such times one must expect the unexpected.

Afterword to the Third German Paperback Edition

My colleague Ernst Badian, one of the greatest prosopographical experts on the late republic, has pointed out some omissions in earlier editions of this work.

It is known that in his youth Caesar composed a poem in praise of Heracles and a tragedy on the theme of Oedipus, as well as a collection of maxims. Badian finds it significant that at an early age Caesar addressed the themes of achievement and fate. We know too that after an accident Caesar took to reciting a short magic formula three times before setting out on a journey, so that it might pass off safely. Badian observes that this is welcome evidence of a superstition to which even Caesar subscribed.

Badian also draws attention to Caesar's extraordinary expenditure. For the site of the Forum Iulium he paid over 100 million sestertii. He also paid L. Paullus 36 million for his political support. No precise figures can be put on the other bribes that he paid, to Curio and many others, or on the gifts and loans he made, but they must have been huge. In addition to his munificent gifts to his soldiers and officers at the start of the civil war and on the occasion of his triumphs, he made large donations to the urban plebs and spent unprecedented sums on feasts and games. Yet this is the man who in 62, before departing for Gaul, was so deeply in debt that he almost had to pledge his equipment as collateral. Indeed, as Badian writes, Caesar must have been the biggest robber (and exploiter) in contemporary Rome. Badian even goes so far as to wonder whether Caesar's rapacity was not unique in history.

I should also have paid some attention to the inscription from Taranto published by L. Gasperini. If the text is correctly restored, it designates Caesar as *dictator rei publicae constituendae*. This would mean that he, like Sulla, really had been charged with the restoration of the commonwealth. I now think it likely that Caesar was empowered not merely to act as he saw fit, but to reconstitute the republic, and that ten years were mutually agreed as the time he would need to perform the task. This may have been the only way to justify Caesar's long-term dictatorship; or at any rate, to lay such a charge upon the dictator was the best way to make it tolerable. Yet this does not alter the fact that, apart from his organizing abilities, Caesar had no qualification for the task. The evidence of the inscription is therefore interesting – and ought not to have been omitted – but it does not essentially change our understanding of Caesar's situation after the civil war.

I also failed to mention the fact that Caesar appointed Octavius to the post of *magister equitus*. This shows that the dictator wished to give his grandnephew, if only temporarily, an important military function that conferred a special distinction and would at the same time contribute to his training, though it is hard to see the appointment as evidence that Octavius was being groomed to succeed Caesar as monarch. In the following year the office was to be taken over by Domitius, the consul of 53.

Badian's comments are published in *Gnomon* 62 (1990), 22ff. I should like to record my gratitude to him.

INDEX

INDEX